# NEW FORMS OF CARBON

Nanocarbons

# NEW FORMS OF CARBON

## Nanocarbons

*Edited*

**Aneeya Kumar Samantara, PhD**
**Satyajit Ratha, PhD**

APPLE ACADEMIC PRESS

First edition published 2024

**Apple Academic Press Inc.**
1265 Goldenrod Circle, NE,
Palm Bay, FL 32905 USA

760 Laurentian Drive, Unit 19,
Burlington, ON L7N 0A4, CANADA

**CRC Press**
2385 NW Executive Center Drive,
Suite 320, Boca Raton FL 33431

4 Park Square, Milton Park,
Abingdon, Oxon, OX14 4RN UK



**Library and Archives Canada Cataloguing in Publication**

Title: New forms of carbon : nanocarbons / edited Aneeya Kumar Samantara, PhD, Satyajit Ratha, PhD.
Other titles: Nanocarbons
Names: Samantara, Aneeya Kumar, editor. | Ratha, Satyajit, editor.
Description: First edition. | Includes bibliographical references and index.
Identifiers: Canadiana (print) 20230598633 | Canadiana (ebook) 20230598668 | ISBN 9781774912799 (hardcover) | ISBN 9781774912805 (softcover) | ISBN 9781003376460 (ebook)
Subjects: LCSH: Carbon.
Classification: LCC TP245.C4 N49 2024 | DDC 662/.93—dc23

**Library of Congress Cataloging-in-Publication Data**

CIP data on file with US Library of Congress

ISBN: 978-1-77491-279-9 (hbk)
ISBN: 978-1-77491-280-5 (pbk)
ISBN: 978-1-00337-646-0 (ebk)

# Dedication

**Dr. Aneeya K. Samantara** dedicates this work to his daughter Aayra

**Dr. Satyajit Ratha** dedicates this work to his parents, Mrs. Prabhati Ratha and Mr. Sanjaya Kumar Ratha

# About the Editors

**Aneeya Kumar Samantara, PhD**
*Postdoctorate Fellow, National Institute of Science Education and Research, Odisha, India*

Aneeya Kumar Samantara, PhD, is a postdoctorate fellow at the National Institute of Science Education and Research, Odisha, India. He did his PhD in Chemistry at the CSIR-Institute of Minerals and Materials Technology, Odisha, India. Before joining PhD, he completed his MSc degree in Advanced Organic Chemistry at Ravenshaw University, Cuttack, Odisha, and MPhil in Chemistry at Utkal University. Dr. Samantara's research interest includes the synthesis of metal and carbon nanostructured particles for electrochemical energy storage/conversion and sensing applications. He has authored and coauthored about 30 peer-reviewed international journals, six book chapters, and has edited/authored 13 books for reputed publishing houses. For his outstanding research activity, he has received awards at many national and international forums.

**Satyajit Ratha, PhD**
*Researcher, Indian Institute of Technology Bhubaneshwar, Bhubaneshwar, India*

Satyajit Ratha, PhD, pursued his PhD at the Indian Institute of Technology Bhubaneshwar, Bhubaneswar, India. Prior to joining IIT Bhubaneswar, he received his BSc from Utkal University in 2008 and MSc from Ravenshaw University in 2010. Dr Ratha's research interests include two-dimensional semiconductors, nanostructure synthesis, applications, energy storage devices, and supercapacitors. He has authored and coauthored about 28 peer reviewed international journals, eight books and one book chapter.

# Contents

# Contributors

**Chinmayee Acharya**
CSIR—Institute of Minerals and Materials Technology, Bhubaneswar, Odisha, India
North Orissa University, Baripada, India

**Anil Arya**
Department of Physics, Kurukshetra University, Kurukshetra, Haryana, India

**Madhabi Madhusmita Bhanjadeo**
Department of Biochemistry, School of Life Sciences, Ravenshaw University, Cuttack, Odisha, India

**Avijit Biswal**
College of Engineering and Technology (Autonomous) Bhubaneswar, Odisha, India

**Dipti P. Das**
Academy of Scientific and Innovative Research, Ghaziabad, Uttar Pradesh, India
Central Characterization Department, CSIR—Institute of Minerals and Materials Technology, Bhubaneswar, Odisha, India

**Bibhuti Bhusan Jena**
Department of Chemistry, Ravenshaw University, Cuttack, Odisha, India

**Sushri Soumya Jena**
School of Basic Sciences, Indian Institute of Technology Bhubaneswar, Bhubaneswar, India

**Vijay Kumar**
Department of Physics, Institute of Integrated and Honors Studies (IIHS), Kurukshetra University, Kurukshetra, Haryana, India

**Anoop Kumar Kushwaha**
School of Basic Sciences, Indian Institute of Technology Bhubaneswar, Bhubaneswar, India

**Sanghamitra Mohapatra**
CSIR—Institute of Minerals and Materials Technology, Bhubaneswar, Odisha, India
AcSIR—Academy of Scientific and Industrial Research, New Delhi, India

**Jagannath Panda**
CSIR—Institute of Minerals and Materials Technology, Bhubaneswar, Odisha, India

**Prasanna Kumar Panda**
CSIR—Institute of Minerals and Materials Technology, Bhubaneswar, Odisha, India

**Tanaswini Patra**
School of Applied Sciences, KIIT Deemed to be University, Bhubaneswar, Odisha, India

**Manas R. Pattanayak**
Syngene International LTD, Medchal, Hyderabad, India

**Satyajit Ratha**
School of Basic Sciences, Indian Institute of Technology Bhubaneswar, Bhubaneswar, Odisha India

**Mihir Ranjan Sahoo**
School of Basic Sciences, Indian Institute of Technology Bhubaneswar, Bhubaneswar, Odisha, India

**Rojalin Sahu**
School of Applied Sciences, KIIT Deemed to be University, Bhubaneswar, Odisha, India

**Alaka Samal**
Academy of Scientific and Innovative Research, Ghaziabad, Uttar Pradesh, India
Central Characterization Department, CSIR-Institute of Minerals and Materials Technology, Bhubaneswar, Odisha, India

**Rashmi Rekha Samal**
Academy of Scientific & Innovative research (AcSIR), New Delhi, India
Environment & Sustainability Department, CSIR—Institute of Minerals and Materials Technology, Bhubaneswar, Odisha, India

**Aneeya K. Samantara**
School of Chemical Sciences, National Institute of Science Education and Research, Bhubaneswar, Odisha, India

**Annu Sharma**
Department of Physics, Kurukshetra University, Kurukshetra, Haryana, India

**A. L. Sharma**
Department of Physics, Central University of Punjab, Bathinda, Punjab, India

**Bankim Chandra Tripathy**
CSIR—Institute of Minerals and Materials Technology, Bhubaneswar, Odisha, India

# Abbreviations

| | |
|---|---|
| AC | activated charcoal |
| ACFs | activated carbon fibers |
| AEA | adiabatic electron affinity |
| AEC | anion exchange capacity |
| AR | androgen receptor |
| ASSC | an asymmetric supercapacitor |
| BBB | blood–brain barrier |
| BNCT | boron neutron capture therapy |
| CEC | cation exchange capacity |
| CISD | configuration-interaction singles and doubles |
| CNT | carbon nanotube |
| CO | carbon monoxide |
| COX | cyclooxygenase |
| CQDs | carbon quantum dots |
| CTCs | circulating tumor cells |
| CV | cyclic voltammetry |
| CVD | chemical vapor deposition |
| DFTB | density functional tight binding |
| DMF | N, N-dimethyl formamide |
| DOS | density of states |
| EAC | extruded activated carbon |
| EDLC | electric double-layer capacitor |
| EDS | energy-dispersive spectroscopy |
| EGFR | epidermal growth factor receptor |
| ER | estrogen receptor |
| ESR | equivalent series resistance |
| FMO | frontier molecular orbital |
| FR | folate receptor |
| FVP | flash vacuum pyrolysis |
| FWS | fullerene water suspension |
| GCC | ground cherry calyces |
| GCD | galvanostatic charge/discharge |
| GNS | graphene nanosheets |

| | |
|---|---|
| GNTs | golden nanotubes |
| GO | graphene oxide |
| GQDs | graphene quantum dots |
| HF | Hartree–Fock |
| HIF | hypoxia-inducible factor |
| HOMO | highest occupied molecular orbital |
| HPC | honeycomb-like porous carbon |
| HPCs | hierarchically porous carbons |
| IBI | International Biochar Initiative |
| LBD | ligand-binding domain |
| LUMO | lowest unoccupied molecular orbital |
| MB | methylene blue |
| MRI | magnetic resonance imaging |
| MSC | mesenchymal stem cells |
| MWCNTs | multiple-walled carbon nanotubes |
| NIR | near-infrared |
| NPs | nanoparticles |
| NSAIADs | nonsteroidal anti-inflammatory drugs |
| nZVI | nanoscale zero-valent iron |
| OFTETS | organic field effect transistors |
| OPVs | organic photovoltaics |
| PA | photoacoustic |
| PAC | powdered activated carbon |
| PAHs | poly-aromatic hydrocarbons |
| PC | pseudo-capacitor |
| PCB | polychlorinated biphenyles |
| PDT | photodynamic therapy |
| PTE | photothermal–electrical |
| PTT | photothermal therapy |
| PU | polyurethane |
| PVA | polyvinyl alcohol |
| QDs | quantum dots |
| RARs | retinoic acid receptors |
| RH | rice husk |
| RO | reverse osmosis |
| ROS | reactive oxygen species |
| SC | supercapacitor |
| SEM | scanning electron microscopy |

| | |
|---|---|
| SiC | silicon carbide |
| SRF | senior research fellowship |
| SSA | specific surface area |
| SSC | symmetric supercapacitor |
| SWCNTs | single-walled carbon nanotubes |
| TEM | transmission electron microscopy |
| TTR | transthyretin |
| UV | ultraviolet |
| VDE | vertical detachment energy |
| VOC | volatile organic compounds |
| WBI | Wiberg bond indices |
| XPS | X-ray photoelectron spectroscopy |
| XRD | X-ray diffraction |
| ZVI | zero-valent iron |
| 0D | zero-dimensional |

# Preface

Carbon, being the second most abundant element in the human body and a natural building block for almost all the organic molecules, has been the topic of both discussion and debate since ancient times upto the modern era. The vast range of compounds that carbon forms due to its catenation property is second to none of the elements in the modern periodic table. Starting from the black colored charcoal to grayish graphite and mesmerizingly transparent diamond, carbon has been the cynosure of myriad of applications ranging from catalysis chemistry to medicinal biology and applied physics. It is popular among the scientific community in the form of activated charcoal/carbon and the potential of this ancient and faithful element was not realized until the advent of the buckminsterfullerene in the year 1985. Since then, the discovery of carbon nanotube (1991) and graphene (2010) has catapulted the research interest on the carbon-derived allotropes. These forms of carbon have at least one dimension falling in the nanoscale order and are referred to as nanocarbons.

Often credited for the discovery of these wonderful carbon structures, nanotechnology has played a phenomenal role in providing us just the set of tools and techniques to gain significant insights for carbon-based research. However, as we come to know one aspect of the element carbon, it surprises us with a new set of unheralded characteristics. Nevertheless, these new forms of carbon have the potential to revolutionize many aspects of research including drug delivery, catalysis, energy conversion and storage, high strength physics, structural engineering, and so on.

This book provides a detailed overview of some of the fundamental aspects of carbon and its allotropic forms, their preparation techniques, and application in several fields of interest. The content of this book has been meticulously arranged to seamlessly guide the readers through the transformation of carbon from its traditional form to the present-day structure(s). We hope that this book will cater to the needs of students at both undergraduate and graduate levels and researchers who are interested in the new forms of carbon.

# Preface

# Summary

The physicochemical properties possessed by carbon and its various allotropes have fascinated the scientific community since a very long time. Incidentally, nanotechnology has revamped the carbon chemistry through sophisticated techniques to carve out the best possible structures, such as fullerenes, graphene, and carbon nanotubes. These carbon structures have redefined the carbon chemistry as well as many of the key research areas, such as medical health and diagnostics, opto-electronics, semiconductor physics, energy storage/conversion, and so forth. There are a large number of futuristic applications and unlimited possibilities with these new forms of carbon.

This book is divided into six chapters and presents a vivid detail of the origin of these new forms of carbon and some of their important characteristics including their application in several fields of interest. Chapter 1 explains the unprecedented physicochemical properties of the two-dimensional carbon structure, graphene. The authors have discussed both synthesis and characterization of graphene including some of its key applications such as water treatment, energy, and corrosion-resistant technologies. Chapter 2 summarizes the application of carbon in one of the futuristic electrical energy storage devices, that is, supercapacitors. In this chapter, biowastes have been effectively treated to yield the required form of carbon through affordable techniques. In Chapter 3, atomic carbon clusters have been introduced in the context of existing cluster research that are mostly based on metal-based atomic clusters. The authors have discussed few organic carbon clusters of importance and the effect of doping on these clusters using different dopant atoms such as phosphorous, gold, boron, silicon and others. The authors suggested that similar to the metallic clusters, these carbon-based clusters and their doped counterparts could prove their potential in several electrochemical and electronic applications. Chapter 4 deals with the synthesis and application of fullerene type of carbon structures. The authors have provided a broad range of applications which include water treatment, energy storage/conversion, biomedical, and pharmaceutical applications that show the huge potential of fullerene. Chapter 5 discusses the production and activation of biochar to check/treat both the organic and inorganic pollutants. A detailed study on the modification of biochars have also been provided, where several methods, such as mineral impregnation, apourdametal assistance, and surface oxidation/reduction have been implemented to modify the biochar to put it to effective use. The concluding Chapter 6 deals with the preparation and properties of activated carbon. A brief account of activated carbon in terms of its activation processes, classification, and adsorption properties have been provided in detail. We believe that the book in its present form would provide an overall picture of the carbon chemistry and the rich physicochemical aspects of the new forms of carbon to the readers, and would be of some use to students, researchers, and academicians who would be interested in these carbon allotropes/structures.

# CHAPTER 1

# Graphene: A New Form of Carbon for Future Sustainability

ALAKA SAMAL[1,2] and DIPTI P. DAS[1,2]

[1]*Academy of Scientific and Innovative Research, Ghaziabad, Uttar Pradesh, India*

[2]*Central Characterization Department, CSIR-Institute of Minerals and Materials Technology, Odisha, India*

## ABSTRACT

A well-known and documented carbon allotrope, "graphene," has a hexagonal carbon arrangement in the route framework. Graphene also has an unusual behavior not really identical to the conductor or insulator, which would be the energy band with a "Dirac cone" having a zero band gap value. Graphene has a typical behavior of very high mechanical strength as well as remarkably transparent with very high electrical and thermal conductivity. After the breakthrough of the single atomic thick graphene, the graphene-hybridized semiconductor scientific field has grown well over the last decade, with great results and feasible aspects of life for the future. The graphene-derived semiconductor photocatalysis has a wide range of applications in scientific fields, from energy generation to storage, including environmental cleanup, electronic equipment, and also in the desalinization of ocean water. Graphene-centered nanocomposite formulation has indeed attracted specific attention in applied science research work. The need for more drinkable water for countless millions of thirsty people could also be mitigated in the future with graphene-customized

New Forms of Carbon: Nanocarbons. Aneeya Kumar Samantara & Satyajit Ratha (Eds.)

technological advances. For this whole purpose, the water desalination of seawater and the decontamination of wastewater have been examined conspicuously with a sound result. Again, the manufacturing of $H_2$ energy with water, graphene-derived photocatalysts, has been used extensively until now. The existence of graphene shows a significant improvement in the material's specific surface area and contributes to the enhanced corrosion protection and conductivity applications without sacrificing the characteristics or composition of the material. This new strategy creates a new way to easy manufacturing of sophisticated surface coatings that can be used to develop thermal exchangers and biocompatible materials. This chapter is predicated mostly on the literature review and discussion of recent graphene-derived semiconductor implementations and advancements. The numerous graphene-based photocatalytic synthesis procedures are briefly at the end elaborated with the advantages and disadvantages. The innumerable applications of graphene-derived architectures were then rigorously assessed, including those of organic/inorganic water contamination degradation/mineralization, $H_2/O_2$ evolution by water photosplitting, energy storage application, coating applications, and so forth. Furthermore, the analysis discusses the future of graphene technology and its impoverishment in an exact manner such that a marvelous graphene technology could be accomplished in the immediate future.

## 1.1 INTRODUCTION

Owing to the remarkable properties of graphene as an outstanding material (such as high mobility of charge carriers called mass-less Dirac fermions, theoretical surface area, thermal conductivity, flexibility, transparency, inexpensive, and bearing very lightweight), it has become the most studied material in the last few years (Bolotin et al., 2011a, b; Du et al., 2008; Peigney et al., 2001; Balandin et al., 2008; Wang et al., 2009; Guinea et al., 2010; Du et al., 2009; Bolotin et al., 2011a, b). Due to its adaptable nature, it could be a solution starting from super-small computers to high-capacity batteries. For the improvement of spatial separation and transfer of photoinduced charge carriers, in that way increasing the activity of semiconductors for photocatalytic reactions, graphene has been used immensely for years. Copious numbers of studies also confirm about the reliability of graphene in the charge carriers separation and photocatalytic activity improvisation with graphene (Xiang et al., 2012; Tu et al., 2013; Lightcap et al., 2010; Ng et al., 2010a, b; Zhang et al., 2014; Wang et al., 2014; An et al., 2014; Moon et al., 2014). The two-dimensional hexagonal sheet of graphene also provides an obvious support matrix to the semiconductor positioned on it. This further increases the surface area of photocatalyst with advancement in adsorption of the organic molecule in the organic pollutant degradation method from water (Han et al., 2012; Wang et al., 2014; Som et al., 2014). Graphene-based photocatalytic approach has given tremendous result in improving the photocatalytic performance of the semiconductor particles along with enhancement of light absorption capacity. So far, many graphene-based semiconductor photocatalysts have been introduced in the field with an improved photocatalytic activity *Ca.* RGO–ZnO, RGO–$SnO_2$, graphene–$MnO_2$, $Mfe_2O_4$/graphene, graphene–$TiO_2$, graphene–$Cu_2O$, graphene–$Fe_2O_3$, graphene–NiO, graphene–$WO_3$, graphene–ZnS, graphene–CdS, graphene–$MoS_2$, and others (Akhavan, 2011; Yang et al., 2013; Qu et al., 2014; Yao et al., 2014; Bai et al., 2012; Xiang et al., 2012; Yang et al., 2006; Yao et al., 2010; Xu et al., 2008; Morishige and Hamada, 2005; Guo et al., 2012; Hu et al., 2011; Cao et al., 2010; Nethravathi et al., 2009; Li et al., 2011a, b).

Chemically derived graphene oxide (GO) is now among the one specific branch of graphene research getting attention (Dreyer et al., 2010; Yin et al., 2014). GO possesses unique properties of graphene and also some extra characteristics like hydrophilicity providing platform for nucleation and growth of the nanoparticles. The structure of GO represents edges with carboxylic acid group and at the basal plane hydroxyl/epoxide groups (Dreyer et al., 2010). The functional groups present on the GO give a base for the directional growth of the nanoparticles with a proper distribution. Furthermore, the physico/chemical properties and the band gap of GO can be tuned by engineering its atomic and chemical structures (Hatakeyama et al., 2014). Several reviews have been also published on this particular topic enlightening concerning the incomparable performance of graphene (Sun et al., 2011; Huang et al., 2011; Zhang et al., 2015; Yang et al., 2014; Zhang et al., 2012; Yang et al., 2013).

Due to the rising cogent call for environmental and energy remediation, the need of an amazing material that could solve many purposes at an ease is mandatory. Graphene acts as a versatile actor to solve many problems involving energy and environment. Graphene not only solves the water-quality issue by decontaminating the polluted water but also helps in the lack of energy for the future generation by producing $H_2/O_2$ fuel (Zhao et al., 2017). The process of photocatalysis uses the undiminishable source, the Sun, to vitalize the energy from water fission and also encourages the decontamination of water with the help of a photocatalyst. Graphene-based photocatalyst has a well-known craze among material scientists. Also, graphene is applied in the desalination process of seawater to yield drinkable water for the living organism (Homaeigohar and Elbahri, 2017). Along with this, graphene has been also used in the $CO_2$ reduction reaction processes for simultaneously solving energy and global warming problems (Low et al., 2015).

Moreover, the unique features of graphene (like those of chemical inertness, extraordinary heat and chemical stability, outstanding versatility, visible optical transparency, and molecular impermeability even as small as helium) make it the most conducive to passive-layer formation, particularly in the natural environment, for protecting against metal oxidation and corrosion. It actually forms the natural barrier of diffusion that separates the guarded metal/substrate from the reactants physically due to the hydrophobic nature of graphene. Thus, graphene has indeed been reported as a corrosion protection material by the researcher Böhm et al. (2014).

Day by day, graphene-based semiconductor articles/reviews are published in numerous amounts in a way that there become saturation in the articles based on graphene. So, it is a trivial task to cover all the updates about the new mark and applications of graphene. Indeed, we have tried to cover most of the new articles regarding graphene-based materials and its applications in various fields in this chapter. This chapter here describes why the use of graphene is important and what are the benefits we could get from graphene-based materials. After that, this chapter describes the synthesis of various graphene-based semiconductors' importance and application. Afterward, the summary and future prospects have been discussed in the field of graphene materials.

## 1.2 CHARACTERIZATION OF GRAPHENE

Far too many scientists have recorded the electronic, mechanical, and optical properties of such graphene that can actually make this a really useful material. Figure 1.1 depicts the extraordinary characteristics of 2D hexagonal honeycomb-like structure. It is indeed a zero band gap conductive material, because its conduction and valence bands meet up in Dirac points, and it has extremely high conductance too. In graphene hexagonal gigantic 2D structure, each atom of carbon is connected by $sp^2$-hybridization with three other adjacent carbon atoms. Maybe this actually makes one electron free for electronic conduction in the third dimension. Due to its surface structures and excellent conductive properties, graphene is being used in various electronic applications currently.

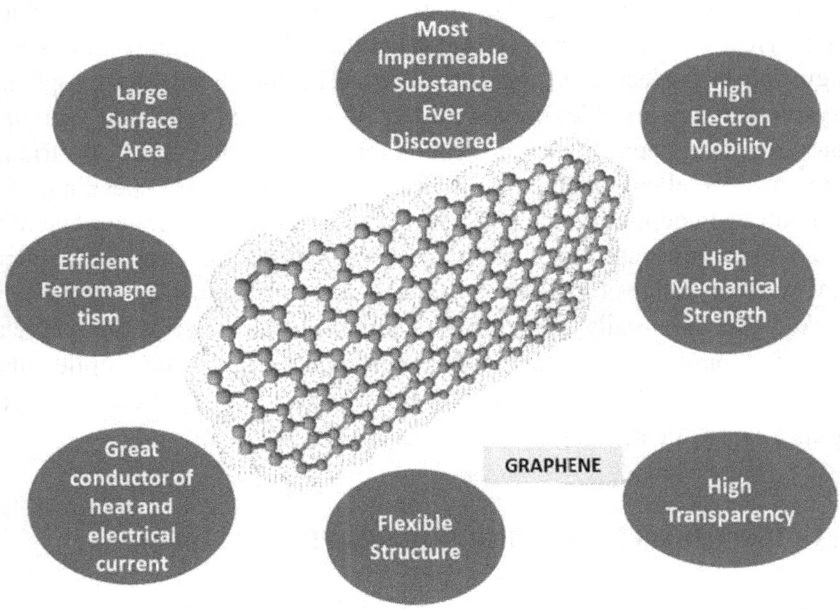


**FIGURE 1.1** Graphene structure with its unique characteristics.

At room temperature, the electronic mobility of graphene is very high and almost independent of temperature. Such extreme mobile electrons are apparently called π-electrons that are positioned above and below the graphene sheet. Basically, the bonding and antibonding of these π-orbitals dictate the great electronic properties of graphene. Graphene is the most robust substance ever really examined with a tensile strength of 130 GPa and Young's module of 1 TPa. Other than this, graphene is incredibly light in weight and is a soft material harder than that of a diamond. Graphene is thought to have been the toughest, 200 times more powerful even than steel material, yet uncovered. It is exceptionally rigid as well as bendable (e.g., rubber). This is because further flat planes of carbon atoms of graphene can quickly and easily wiggle even without the separation of the atoms. Anything that is just one-atom thick is supposed to also be fairly light weight. Thus, this graphene really would apparently cover a sports field only with less than one teaspoon. Graphene is way better to transport high temperature, which has very extremely high thermal conductivity than almost any other type of material. Once again, this most probably uncovers the advantages of graphene by using composite materials that could be used to increase other materials' thermal resistance or conductivity. Heat-conducting materials also conduct electricity quite well, since they transport electricity using electrons. The flat, hexagonal graphene crystal structure provides comparatively little electron resistance that zips throughout it speedily with no trouble and carries additional electricity than even excellent conductors like copper. Evolutionarily speaking, this should conclude that perhaps graphene in electrons has a longer actually mean free route than just about any other material that can go beyond that without collapsing. Graphene sheets have carbon atoms so densely packed that they really work just like superfine atomic grids that prevent the development of other components on their surface. This implies that graphene is indeed good for trapping gases and also can be used in encouraging new applications in hydrogen storage.

## 1.3 SYNTHESIS PROCEDURES OF GRAPHENE

Different techniques for graphene synthesis have been developed in recent years. The most widely utilized techniques today are thermal chemical vapor deposition (CVD) (Böhm et al., 2014), chemical exfoliation (Allen et al., 2010), mechanical cleaving (Novoselov et al., 2004), and chemical synthesis (Park and Ruoff, 2009). Another technique, including microwave (Jiao et al., 2010) and unzipping of nanotubes (Xin et al., 2010), is also reported. CVD, mechanical exfoliation, and synthesis by chemical approach are still the most reliable methods. The graphene layers are first prepared by the mechanical exfoliation method using a tape called "scotch tape" method by R. Ruoff and group (Lu et al., 1999). The graphene layer flakes were exfoliated from the graphite mechanically followed by drying on a silicon wafer via a dry deposition method. The acceptable situation in this process is that graphene can be pelated layer after layer from the bulk graphite. It is again a mechanical issue to counteract the van der Waals attraction between the layers of graphite, peel the layer, and then accomplish graphene structure. It is important to note that such purely mechanical paths are preconditions for the development of graphene to all previously reported exfoliation strategies. This can be predicted too that regulated graphite exfoliation is indeed possible by customizing the mechanical paths to achieve high-quality graphene with a high efficiency. Ultrasonication exfoliation is a liquid process exfoliation that produces large numbers of graphene sheets. In 2014, Coleman's group first time revealed a higher yield of graphene by the liquid-phase exfoliation process (Paton et al., 2014). One advantage of such an approach is that it seems quite pretty easy to produce graphene. By far, the most deficient problem is the incredibly low concentration of graphene, which is not really economical. The whole exfoliation technique does not really make it easy to achieve larger quantities of graphene, even bringing the inadequacy of feasible flakes into account. The problem with this process is very low, but graphene flakes must be figured at the surface of the substructure, which is exhaustive to use. But here the graphene quality is very high without any defects. Another method that involves graphite oxide manufacturing from graphite by a chemical exfoliation method is named as Hummers method (Samal and Das, 2018). By oxidizing the graphite layers using a Hummers method, graphite is chemically translated into an intermediate water-dispersible substance called graphite oxide. Graphite oxide is indeed a sheet-layered stack that exfoliates in the mechanical force copiously. The low cost and scalability are the main advantages of this procedure. But current synthesis relies on graphite reactions mostly with strong mixed oxidants that are at risk of explosion (Songfeng et al., 2018).

Electrochemical procedures have also been acquired lately and used popularly to produce graphene because of the ecological quality, high efficiency, and reduced cost (Yang et al., 2016). Again, it has been shown that the electrochemical exfoliation of graphite foil could also reproduce almost pristine graphene in a few seconds (Yang et al., 2015). Driven by such achievements, many scientists also attempted to synthesize GO with electrochemical oxidation of diverse graphitic materials such as pencil core (Liu et al., 2013), graphite rod (Parvez et al., 2016), and graphite flakes (Gurzęda et al., 2016; Yu et al., 2016). Moreover, the CVD technique is considered to be a more promising way for the scalable synthesis of high-quality and single-layer graphene (Yu et al., 2008). The CVD methodology is used to synthesize graphene in the decomposition of carbon source molecules. Due to the high segregation of carbon atoms on the metallic substrates (i.e., Ni foil), the graphene structure included about 35 graphene layers was synthesized. The method has been further established to accomplish control growth on the Ni-deposited $SiO_2$/Si and Ni film for single-layer, two-layer, and few-layer graphene (Yu et al., 2008; Kalita et al., 2010). But in this process, the major challenge is the removal and transfer of the as-developed graphene sheet on an isolating substrate for the characterization and application. Well into the later point, CVD graphene has been transferred to an arbitrary substrate without distressing its intrinsic characteristics (Reina et al., 2009). CVD can be transferred through wet or dry graphic catalytic layer graph transfer onto the desired surface of the substratum. The process of CVD was considered to grow single-glass large-size crystals, monolayers, and two-layer graphs in high quality. In this perspective, graphene nucleation and growth mechanism on a catalytic substrate play a key role in the formation of large domains and a number of layers.

Among the most acclaimed techniques of graphene synthesis is epitaxial thermal growth on a single-crystalline silicon carbide (SiC) surface. Graphene growth on SiC productivity is envisioned as a highly promising method for large-scale graphene production along with the practical application in electronics. The graphene growth rate on SiC depends on the particular polar SiC crystal face (Low et al., 2012). In unzipping a carbon nanotube to get graphene nanoribbons, the chemical and plasma-etched methods have been used by the researchers. Graphene nanoribbon establishes a thin extended graphene strip that shows straight edges. The production of multilayer graphene or single-layer graphene entirely depends on the precursor wall thickness. In this method, researchers have used a multiwalled nanotube on a Si substrate pretreated 3-aminopropyltriethoxysilane to get high-quality graphene nanoribbons from it with a step height of 0.8–2 nm (Jiao et al., 2009). Some other procedure to open the multiwalled nanotubes is the use of electrical field to get graphene nanoribbons (Bhuyan et al., 2016). A single multiwalled nanotube with a tungsten electrode was applied with an electrical field and it was portrayed that perhaps the semicontact multiwalled nanotubes' end had begun unwrapping and graphene nanoribbon emerged. The graphene nanoribbon manufacturing process by this ensures highly clean, faultless, monitored sequencing for optimized equipment in electronic components.

## 1.4 APPLICATIONS OF GRAPHENE

### *1.4.1 DESALINATION OF WATER*

For the cause of industrialization and urbanization, human being is facing the consequences of water shortage for the last many years. Due to the population explosion, the available fresh water is insufficient for mankind. Again, the fresh-water sources are getting contaminated day by day due to increasing industrialization. For this, seawater desalination would be the optimum solution for all the above-cited problems without provoking the ecosystem. For this, the technology consuming less cost and energy is necessary for water purification. As of not long ago, multistage flash distillation and reverse osmosis (RO) methodology has been utilized basically for seawater desalination (Xu et al., 2016). In the RO procedure, the core of RO is a semipermeable layer that aides in the detachment of unadulterated water from seawater. That is the reason RO membrane ought to be stronger, more slender and artificially safe than the polyamide dynamic layers in thin-film composite RO layers. Because of its one-of-a-kind properties, graphene could be viewed as a definitive RO membrane. It has been reported that GO indicates antimicrobial properties, in this manner bringing down film biofouling, subsequently enhancing the layer lifetime and energy utilization of the water decontamination forms (Goh and Ismail, 2015; Mahmoud et al., 2015). For such reasons, broad research is as of now being completed to understand its potential as a cutting-edge desalination film (Cohen-Tanugi and Grossman, 2015). As indicated by the literatures, nanoporous graphene layers with the quick stream rate of water to the atomic thickness of the graphene film can effectively desalinate water (Wang and Karnik, 2012). As indicated by another investigation published in *Nature*, the desalination of ocean water to drinkable water may soon have the capacity to be done effectively with a graphene channel (Abraham et al., 2017). Here, the modified graphene-oxide membrane sieve acts as a filter with tiny nanometer-wide holes drilled into the sieve. The filter retains the salt molecules as the water flows through the sieve. Actually, the graphene membrane sieve allows certain molecules to go through keeping other molecules out which effectively filters out the salt and also extremely polluting organic dye molecules. But this process is not able to filter out the entire salt molecules. Further research work is needed to improve the sieve quality to filter out the seawater to pure drinkable water with an inexpensive and scalable method. The researchers hope that the sieve will act as a model for future graphene-oxide sieves with holes of different sizes to filter molecules larger or smaller than salt. The sieve would require even more testing, including durability testing to excess seawater exposure, before being used in an industrial setting. For the first time, Huang and Feng (2018) have fabricated novel GO/polyimide hollow fiber films by direct spinning of a GO/PI suspension by means of a coaxial two-capillary spinning technique. The GO/PI hollow fiber film performances were assessed for desalination of ocean water by pervaporation strategy. Here, the GO/polyimide hollow fiber layer showed great water penetrability and salt dismissal for desalination of various concentrations of seawater with a high strength with quite a while run estimation. The ion rejection of the majority of the ions exhibits in water-like $Na^+$, $Cl^-$, $Ca^{2+}$, $K^+$, $F^-$, $Mg^{2+}$, and $PO_3^-$ is 99.9%, 99.9%, 99.9%, 99.8%, 99.8%, 99.9%, and 99.8%, respectively, through the film demonstrating a magnificent outcome for water refinement. They have depicted the superb desalination execution of the GO/polyimide hollow fiber layer with a lot of hydrophilic locales that encourage sorption and dispersion of water molecules through the films which are better than traditional zeolite layers. This high partition exhibitions joined with high steadiness prescribe the created GO/polyimide hollow fiber films to guarantee for the seawater desalination demonstrating the colossal execution of graphene subsidiary toward future water lack for billions of individuals around the globe.

There are many endeavors toward the improvement of GO-based films for water desalination by other researchers (Hu and Mi, 2013; Han et al., 2013; Joshi et al., 2014; Sun et al., 2014; Sun et al., 2013; Nicolai et al., 2014; Surwade et al., 2015; Hega and Zou, 2015). For the time, the development of technology for the fabrication of large–size, high-quality graphene is mandatory for the molecular and ionic sieving membrane. It was described by Surwade et al. (2015) experimentally, the possibility of desalination utilizing single-layer graphene with a couple of controllable openings. You et al. (2016) have made incredible endeavor to advance more desalination tests utilizing the layers for the extensive scaled desalination with a few conceivable outcomes with an expectation of gigantic extent of research on using single layer with defects made artificially as pores for seawater desalination. The group has done brief examinations between nanoporous graphene and GO layers utilized for desalination. Pristine graphene is to a great degree impermeable to fluid or gases, along these lines as indicated by the group to make the water pathways, the stacking of GO sheets or to generate nanopores in monolayer graphene are the two conceivable outcomes. In the figure, they have demonstrated the plan of two conceivable outcomes.

### *1.4.2 WATER DECONTAMINATION*

The chemical wastes from industries badly impact the water bodies. Particularly, the toxic organic pollutants coming from the textile industry contributes a huge share to the water body pollution. These organic pollutants can be removed from the water bodies through various chemical, biological, and physical technologies, that is, advanced oxidation processes, photocatalysis, sonolysis, ozonation, including the Fenton reaction, and combinations of these techniques (Chong et al., 2010; Bremner et al., 2009). The basic fundamental of advanced oxidation process including photocatalysis based on the growth of radicals and their interaction with pollutants (organic molecules) in presence of light lead to degradation. The photocatalysis process has drawn great attention for the removal of organic pollutants fruitfully. The eye-catching application of graphene in photocatalytic organic degradation came into the performance due to the huge limitations in the photocatalysts like $TiO_2$ (having wide band gap and high electron–hole recombination character), CdS (highly photocorrosive and restricted photocatalytic action due to the high recombination rate of photogenerated electron–hole pairs), and the use of highly toxic and costly noble metals in the semiconductor photocatalysts. Also, the most downside of the light-sensitive semiconductor material is the lack of visible light utilization. In order to improve the photocatalytic efficacy, these semiconductor photocatalysts (i.e., $TiO_2$) coupled with noble-metal nanostructures (i.e., Ru, Rh, Pd, Pt, etc.). Indeed, this combination has a great impact on the improvisation of the efficiency of the photocatalyst but the actual high cost of the technology (the noble metals) limits the practical application (Yu et al., 2016; Li et al., 2009). Due to the slow kinetics, the photocatalysis reaction has not been broadly connected to treat industrial wastewater (Linsebigler et al., 1995). Photocatalytic decomposition of a variety of organic species has been broadly examined and established for photocatalysis technology (Ullah et al., 2012).

After so many trials, graphene has been used as an efficient cocatalyst, sensitizer, and Z-scheme mediator in the field of photocatalytic degradation of pollutants. Photocatalytic water treatment causes photodegradation of organic matter such as alkanes, lipids, organic acids, aromatics, and so on. These molecules get fragmented by oxidation reaction with charge carriers to low atomic weight intermediates and in the end into $CO_2$, $H_2O$, and other anions with radiation (Ullah et al., 2012). Yeh et al. (2011) have contemplated GO as a semiconductor demonstrating photocatalytic reactivity. Graphene-based semiconductor photocatalysts have been broadly connected for the photocatalytic debasement of organic contaminations. Various endeavors in the utilization of GO (oxidized form of graphene) or reduced graphene oxide (RGO; reduced form of GO) for the change of $TiO_2$ for photocatalytic debasement of organic toxins (Gao et al., 2009; Fujishima et al., 2008). Although $TiO_2$ is a standout photocatalyst in the degradation of organic pollutants, but the high recombination of charge carrier hinders the utilization practically. Lui et al. (2013) explained the decomposition of methylene blue dye (MB) with synthesized hierarchical $TiO_2$. The enhanced activity of the photocatalyst with graphene modification explained by the excellent transport of the charge carriers due to the great electron conductivity of graphene that suppresses the recombination of the electron–hole pair leading to enhanced degradation of MB. Muthirulan et al. (2014) used a one-step effective route and applied cost-effective strategy for the synthesis of graphene–$TiO_2$ nanocomposite. Furthermore, the group has studied the photo-oxidation of AO7 under UV irradiation following the pseudo-first-order kinetics as professed by the Langmuir–Hinshelwood model. It has likewise been explained that the productivity of graphene-based $TiO_2$ is more than pristine $TiO_2$ which is because of the restraint of the charge carrier recombination. Zhang et al. (2010) demonstrate that high adsorption material will bolster a higher organization of reactants close to the $TiO_2$. Leary and Westwood (2011) observed graphene sheets acting as flawless electron sinks or electron transfer bridges.

As described by Zhang and group (2011), the enhanced photocatalytic action of the graphene–$TiO_2$ is because of its high absorptivity of pollutants, good-charge transportation, and partition which is not found in other

$TiO_2$/carbon composites. In the photodegradation of MB, P25-GR indicated a huge change compared with the bare P25. Besides, $TiO_2$–graphene composites, other composites of semiconductor, and graphene photocatalysts have been reported earlier for the degradation of organic pollutants in water such as $SnO_2$–graphene, ZnO–graphene, $Bi_2WO_6$–graphene, Ag/AgCl/GO, Ag/AgBr/GO, $ZnFe_2O_4$–graphene, $InNbO_4$–graphene, and g-$Bi_2MoO_6$/graphene (Li et al., 2011a, b; Gao et al., 2011; Zhang et al., 2011; Zhu et al., 2011; Fu and Wang, 2011; Zhang et al., 2011; Zhou et al., 2011; Yi et al., 2010; Liang et al., 2012). Another eye-catching visible light semiconductor is $Ag_3PO_4$ which is very much attractive for the decomposition of many organic dye pollutants. $Ag_3PO_4$ has very high quantum efficiency toward $O_2$ evolution and is many times quicker than $BiVO_4$ and $TiO_{2-x}N_x$, toward the decomposition of MB (Liang et al., 2012). $Ag_3PO_4$ is a great efficient catalyst for organic pollutant mineralization under visible light, but photocorrosive nature hinders its stability and recyclability. Graphene played a great role in the enhancement of reactivity and stability of $Ag_3PO_4$ photocatalysts significantly. Yang et al. (2013) fabricated GO–$Ag_3PO_4$ by methods for a two-phase nucleation development process that included GO sheets as a capping agent and substrates. In another work, Dong et al. (2013) fabricated $Ag_3PO_4$/GO by methods of liquid-phase deposition. Yang et al. (2013) created $Ag_3PO_4$/reduced GO nanocomposites by methods for a simple chemical precipitation approach in *N,N*-dimethyl formamide solvent. Liu et al. (2012) fabricated $Ag_3PO_4$/graphene composites by hydrothermal method. Jiang et al. (2013a, b) arranged GO-enwrapped $Ag_3PO_4$ composites through an ion-exchange method for $CH_3COOAg$ and $Na_2HPO_4$ in presence of GO sheets. Our research group has designed the unique Z-scheme RGO–$Ag_3PO_4$ heterostructure by means of a novel photoreduction method for the first time (Samal et al., 2016a, b). The resultant heterostructure shows better photomineralization of textile colors and protection from photocorrosion with respect to its counterpart. These heterostructures not just fundamentally upgrade the activity regarding mineralization of colors and hydrogen production yet in addition radically enhances its stability. Additionally in our laboratory, we have developed RGO/$Ag_3VO_4$ nanocomposites by methods for a photoreduction technique utilizing different sacrificial agents for photohydroxylation of phenol without a hydroxylating species and photodegradation of textile colors under visible light irradiation (Samal et al., 2016a, b).

Likewise, we have also developed graphene-based flowers for photocatalytic decolorization and $H_2$ evolution (Samal et al., 2016a, b). Figure 1.2 describes beautiful morphologies of as-synthesized $Co_3(PO_4)_2$ and 12 RGO–$Co_3(PO_4)_2$ investigated by FESEM. Figure 1.2a shows the archetypical 3D flower-like structures, along with RGO sheets in the case of 12 RGO–$Co_3(PO_4)_2$ (Fig 1.2b and c). The $Co_3(PO_4)_2$ plates with thicknesses of 50–100 nm were associated with each other to resemble 3D flowers (Fig. 1.2a). These flowers have hierarchical structures with high surface-to-volume ratios. The XRD pattern reveals that as-synthesized $Co_3(PO_4)_2$ is well crystallized (Fig. 1.2d). The considerable photoactivity is due to the staggered type II heterojunction system with RGO which prompts the improvement of photoactivity under visible light irradiation.

**FIGURE 1.2** FESEM micrographs of $Co_3(PO_4)_2$ (a) and 12 RGO–$Co_3(PO_4)_2$ (b and c); (d) XRD patterns of the xRGO–$Co_3(PO_4)_2$ 3D flowers.
*Source:* Reprinted with permission from Samal et al., 2016b. Copyright © 2016 Wiley-VCH Verlag GmbH & Co. KGaA, Weinheim.

Reduced GO-based silver nanoparticle-containing composite were synthesized by Jiao et al. (2015a, b) has by in-situ process through the concurrent reduction of GO and noble metal precursors inside the GO gel lattice. The researchers portrayed practical utilizations of the reduced GO-based nanoparticle-containing composite hydrogels for organic dye color evacuation and wastewater treatment. Figure 1.3 demonstrates the morphologies of RGO and Ag nanoparticles containing RGO-based mixture gels. Both SEM and TEM images of RGO hydrogel (Fig. 1.3a, d) demonstrate that the GO sheets were cross-connected in the permeable poly(ethyleneimine) systems. The examination with RGO-based gels and silver nanoparticle-containing RGO-based composite gels uncover that the synthesized silver nanoparticles were homogeneously distributed on the surface of the RGO nanosheets to acquire a gel-based nanocomposite framework (Xiong et al., 2011). Xiong et al. (2011) portrayed that RGO all by oneself can degrade Rhodamine B (RhB) under visible-light irradiation, yet with an excessively moderate outlay. So they accomplished copper-ion-modified RGO integration materials by an immersion method, which showed a speedier photocatalytic response rate and better mineralization under visible-light irradiation than gold-modified RGO. They explained the whole degradation of dye is due to the reactive oxygen species as follows. Firstly, RhB was from the ground up excited to $RhB^*$, trailed by electron exchange to the RGO, at that point the photoelectrons were caught by surface-adsorbed $Cu^{2+}$ particles to shape $Cu^+$ species, which were immediately oxidized to $Cu^{2+}$ by $O_2$. The $RhB^{\cdot+}$ radical was degraded itself or individually formed reactive oxygen species. At that connect after the reversible Cu species go about as an electron communicator, which passes the electrons from the RGO surface to the $O_2$, subsequently prompting the consistent age of reactive oxygen species for RhB degradation. This reactive oxygen species formation makes the present RGO–Cu integration a great photocatalyst for the degradation of dye pollutants under visible-light irradiation.

In recent times, many articles have been published on new graphene-modified composites for the waste-water treatment application. Jiao et al. (2015) have designed RGO/chitosan/silver nanoparticle combinations via self-assembly practice and simultaneous reduction of chitosan molecules by the whole GO. And, the composites performed as a very good photocatalyst for the degradation of RhB and MB dye and even for dye mixture prove the potential of the GO composite materials for wastewater treatment. Shanmugam et al. (2015) synthesized graphene–$V_2O_5$ rods by sequence-mixing method and the fusion indicated incomprehensible photocatalytic reaction for the degradation of MB color by all of the coordinate sunlight illumination contrasted with UV and visible-light sources. Gan et al. (2014) revealed the photothermal impact of graphene-based nanocomposites, which is surely a valuable photocatalytic development to organic pollutants. Chen et al. (2013a, b) additionally unified the GO–chitosan building a whole hydrogel and express it as ecstatic range absorbents for water cleansing. As a modern class of QD materials, graphene quantum dots have right urgently turned facing a dynamic zone of the incredible draw not quite the alike as harmful and flimsy ordinary semiconductor QDs. Additionally, GQDs are consequently benevolent and greatly hearty against ferocious synthetic chemical assaults or solid UV-light illumination. Pan et al. (2015) synthesized graphene quantum dots $TiO_2$ nanoparticles which have extended light absorption from UV to visible range. Here, the arrangement of heterojunctions between $TiO_2$ nanoparticles and GQDs in the composite assumes a key part in the improved noticeable light reactant action. Additionally, Singh et al. (2017) influenced a huge new approach for the natural utilization of

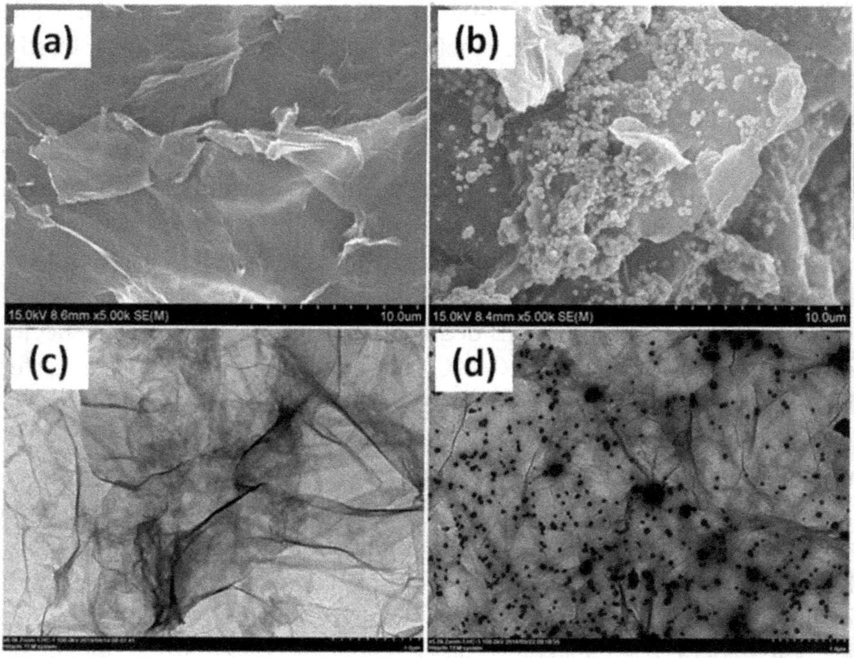


**FIGURE 1.3** SEM and TEM images for the lyophilized RGO/PEI hydrogel (a, c) and RGO/PEI/Ag hydrogel (b, d).

*Source:* Reprinted with permission From Jiao et al., 2015. Copyright © 2015, American Chemical Society pollutant soot to residue by magnificently using a contamination material for degradation of the other pollutant material. The group has done isolation of two-dimensional graphitic materials as water-dissolvable graphene nanosheets (GNS) from dirty–dangerous dark pollutant particulate matter such as dark carbon from petroleum ash. These graphene sheets additionally utilized for the photocatalytic degradation of dangerous dye color, for example, MB affected by visible light illumination. Curiously, this new pollutant catalyst demonstrated extraordinary movement almost 11 times quicker debasement rate inside ~90 min of visible light irradiation than insoluble graphene sheet. The group additionally detailed investigation comparison of the photocatalytic degradation conduct of insoluble GNS and water-soluble GNS on the broadly utilized organic dye color as MB under the visible light illumination utilizing a 60-W tungsten bulb. In Figure 1.4, the component of the dynamic debasement of color was depicted by the surface adsorption/interactions of MB on water-soluble GNS toward the higher side when contrasted with graphene sheets which essentially quickened holding cooperation of MB with surface carboxylic acid groups of the dye. This procedure starts the debasement of the dye in nearness of wsGNS. Alongside this, the high thickness functionalization of wsGNS with negative charges prompts upgraded spatial partition between the highest occupied molecular orbital and lowest unoccupied molecular orbital (HOMO–LUMO) alongside generation of oxygen radicals and thusly enhancing the proficiency of photocatalytic degradation of organic dyes.

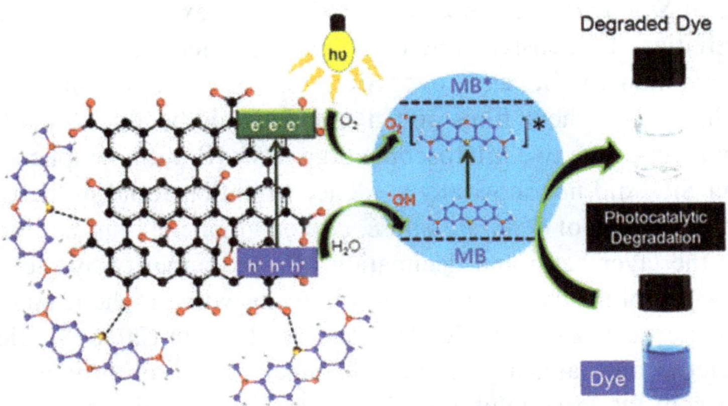


**FIGURE 1.4** Schematic diagram showing MB photodegradation.

*Source:* Reprinted with permission from Singh et al., 2017. Copyright © 2017, American Chemical Society

Apart from organic pollutants, heavy metal ions such as bivalent cadmium [Cd(II)], bivalent lead [Pb(II)], and hexavalent chromium [Cr(VI)] are very harmful toxic inorganic pollutants in water bodies. Zhang et al. (2013) derived RGO incorporated with commercial ZnO particles by a solvothermal reaction to form RGO–ZnO composites for photocatalytic degradation of Cr(VI) in water under UV light irradiation. ZnO is a well-known photocorrosive photocatalyst, but the RGO–ZnO composite exhibits enhanced photoactivity as compared to blank ZnO sample. And also, RGO–ZnO shows great recycling activity test without photocorrosion of ZnO. The enhanced photocatalytic performance of RGO–ZnO composite is credited to the effective hybridization between ZnO and RGO with a π-conjugative 2D framework. The RGO productively isolates the photogenerated charge transporters by decreasing the enactment of surface oxygen atoms in ZnO, accordingly upgrading the photoactivity and photostability of RGO–ZnO composites. Additionally, graphene sheets assume a key part in advancing the photocatalytic purification adequacy of graphene-based composites. For instance, Akhavan (2009) has arranged RGO–$TiO_2$ thin films through deposition of GO sheets on $TiO_2$ thin films took after by annealing and a UV−vis light-assisted reduction process and used them for photo-inactivation of *Escherichia coli* in a watery arrangement under sunlight illumination. Also, Cao and coworkers (2013) have reported the preparation of RGO–$TiO_2$ composites through the redox response among $TiCl_3$ and GO for photocatalytic inactivation of *E. coli*. Recently, Li et al. (2014) have studied the antibacterial activity by the use of *Staphylococcus aureus* and *E. coli* to examine the antibacterial activities of extensive zone monolayer graphene film on conductor Cu, semiconductor Ge, and cover $SiO_2$. The outcomes obtained by him show that the graphene films on Cu and Ge could be able to inhibit the development of the two microscopic organisms. But, the graphene insulator on $SiO_2$ did not show any evidence of antibacterial property. The SEM morphology of *S. aureus* and *E. coli* on graphene films additionally affirms the layer harm and decimation of layer integrity by the immediate contact of the two microscopic organisms with graphene on Cu and Ge in the article (Li et al., 2014). Nanda and group (2016) studied the antibacterial mechanism of GO at the subatomic level through Raman spectroscopy at least inhibitory focus (MIC) by utilizing cultures of *E. coli* and *Enterococcus faecalis.* They described the mechanism that the GO infiltrates cell layer of *E. coli* through its sharp edges. In case of *E. faecalis*, clear membrane damages are observed. In addition, they also observed that when the concentration of GO is increased, debasement of the external and the internal cell films of microscopic organisms increases. A thorough study by Chong et al. (2017) explained the underlying mechanism of enhancement of antibacterial activity of GO by simulated sunlight. They observed in their experiment that the singlet oxygen generated by GO is responsible for slight oxidation of antioxidant biomolecules. Be that as it may, the fundamental driver of oxidation they found was the light-actuated electron–hole pairs evolved on the GO surface.

### *1.4.3 ENERGY APPLICATIONS*

Owing to the exhaustion of the world energy store and the requirement of cleaner environment, the eagerness for elective and renewable fuels has extended rapidly around the world. Hydrogen vitality has been perceived as one of these potential energy vectors. It has been utilized broadly as a fuel. Photocatalytic splitting of water by solar power presumed to be efficient one for hydrogen generation. Different kinds of water-splitting photocatalysts have been investigated so far. For the first time, Fujishima and Honda reported hydrogen evolution from photoelectrochemical water splitting in 1972. $TiO_2$, having extraordinary properties, has been employed as efficient photocatalyst for water splitting under UV light illumination; however, it can use just 4% of the entire solar light and has low quantum efficiency. The wide band gap of $TiO_2$ was incorporated with GO which has been very effective in separating charges on $TiO_2$. Also, there happens to be the formation of Ti–O–C bonding with unpaired π-electrons of GO and surface Ti atoms of $TiO_2$ which extend the light absorption range of $TiO_2$ (Zhang et al., 2010). Also, GO can form p–n heterojunction with *n*-type ZnO, $TiO_2$, and so on for visible light absorption [148]. The GO as electron sink is responsible for the higher photocatalytic activity of GO–$TiO_2$ composites in $H_2$ evolution from water. The wide band gap catalyst being UV light active has the limitation to harness visible spectrum. Narrow band gap photocatalysts like CdS individually cannot be utilized as photocatalyst for water splitting practically as it tends to get decompose to metallic form under photoirradiation, and also bare CdS has a tendency to get aggregate to decrease the surface area having a higher recombination possibility. GO not only hindered the aggregation and recombination owing to its large 2D structure but also promoted the photocatalytic activity of CdS. So GO-based CdS photocatalysts have been reported of having quantum efficiency of 22.5% at 420 nm (Li et al., 2011a, b). Instead of being cocatalyst or sensitizer for other catalysts, GO itself has the calibration with so many flexibility. GO/RGO has been used directly as photocatalyst for water splitting by researchers (Yeh et al., 2011). Jiang et al. (2013a, b) investigated the redox energy levels of GO with respect to water oxidation and reduction potential levels. It has been observed that all five GOs have appropriate band gap and ideal band positions for water splitting under visible-light illumination.

Yeh and coworkers (2013) changed the conductivity type of GO from p-type to n-type by introducing amino and amide groups on the GO surface which is able to catalyze the $H_2$ and $O_2$ evolutions from water simultaneously. In another report, the same group has presented the hydrogen generation by utilizing RGO with a band gap varying from 2.4 to 4.3 eV with no cocatalyst (Yeh et al., 2010). GO or GO-based photocatalysts have the ability to split water and also separate the electron–hole during the photocatalytic reaction. Furthermore, Wang et al. (2017) carried out the DFT calculations of RGO/Pt–$TiO_2$ composite and observed a reduced band gap from 2.88 eV to 2.76 eV. They also determined that the unique energy levels situated at $TiO_2$ band gap were packed with C2p orbitals of graphene, which brought about energized electrons in $TiO_2$ effectively exchanging to graphene to achieve good promotion in photoactivity toward $H_2$ production which is 81 times better than bare $TiO_2$. Recently, our group has developed p–n heterojunction using *p*-type RGO and *n*-type $Cu_3(PO_4)_2$ by obtaining the methods involving solid-state as well as liquid-state reactions (Samal and Das, 2018). This heterojunction catalyzes the water splitting efficiently and produces $H_2$ under visible light irradiation. Here, the RGO/copper phosphate photocatalyst has a unique rod, flower, caramel-treat-like morphology and showed enhanced activity for $H_2$ generation under visible light irradiation. Among them, maximum charge carrier separation was achieved in case of rod-like structure which showed 7500 μmol/h/g amount of $H_2$. This work presents the controlled morphology growth of photocatalyst for hydrogen evolution through water splitting.

Graphene has also been used to convert $CO_2$ into hydrocarbon fuel. Liang et al. (2011) explored the impact of structural defect of graphene on the photocatalytic reduction of $CO_2$. Also, an improvement was observed (~sevenfold) in the photoreduction of $CO_2$ by using $TiO_2$/graphene nanocomposites as compared to pure $TiO_2$. Tu et al. (2012) reported the fabrication of robust hollow spheres by a layer-by-layer deposition method for efficient photocatalytic application toward $CO_2$ conversion. They compared the photocatalytic activity of commercial P25 and $TiO_2$/graphene nanocomposites for photoreduction of $CO_2$, where nine times enhanced photoactivity was achieved in presence of graphene. This is because of the ultrathin nature of nanosheets which allows fast movement of charge carriers onto the surface of photocatalyst thereby achieving spatial charge separation. Tan et al. (2013) have synthesized rGO/$TiO_2$ by simple solvothermal method to get the hybrid nanocrystals and evaluated the photocatalytic activity of the prepared rGO/$TiO_2$ nanocomposites in the reduction of $CO_2$. The observed result confirms the advanced photocatalytic activity of the rGO/$TiO_2$ as compared to pristine graphite oxide and pure anatase $TiO_2$.

Along with other energy applications, RGO, among graphene family, has been most commonly used as material in lithium-ion battery research. In the course of the most recent two decades, lithium-ion batteries have turned out to be the most exceptionally used energy storage systems in cell phones, energy-effective transportation, and energy-storage devices. Cheng (2017) developed "Magic G" having high surface area and electrical conductivity, used for both anode and cathode materials of Li-ion batteries. This Magic G is a honeycomb-like porous graphene sponge that increases the rate capacity and cyclability of lithium-ion batteries. The cell with Magic G-added substance demonstrated an extraordinary improvement of capacity retention in high-rate charging, discharging, and cycling and is a promising added substance material for cutting-edge lithium-particle batteries.

### *1.4.4 CORROSION-RESISTANT COATING*

Graphene is used to create corrosion-resistive coatings which might safeguard the important building as well as machine components from corrosion. Owing to outstanding characteristics of graphene, the quality of coating could be significantly improved. Graphene is quite suitable for top-quality nanocomposite coating material with a high-efficiency filler matrix. Moreover, this has been used as high-quality coating material with the formation of graphene–nanoparticles composites. This means that graphene can become a protector of scratch or other physical damage to a substrate. Graphene has been shown to have an effective barrier to corrosion since it has been labeled indestructible under the environments in which certain substrates' chemical reactions may happen. It has also been observed that the graphene can be used as a corrosion protector of steel by growing graphene flakes directly onto a stainless steel surface. Ludovi's work (2015) uses graphene to repel water from the surface of stainless steel fibers with its natural hydrophobic characteristics. Additionally, the electric characteristics of graphene lower the chance of redox reactions on the surface discouraging from corrosion. Graphene nanofillers seem to be distinctive because they can boost the characteristics even at minimum loads of protective coatings. In glass corrosion protection, the coating material should be strong, transparent, very thin, and chemically inert. Graphene coating on glass materials has been evidently impeccable with all these extraordinary characteristics. The GO-based polyethylene imide coating for corrosion barrier was investigated by the researchers; GO was produced in accordance with the modified Hummers method by oxidization and exfoliation of graphite (Yu et al., 2012). The coating was made on a polyethylene imide film with GO through the layer-by-layer method. Wang et al. (2011) have noted that the capability and cycling stability of rechargeable lithium–sulfur battery cathode material can be modified by using GO-enveloped sulfur particulate material. The result was a high and stable composite having capability with approximately 600 mA h/g over 100 cycles. Liao et al. (2012) using in-situ plate synthesized graphene-reinforced polyurethane acrylate coating. Their findings illustrated that the composite electrical conductivity increased as the graphene amount increased in the synthesized composite material.

## 1.5 SUMMARY AND FUTURE PERSPECTIVES

From this chapter, it could be derived that graphene is the optimum future solution for energy and environmental problems. The high light absorption and antirecombination property of graphene lead the photocatalytic field applications. Graphene could be used in water purification, water desalination, $H_2$ evolution, $CO_2$ reduction, and also in lithium batteries.

Particularly, GO band gap could be varied to achieve an efficient photocatalyst for a practical approach. In the matter of photocatalytic water splitting, the flexible band gap of GO should be used efficiently and prominently to get better solar-to-hydrogen conversion efficiency. Graphene is one of the cheapest and abundant solutions for future perspectives. Generally, at the point when photocatalyst nanoparticles are stacked on graphene sheets just a little division region of photocatalyst get in contact with the graphene sheet, so the efficient electron is not able to get transferred to the reaction-site properly to catalyze the reaction efficiently in graphene-based applications. The photocatalytic activities of the photocatalysts are increased by the light absorption/ band edge alternation ought to be portrayed conspicuously as well. To date, a cost-effective, feasible, and industrial approach has not yet been developed for the production of graphene.

## ACKNOWLEDGMENTS

The authors are thankful to Prof. Suddhasatwa Basu, Director, CSIR-IMMT for giving permission to publish the book chapter. Alaka Samal is thankful to CSIR for the grant of senior research fellowship.

## KEYWORDS

- **graphene**
- **GO membranes**
- **water decontamination**
- **energy applications**

## REFERENCES

Abraham, J.; Vasu, K. S.; Williams, C. D.; Gopinadhan, K.; Su, Y.; Cherian, C. T.; Dix, J.; Prestat, E.; Haigh, S. J.; Grigorieva, I. V.; Carbone, P.; Geim, A. K.; Nair, R. R. Tunable Sieving of Ions Using Graphene Oxide Membranes. *Nat. Nanotechnol.* **2017,** *12*, 546–550.

Akhavan, O. Graphene Nanomesh by ZnO Nanorod Photocatalysts. *ACS Nano* **2010,** *4*, 4174–4180.

Akhavan, O.; Ghaderi, E. Photocatalytic Reduction of Graphene Oxide Nanosheets on TiO2 Thin Film for Photo Inactivation of Bacteria in Solar Light Irradiation, *J. Phys. Chem. C* **2009,** *113*, 20214–20220.

Akhavan, O.; Photocatalytic Reduction of Graphene Oxides Hybridized by ZnO Nanoparticles in Ethanol. *Carbon* **2011,** *49*, 11–18.

Allen, M. J.; Tung, V. C.; Kaner, R. B. Honeycomb Carbon: A Review of Graphene. *Chem. Rev.* **2010,** *110*, 132–145.

An, X.; Li, K.; Tang, J. $Cu_2O$/Reduced Graphene Oxide Composites for the Photocatalytic Conversion of $CO_2$. *Chem Sus Chem* **2014,** *7*, 1086–1093.

Aneja, K. S.; Mallik, H. L.; Böhm, Khanna, A. S.; Böhm, S. Functionalised Graphene as a Barrier Against Corrosion. *Flat Chem.* **2017,** *1*, 11–19.

Bai, S.; Shen, X.; Zhong, X.; Liu, Y.; Zhu, G.; Xu, X.; Chen, K. One-Pot Solvothermal Preparation of Magnetic Reduced Graphene Oxide-Ferrite Hybrids for Organic Dye Removal. *Carbon* **2012,** *50*, 2337–2346.

Balandin, A. A.; Ghosh, S.; Bao, W. Z.; Calizo, I.; Tewelddbrhan, D.; Miao, F.; Lau, C. N. Superior Thermal Conductivity of Single-Layer Graphene. *Nano Lett.* **2008,** *8*, 902–907.

Bhuyan, S. A. Md.; Uddin, N. Md.; Islam M. Md.; Bipasha, F. A.; Hossain, S. S. Synthesis of Graphene. *Int Nano Lett.* **2016,** *6*, 65–83.

Böhm, S. et al. Graphene against corrosion. *Nat. Nanotechnol.* **2014,** *9*, 741–742.

Bolotin, K. I.; Ghahari, F.; Shulman, M. D.; Stormer, H. L.; Kim, P. Observation of the Fractional Quantum Hall Effect in Graphene. *Nature* **2011a,** *462*, 196–199.

Bolotin, K. I.; Sikes, K. J.; Jiang, Z.; Klima, M.; Fudenberg, Hone, J.; Kim, P.; Stormer, H. L. Ultrahigh Electron Mobility in Suspended Graphene. *Sol. State Commun.* **2011b,** *146*, 351–355.

Bremner, D. H.; Molina, R.; Martınez, F.; Melero, J. A.; Segura, Y. Degradation of Phenolic Aqueous Solutions by High Frequency Sono-Fenton Systems (US–$Fe_2O_3$/SBA-15–$H_2O_2$), *Appl. Catal. B* **2009,** *90*, 380–388.

Cao, A. N.; Liu, Z.; Chu, S. S.; Wu, M. H.; Ye, Z. M.; Cai, Z. W.; Chang, Y. L.; Wang, S. F.; Gong, Q. H.; Liu, Y. F. A Facile One-step Method to Produce Graphene–CdS Quantum Dot Nanocomposites as Promising Optoelectronic Materials. *Adv. Mater.* **2010,** *22*, 103–106.

Cao, B.; Cao, S.; Dong, P.; Gao, J.; Wang, J. High Antibacterial Activity of Ultrafine $TiO_2$/Graphene Sheets Nanocomposites Under Visible Light Irradiation, *Mater. Lett.* **2013,** *93*, 349–352.

Chen, G.; Sun, M.; Wei, Q.; Zhang, Y.; Zhu, B.; Du, B. Ag3PO4/Graphene-Oxide Composite with Remarkably Enhanced Visible-Light-Driven Photocatalytic Activity Toward Dyes in Water. *J. Hazard. Mater.* **2013a,** *244–245*, 86–93.

Chen, Y.; Chen, L.; Bai, H.; Li, L. Graphene Oxide–Chitosan Composite Hydrogels as Broad-Spectrum Adsorbents for Water Purification. *J. Mater. Chem. A* **2013b,** *1*, 1992–2001.

Cheng, Q. Porous Graphene Sponge Additives for Lithium Ion Batteries with Excellent Rate Capability, *Scientific Reports* **2017,** *7*, 925.

Chong, M. N.; Chow, C. W.; Saint, C. Recent Developments in Photocatalytic Water Treatment Technology: A Review. *Water Res.* **2010,** *44*, 2997–3027.

Chong, Y.; Ge, C.; Fang, G.; Wu, R.; Zhang, H.; Chai, Z.; Chen, C.; Yin, J.-J. Light-Enhanced Antibacterial Activity of Graphene Oxide, Mainly via Accelerated Electron Transfer. *Environ. Sci. Technol.* **2017,** *51*, 10154–10161.

Cohen-Tanugi, D.; Grossman, J. C. Nanoporous Graphene as a Reverse Osmosis Membrane Recent Insights from Theory and Simulation. *Desalination* **2015,** *366*, 59–70.

Das, D. P.; Samal, A.; Das, J.; Dash, A.; Gupta, H. One-Pot Fabrication of RGO–$Ag_3VO_4$ Nanocomposites by In Situ Photo Reduction Using Different Sacrificial Agents: High Selectivity Toward Catechol Synthesis. *Photochem. Photobiol.* **2014,** *90*, 57–65.

Dong, P. Y.; Wang, Y. H.; Cao, B. C.; Xin, S. Y.; Guo, L. N.; Zhang, J.; Li, F. H. Ag3PO4/Reduced Graphite Oxide Sheets Nanocomposites with Highly Enhanced Visible Light Photocatalytic Activity and Stability. *Appl. Catal. B* **2013,** *132–133*, 45–53.

Dreyer, D. R.; Park, S.; Bielawski, C. W.; Ruoff, R. S. The Chemistry of Graphene Oxide. *Chem. Soc. Rev.* **2010,** *39*, 228–240.

Du, X.; Skachko, I.; Barker, A.; Andrei, E. Y. Approaching Ballistic Transport in Suspended Graphene. *Nat. Nanotech.* **2008,** *3*, 491–495.

Du, X.; Skachko, I.; Duerr, F.; Luican, A.; Andrei, E. Y. Fractional Quantum Hall Effect and Insulating Phase of Dirac Electrons in Graphene. *Nature* **2009,** *462*, 192–195.

Eder, D.; Motta, M.; Windle, A. H. Iron-doped Pt–$TiO_2$ Nanotubes for Photo-Catalytic Water Splitting. *Nanotechnology* **2009,** *20*, 055602.

Fu, Y. S.; Wang, X. Magnetically Separable ZnFe2O4–Graphene Catalyst and Its High Photocatalytic Performance Under Visible Light Irradiation. *Ind. Eng. Chem. Res.* **2011,** *50*, 7210–7218.

Fujishima, A.; Honda, K. Electrochemical Photolysis of Water at a Semiconductor Electrode. *Nature* **1972,** *238*, 37–38.

Fujishima, A.; Zhang, X.; Tryk, D. A. $TiO_2$ Photocatalysis and Related Surface Phenomena. *Surf. Sci. Rep.* **2008,** *63*, 515–582.

Gan, Z.; Wu, X.; Meng, M.; Zhu, X.; Yang, L.; Chu, P. K. Photothermal Contribution to Enhanced Photocatalytic Performance of Graphene-Based Nanocomposites. *ACS Nano* **2014,** *8*, 9304–9310.

Gao, B.; Chen, G. Z.; Puma, G. L. Carbon Nanotubes/Titanium Dioxide (CNTs/$TiO_2$) Nanocomposites Prepared by Conventional and Novel Surfactant Wrapping Sol–Gel Methods Exhibiting Enhanced Photocatalytic Activity. *Appl. Catal. B: Environ.* **2009,** *89*, 503–509.

Gao, E. P.; Wang,W. Z.; Shang, M.; Xu, J. H. Synthesis and Enhanced Photocatalytic Performance of Graphene-$Bi_2WO_6$ Composite. *Phys. Chem. Chem. Phys.* **2011,** *13*, 2887–2893.

Goh, P.; Ismail, A. Graphene-Based Nanomaterial: The State-of-the-Art Material for Cutting Edge Desalination Technology. *Desalination* **2015,** *356*, 115–128.

Guinea, F.; Katsnelson, M. I.; Geim, A. K. Energy Gaps, Topological Insulator State and Zero-Field Quantum Hall Effect in Graphene by Strain Engineering. *Nat. Phys.* **2010,** *6*, 30–33.

Guo, J. J.; Li, Y.; Zhu, S. M.; Chen, Z. X.; Liu, Q. L.; Zhang, D.; Moonc, W-J.; Song, D-M. Synthesis of $WO_3$@Graphene Composite for Enhanced Photocatalytic Oxygen Evolution from Water. *RSC Adv.* **2012,** *2*, 1356–1363.

Gurzęda, B.; Florczak, P.; Kempinski, M.; Peplinska, B.; Krawczyk. P.; Jurga, S. Synthesis of Graphite Oxide by Electrochemical Oxidation in Aqueous Perchloric Acid. *Carbon* **2016,** *100*, 540–545.

Han, C.; Yang, M. Q.; Weng, B.; Xu, Y. J. Improving the Photocatalytic Activity and Anti-Photocorrosion of Semiconductor ZnO by Coupling with Versatile Carbon. *Phys. Chem. Chem. Phys.* **2014,** *16*, 16891–16903.

Han, L.; Wang, P.; Dong, S. Progress in Graphene-Based Photoactive Nanocomposites as a Promising Class of Photocatalyst. *Nanoscale* **2012,** *4*, 5814–5825.

Han, Y.; Xu, Z.; Gao, C. Ultrathin Graphene Nanofiltration Membrane for Water Purification. *Adv. Funct. Mater.* **2013,** *23*, 3693–3700.

Hatakeyama, K.; Tateishi, H.; Taniguchi, T.; Koinuma, M.; Kida, T.; Hayami, S.; Yokoi, H. Matsumoto, Y. Tunable Graphene Oxide Proton/Electron Mixed Conductor that Functions at Room Temperature. *Chem. Mater.* **2014,** *26*, 5598–5604.

Hega, H. M.; Zou, L. Graphene Oxide-Assisted Membranes: Fabrication and Potential Applications in Desalination and Water Purification, *J. Membr. Sci.* **2015,** *484*, 95–106.

Homaeigohar, S.; Elbahri, M. Graphene Membranes for Water Desalination. *NPG Asia Mat.* **2017,** *9*, e427.

Hu, H.; Wang, X.; Liu, F.; Wang, J.; Xu, C.; Rapid Microwave-Assisted Synthesis of Graphene Nanosheets–Zinc Nanocomposites: Optical and Photocatalytic Properties. *Synth. Met.* **2011,** *161*, 404–410.

Hu, M.; Mi, B. Enabling Graphene Oxide Nanosheets as Water Separation Membranes, *Environ. Sci. Technol.* **2013,** *47*, 3715–3723.

Huang, X.; Yin, Z.; Wu, S.; Qi, X.; He, Q.; Zhang, Q.; Yan, Q.; Boey, F.; Zhang, H. Graphene Based Materials: Synthesis, Characterization, Properties, and Applications. *Small* **2011,** *7*, 1876–1902.

Huanga, A.; Feng, B. Synthesis of Novel Graphene Oxide-Polyimide Hollow Fiber Membranes for Seawater Desalination. *J. Membrane Sci.* **2018,** *548*, 59–65.

Jia, L.; Wang, D. H.; Huang, Y. X.; Xu, A. W.; Yu, H. Q. Highly Durable N-Doped Graphene/CdS Nanocomposites with Enhanced Photocatalytic Hydrogen Evolution from Water under Visible Light Irradiation. *J. Phys. Chem. C* **2011,** *115*, 11466–11473.

Jiang, B. J.; Wang, Y. H.; Wang, J. Q.; Tiang, C. G.; Li, W. J.; Feng, Q. M.; Pan, Q. J.; Fu, H. G. In Situ Fabrication of Ag/$Ag_3PO_4$/Graphene Triple Heterostructure Visible-Light Photocatalyst Through Graphene-Assisted Reduction Strategy, *Chem. Cat. Chem.* **2013a,** *5*, 1359–1367.

Jiang, X.; Nisar, J.; Pathak, B.; Zhao, J.; Ahuja, R. Graphene Oxide as a Chemically Tunable 2-D Material for Visible-Light Photocatalyst Applications. *J. Catal.* **2013b,** *299*, 204–209.

Jiao, L. Y.; Wang, X. R.; Diankov, G.; Wang, H. L.; Dai, H. J. Facile Synthesis of High-Quality Graphene Nanoribbons. *Nat. Nanotechnol.* **2010,** *5*, 321–325.

Jiao, L. Y.; Zhang, L.; Wang, X. R.; Diankov, G.; Dai, H. J. Narrow Graphene Nanoribbons from Carbon Nanotubes. *Nature* **2009,** *458*, 877–880.

Jiao, T.; Guo, H.; Zhang, Q.; Peng, Q.; Tang, Y.; Yan, X.; Li, B. Reduced Graphene Oxide-Based Silver Nanoparticle-Containing Composite Hydrogel as Highly Efficient Dye Catalysts for Wastewater Treatment. *Sci. Rep.* **2015a,** *5*, 11873.

Jiao, T.; Zhao, H.; Zhou, J.; Zhang, Q.; Luo, X.; Hu, J.; Peng, Q.; Yan, X. Self-Assembly Reduced Graphene Oxide Nanosheet Hydrogel Fabrication by Anchorage of Chitosan/

Silver and Its Potential Efficient Application toward Dye Degradation for Wastewater Treatments. *ACS Sust. Chem. Eng.* **2015b,** *3*, 3130–3139.

Joshi, R. K.; Carbone, P.; Wang, F. C.; Kravets, V. G.; Su, Y.; Grigorieva, I. V.; Wu, H. A.; Geim, A. K.; Nair, R. R. Precise and Ultrafast Molecular Sieving Through Graphene Oxide Membranes. *Science* **2014,** *343*, 752–754.

Kalita, G.; Masahiro, M.; Uchida, H. Few Layers of Graphene as Transparent Electrode from Botanical Derivative Camphor. *Mater. Lett.* **2010,** *64*, 2180–2183.

Leary, R.; Westwood, A. Carbonaceous Nanomaterials for the Enhancement of $TiO_2$ Photocatalysis. *Carbon* **2011,** *49*, 741–772.

Lee, J. M.; Pyun,Y. B.; Yi, J.; Choung, J. W.; Park, W. II. ZnO Nanorod–Graphene Hybrid Architectures for Multifunctional Conductors. *J. Phys. Chem. C* **2009,** *113*, 19134–19138.

Li, B. J.; Cao, H. Q. ZnO@Graphene Composite with Enhanced Performance for the Removal of Dye from Water. *J. Mater. Chem.* **2011,** *21*, 3346–3349.

Li, J.; Wang, G.; Zhu, H.; Zhang, M.; Zheng, X.; Di, Z.; Liu, X.; Wang, X. Antibacterial Activity of Large-Area Monolayer Graphene Film Manipulated by Charge Transfer. *Sci. Rep.* **2014,** *4*, 4359.

Li, Q.; Guo, B.; Yu, J.; Ran, J.; Zhang, B.; Yan, H.; Gong, J. Highly Efficient Visible-Light-Driven Photocatalytic Hydrogen Production of CdS-Cluster-Decorated Graphene Nanosheets. *J. Am. Chem. Soc.* **2011a,** *133*, 10878–10884.

Li, Y. G.; Wang, H. L.; Xie, L. M.; Liang, Y. Y.; Hong, G. S.; Dai, H. J. MoS2 Nanoparticles Grown on Graphene: An Advanced Catalyst for the Hydrogen Evolution Reaction. *J. Am. Chem. Soc.* **2011b,** *133*, 7296–7299.

Li, Y.; Hu, Y.; Peng, S.; Lu, G.; Li, S. Synthesis of CdS Nanorods by an Ethylenediamine Assisted Hydrothermal Method for Photocatalytic Hydrogen Evolution, *J. Phys. Chem. C* **2009,** *113*, 9352–9358.

Liang, Q.; Shi, Y.; Ma, W.; Li, Z.; Yang, X. Enhanced Photocatalytic Activity and Structural Stability by Hybridizing Ag3PO4 Nanospheres with Graphene Oxide Sheets. *Phys. Chem. Chem. Phys.* **2012,** *14*, 15657–15665.

Liang, Y. T.; Vijayan, B. K.; Gray, K. A.; Hersam, M. C. Minimizing Graphene Defects Enhances Titania Nanocomposite-Based Photocatalytic Reduction of $CO_2$ for Improved Solar Fuel Production. *Nano Lett.* **2011,** *11*, 2865–2870.

Liao, K.-H.; Qian, Y.; Macosko, C. W. Ultralow Percolation Graphene/Polyurethane Acrylate Nanocomposites. *Polymer* **2012,** *53*, 3756–3761.

Lightcap, I. V.; Kosel, T. H.; Kamat, P. V. Anchoring Semiconductor and Metal Nanoparticles on a Two-Dimensional Catalyst Mat. Storing and Shuttling Electrons with Reduced Graphene Oxide. *Nano Lett.* **2010,** *10*, 577–583.

Linsebigler, A. L.; Lu, G.; Yates, J. T. Photocatalysis on $TiO_2$ Surfaces: Principles, Mechanisms, and Selected Results. *Chem. Rev.* **1995,** *95*, 735–758.

Liu, J.; Yang, H.; Zhen, S. G.; Poh, C. K.; Chaurasia, A.; Luo, J.; Wu, X.; Yeow, E. K. L.; Sahoo, N. G.; Lin, J.; Shen, Z. A Green Approach to the Synthesis of High-Quality Graphene Oxide Flakes via Electrochemical Exfoliation of Pencil Core. *RSC Adv.* **2013,** *3*, 11745–11750.

Liu, L.; Liu, J. C.; Sun, D. D. Graphene Oxide Enwrapped Ag3PO4 Composite: Towards a Highly Efficient and Stable Visible-Light-Induced Photocatalyst for Water Purification. *Catal. Sci. Technol.* **2012,** *2*, 2525–2532.

Low, J. X.; Yu, J.; Ho, W. Graphene-Based Photocatalysts for $CO_2$ Reduction to Solar Fuel. *J. Phys. Chem. Lett.* **2015,** *6*, 4244–4251.

Low, T.; Perebeinos, V.; Tersoff, J.; Avouris, P. H. Deformation and Scattering in Graphene Over Substrate Steps. *Phys. Rev. Lett.* **2012,** *108*, 096601.

Lu, X. K.; Yu, M. F.; Huang, H.; Ruoff, R. S. Tailoring Graphite with the Goal of Achieving Single Sheets. *Nanotechnology* **1999,** *10*, 269–272.

Ludovi, F.; Dumée, L.; Ziyu, H.; Wang, P.; Xiong, S. J.; Feng, C. F.; Tan, M. Y.; Mikel, F. S.; Gray, D. S.; Pacheco, A.; Hodgson, P.; Majumder, M.; Kong, L. Growth of Nano-Textured Graphene Coatings Across Highly Porous Stainless Steel Supports Towards Corrosion Resistant Coatings. *Carbon* **2015,** *87*, 395–408.

Lui, G.; Liao, J. Y.; Duan, A. Graphene-Wrapped Hierarchical $TiO_2$ Nanoflower Composites with Enhanced Photocatalytic Performance. *J. Mater. Chem. A.* **2013,** *1*, 12255–12262.

Mahmoud, K. A.; Mansoor, B.; Mansour, A.; Khraisheh, M. Functional Graphene Nanosheets: The Next Generation Membranes for Water Desalination. *Desalination* **2015,** *356*, 208–225.

Meidanchi, A.; Akhavan, O. Superparamagnetic Zinc Ferrite Spinel–Graphene Nanostructures for Fast Wastewater Purification. *Carbon* **2014,** *69*, 230–238.

Min,Y. L.; Zhang, K.; Chen, L. H.; Chen,Y. C.; Zhang, Y. G. Ionic Liquid Assisting Synthesis of ZnO/Graphene Heterostructure Photocatalysts with Tunable Photoresponse Properties. *Diam. Relat. Mater.* **2012,** *26*, 32–38.

Moon, G. -H.; Kim, D. -H.; Kim, H. -I.; Bokare, A. D.; Choi, W. Platinum-like Behavior of Reduced Graphene Oxide as a Co-catalyst on $TiO_2$ for the Efficient Photocatalytic Oxidation of Arsenite. *Environ. Sci. Technol. Lett.* **2014,** *1*, 185–190.

Morishige, K.; Hamada, T. Iron Oxide Pillared Graphite. *Langmuir* **2005,** *21*, 6277–6281.

Muthirulan, P.; Devi, C. K. N.; Sundaram, M. M. Fabrication and Characterization of Efficient Hybrid Photocatalysts Based on Titania and Graphene for Acid Orange Seven Dye Degradation under UV Irradiation. *Adv. Mat. Lett.* **2014,** *5*, 163–171.

Nanda, S. S.; Yi, D. K.; Kim, K. Study of Antibacterial Mechanism of Graphene Oxide Using Raman Spectroscopy. *Sci. Rep.* **2016,** *6*, 28443.

Nethravathi, C.; Nisha, T.; Ravishankar, N.; Shivakumara, C.; Rajamathi, M. Graphene–Nanocrystalline Metal Sulphide Composites Produced by a One-Pot Reaction Starting from Graphite Oxide. *Carbon* **2009,** *47*, 2054–2059.

Ng, Y. H.; Iwase, A.; Kudo, A.; Amal, R. Reducing Graphene Oxide on a Visible-Light $BiVO_4$ Photocatalyst for an Enhanced Photoelectrochemical Water Splitting, *J. Phys. Chem. Lett.* **2010a,** *1*, 2607–2612.

Ng, Y. H.; Lightcap, I. V.; Goodwin, K.; Matsumura, M.; Kamat, P. V. To What Extent Do Graphene Scaffolds Improve the Photovoltaic and Photocatalytic Response of $TiO_2$ Nanostructured Films. *J. Phys. Chem. Lett.* **2010b,** *1*, 2222–2227.

Nicolai, A.; Sumpter, B. G.; Meunier, V. Tunable Water Desalination Across Graphene Oxide Framework Membranes. *Phys. Chem. Chem. Phys.* **2014,** *16*, 8646–8654.

Novoselov, K. S.; Geim, A. K.; Morozov, S. V.; Jiang, D.; Zhang, Y.; Dubonos, S. V.; Grigorieva, I. V.; Firsov, A. A. Electric Field Effect in Atomically Thin Carbon Films. *Science* **2004** *306*, 666–669.

Pan, D.; Jiao, J.; Li, Z.; Guo, Y.; Feng, C.; Liu, Y.; Wang, L.; Wu, M. Efficient Separation of Electron–Hole Pairs in Graphene Quantum Dots by $TiO_2$ Heterojunctions for Dye Degradation, *ACS Sust. Chem. Eng.* **2015,** *3*, 2405–2413.

Park, S.; Ruoff, R. S.; Chemical Methods for the Production of Graphenes. *Nat. Nanotechnol.* **2009,** *4*, 217–224.

Parvez, K.; Rincon, R. A.; Weber, N.-E.; Cha, K. C.; Venkataraman, S. S. Onestep Electrochemical Synthesis of Nitrogen and Co-Doped, High-Quality Graphene Oxide. *Chem. Commun.* **2016,** *52*, 5714–5717.

Paton, K. R.; Varrla, E.; Backes, C.; Smith, R. J; Khan, U.; O'Neill, A.; Boland, C.; Lotya, M. Istrate; King, O. M.; Higgins, P. T.; Barwich, Scalable Production of Large Quantities of Defect-Free Few-Layer Graphene by Shear Exfoliation in Liquids. *Nat. Mater.* **2014,** *13*, 624–630.

Peigney, A.; Laurent, C.; Flahaut, E.; Bacsa, R. R.; Rousset, A. Specific Surface Area of Carbon Nanotubes and Bundles of Carbon Nanotubes. *Carbon* **2001,** *39*, 507–514.

Qu, J.; Shi, L.; He, C.; Gao, F.; Li, B.; Zhou, Q.; Hu, H.; Shao, G.; Wang, X.; Qiu, J. Highly Efficient Synthesis of Graphene/$MnO_2$ Hybrids and Their Application for Ultrafast Oxidative Decomposition of Methylene Blue. *Carbon* **2014,** *66*, 485–492.

Reina, A. J.; Ho, X. T.; Nezich, J. D.; Son, H.; Bulovic, V.; Dresselhaus, M. S.; Kong, J. Large Area, Few-Layer Graphene Films on Arbitrary Substrates by Chemical Deposition. *Nano Lett.* **2009,** *9*, 30–35.

Samal, A.; Das, D. P.; Madras, G. Repercussion of Solid State vs. Liquid State Synthesized P-N Heterojunction RGO-Copper Phosphate on Proton Reduction Potential in Water. *Sci. Rep.* **2018,** *8*, 2881.

Samal, A.; Das, D. P.; Nanda, K. K.; Mishra, B. K.; Das, J.; Das, A. Reduced Graphene Oxide–Ag3PO4 Heterostructure: A Direct Z—Scheme Photocatalyst for Augmented Photo Reactivity and Stability. *Chem. Asian J.* **2016a,** *11*, 584–595.

Samal, A.; Swain, S.; Satpati, B.; Das, D. P.; Mishra, B. K. 3D $Co_3(PO_4)_2$–Reduced Graphene Oxide Flowers for Photocatalytic Water Splitting: A Type II Staggered Heterojunction System. *Chem. Sus. Chem* **2016b,** *9*, 3150–3160.

Samal, A; Das, D. P. Transfiguring UV Light Active "Metal Oxides" to Visible Light Active Adation by Reduced Graphene Oxide Hypostatization. *Catalys. Today* **2018,** *300*, 124–135.

Shanmugam, M.; Alsalme, A.; Alghamdi, A.; Jayavel, R. Enhanced Photocatalytic Performance of the Graphene-V2O5 Nanocomposite in the Degradation of Methylene Blue Dye Under Direct Sunlight, *ACS Appl. Mater. Inter.* **2015,** *7*, 14905–14911.

Shi, Y.; Zhou, K.; Wang, B.; Jiang, S.; Qian, X.; Gui, Z.; Yuen, R. K. K.; Hu, Y. Ternary Graphene–$CoFe_2O_4$/CdS Nanohybrids: Preparation and Application as Recyclable Photocatalysts. *J. Mater. Chem. A* **2014,** *2*, 535–544.

Singh, A.; Khare, P.; Verma, S.; Bhati, A.; Sonker, A. K.; Tripathi, K. M.; Sonkar, S. K. Pollutant Soot for Pollutant Dye Degradation: Soluble Graphene Nanosheets for Visible Light Induced Photo degradation of Methylene Blue, *ACS Sust. Chem. Eng.* **2017,** *5*, 8860–8869.

Singh, B. P.; Jena, B. K.; Bhattacharjee, S.; Besra, L. Development of Oxidation and Corrosion Resistance Hydrophobic Graphene Oxide-Polymer Composite Coating on Copper. *Surf. Coat. Technol.* **2013,** *232*, 475–481.

Som, T.; Troppenz, G. V.; Wendt, R.; Wollgarten, M.; Rappich, J.; Emmerling, F.; Rademann, K. Graphene Oxide/a-$Bi_2O_3$ Composites for Visible-Light Photocatalysis, Chemical Catalysis, and Solar Energy Conversion. *Chem Sus Chem* **2014,** *7*, 854–865.

Songfeng, P.; Qinwei, W.; Huang, K.; Cheng, H-M.; Ren, W. Green Synthesis of Graphene Oxide by Seconds Timescale Water Electrolytic Oxidation. *Nat. Commun.* **2018,** 9:145.

Sun, P.; Zheng, F.; Zhu, M.; Song, Z.; Wang, K.; Zhong, M.; Wu, D.; Little, R. B.; Xu, Z; Zhu, H. Selective Trans-Membrane Transport of Alkali and Alkaline Earth Cations Through Graphene Oxide Membranes Based on Cation–π Interactions, *ACS Nano* **2014,** *8*, 850–859.

Sun, P.; Zhu, M.; Wang, K.; Zhong, M.; Wei, J.; Wu, D.; Xu, Z.; Zhu, H. Selective Ion Penetration of Graphene Oxide Membranes, *ACS Nano* **2013,** *7*, 428–437.

Sun, Y.; Wu, Q.; Shi, G. Graphene Based New Energy Materials. *Energy Environ. Sci.* **2011,** *4*, 1113–1132.

Surwade, S. P.; Smirnov, S. N.; Vlassiouk ,I. V.; Unocic, R. R.; Veith,G. M.; Dai, S.; Mahurin, S. M. Water desalination using nanoporous single-layer graphene. *Nat. Nanotechnol.* **2015,** *10*, 459–464.

Tan, L. –L.; Ong, W. –J.; Chai, S. –P.; Mohamed, A. Reduced Graphene Oxide-$TiO_2$ Nanocomposite as a Promising Visible-Light-Active Photocatalyst for the Conversion of Carbon Dioxide. *Nanoscale Res. Lett.* **2013,** *8*, 465.

Tu, W.; Zhou, Y.; Liu, Q.; Tian, Z; Gao, J.; Chen, X.; Zhang, H.; Liu, J.; Zou, Z. Robust Hollow Spheres Consisting of Alternating Titania Nanosheets and Graphene Nanosheets with High Photocatalytic Activity for $CO_2$ Conversion into Renewable Fuels. *Adv. Funct. Mater.* **2012,** *22*, 1215–1221.

Tu, W.; Zhou, Y.; Zou, Z. Versatile Graphene-Promoting Photocatalytic Performance of Semiconductors: Basic Principles, Synthesis, Solar Energy Conversion, and Environmental Applications. *Adv. Funct. Mat.* **2013,** *23*, 4996–5008.

Ullah, R.; Sun, H.; Wang, S. Wet-Chemical Synthesis of $InTaO_4$ for Photocatalytic Decomposition of Organic Contaminants in Air and Water with UV–vis Light. *Ind. Eng. Chem Res.* **2012,** *51*, 1563–1569.

Wang, D. H.; Kou, R.; Choi, D.; Yang, Z. G.; Nie, Z. M.; Li, J.; Saraf, L. V.; Hu, D. H.; Zhang, J. G.; Graff, G. L.; Liu, J.; Pope, M. A.; Aksay, I. A. Ternary Self-Assembly of Ordered Metal Oxide−Graphene Nanocomposites for Electrochemical Energy Storage. *ACS Nano* **2010,** *4*, 1587–1595.

Wang, E. N.; Karnik, R. Water Desalination: Graphene Cleans Up Water. *Nat. Nanotechnol.* **2012,** *7*, 552–554.

Wang, H.; Yang, Y.; Liang, Y.; Robinson, J. T.; Li, Y.; Jackson, A.; Cui, Y.; Dai, H. Graphene-Wrapped Particles as a Rechargeable Lithium-Sulfur Battery Cathode Material with High Capacity and Cycling Stability. *Nano Lett.* **2011,** *11*, 2644–2647.

Wang, P.; Zhan, S.; Xia, Y.; Ma, S.; Zhou, Q.; Li, Y. The fundamental Role and Mechanism of Reduced Graphene Oxide in rGO/Pt-$TiO_2$ Nanocomposite for High-Performance Photocatalytic Water Splitting. *Appl. Cat. B: Environ.* **2017,** *207*, 335–346.

Wang, Y. Q.; Zhang, Z. J.; Zhu, Y.; Li, Z. C.; Vajtai, R.; Ci, L. J.; Ajayan, P. M. Nanostructured $VO_2$ Photocatalysts for Hydrogen Production. *ACS Nano* **2008,** *2*, 1492–1496.

Wang, Y.; Huang, Y.; Song, Y.; Zhang, X. Y.; Ma, Y. F.; Liang, J. J.; Chen, Y. S. Room-Temperature Ferromagnetism of Graphene. *Nano Lett.* **2009,** *9*, 220–224.

Wang, Y.; Yu, J.; Xiao, W.; Li, Q. Microwave-Assisted Hydrothermal Synthesis of Graphene Based Au–$TiO_2$ Photocatalysts for Efficient Visible-Light Hydrogen Production. *J. Mat. Chem. A* **2014,** *2*, 3847–3855.

Wang, B.; Wu, J.; Zhang, N.; Xu, Y. –J. Observing the Role of Graphene in Boosting the Two-Electron Reduction of Oxygen in Graphene–$WO_3$ Nanorod Photocatalysts. *Langmuir* **2014,** *30*, 5574–5584.

Weng, Wang, B.; Yang, M. Q.; Zhang, N.; Xu, Y. J. Toward the Enhanced Photoactivity and Photostability of ZnO Nanospheres via Intimate Surface Coating with Reduced Graphene Oxide. *J. Mater. Chem. A* **2014,** *2*, 9380–9389.

Williams, G.; Kamat, P. V. Graphene–Semiconductor Nanocomposites: Excited-State Interactions Between ZnO Nanoparticles and Graphene Oxide. *Langmuir* **2009,** *25*, 13869–13873.

Wu, J. L.; Bai, S.; Shen, X. P.; Jiang, L. Preparation and Characterization of Graphene/CdS Nanocomposites. *Appl. Surf. Sci.* **2010,** *257*, 747–751.

Xiang, Q. J.; Yu, J. G.; Jaroniec, M. Synergetic Effect of $MoS_2$ and Graphene as Cocatalysts for Enhanced Photocatalytic $H_2$ Production Activity of $TiO_2$ Nanoparticles. *J. Am. Chem. Soc.* **2012,** *134*, 6575–6578.

Xiang, Q.; Yu, J.; Jaroniec, M. Graphene-Based Semiconductor Photocatalysts. *Chem. Soc. Rev.* **2012,** *41*, 782–796.

Xin, G. Q.; Hwang, W.; Kim, N.; Cho, S. M.; Chae, H. A graphene Sheet Exfoliated with Microwave Irradiation and Interlinked by Carbon Nanotubes for High-Performance Transparent Flexible Electrodes. *Nanotechnology* **2010,** *21*, 40.

Xiong, Z.; Zhang, L.; Zhao, X. S. Visible Light Induced Dye Degradation Over Copper Modified Reduced Graphene Oxide. *Chem. Eur. J.* **2011,** *17*, 2428 – 2434.

Xu, C.; Wang, X.; Zhu, J. W. Graphene–Metal Particle Nanocomposites. *J. Phys. Chem. C* **2008,** *112*, 19841–19845.

Xu, K.; Feng, B.; Zhou, C.; Huang, A. Synthesis of Highly Stable Graphene Oxide Membranes on Polydopamine Functionalized Supports for Seawater Desalination. *Chem. Eng. Sci.* **2016,** *146*, 159–165.

Xu, T. G.; Zhang, L. W.; Cheng, H. Y.; Zhu, Y. F. Significantly Enhanced Photocatalytic Performance of ZnO via Graphene Hybridization and the Mechanism Study. *Appl. Catal. B* **2011,** *101*, 382–387.

Yang, A.; Xue, Y.; Zhang, Y.; Zhang, X.; Zhao, H.; Li, X.; He, Y.; Yuan, Z. A Simple One-Pot Synthesis of Graphene Nanosheet/$SnO_2$ Nanoparticle Hybrid Nanocomposites and Their Application for Selective and Sensitive Electrochemical Detection of Dopamine. *J. Mater. Chem. B* **2013,** *1*, 1804–1811.

Yang, H. H.; Guo, L. J.; Yan, W.; Liu, H. T. A Novel Composite Photocatalyst for Water Splitting Hydrogen Production. *J. Power Sources* **2006,** *159*, 1305–1309.

Yang, M.-Q.; Xu, Y.-J. Selective Photoredox Using Graphene-Based Composite Photocatalysts. *Phys. Chem. Chem. Phys.* **2013,** *15*, 19102–19118.

Yang, M.-Q.; Zhang, N.; Pagliaro, M.; Xu, Y.-J. Artificial Photosynthesis Over Graphene–Semiconductor Composites: Are We Getting Better? *Chem. Soc. Rev.* **2014,** *43*, 8240–8254.

Yang, S.; Bruller, S.; Wu, Z. S.; Liu, Z.; Parvez, K.; Dong, R.; Richard, F.; Samori, P.; Feng, X.; Mullen, K. Organic Radical-Assisted Electrochemical Exfoliation for the Scalable Production of High-Quality Graphene. *J. Am. Chem. Soc.* **2015,** *137*, 13927–13932.

Yang, S.; Lohe, M. R.; Muellen, K.; Feng, X. New-Generation Graphene from Electrochemical Approaches: Production and Applications. *Adv. Mater.* **2016,** *28*, 6213–6221.

Yang, X. F.; Cui, H. Y.; Li, Y.; Qin, J. L.; Zhang, R. X.; Tang, H. Fabrication of Ag3PO4-Graphene Composites with Highly Efficient and Stable Visible Light Photocatalytic Performance, *ACS Catal.* **2013,** *3*, 363–369.

Yao, Y.; Cai, Y.; Lu, F.; Wei, F.; Wang, X.; Wang, S. Magnetic Recoverable $MnFe_2O_4$ and $MnFe_2O_4$-Graphene Hybrid as Heterogeneous Catalysts of Peroxymonosulfate Activation for Efficient Degradation of Aqueous Organic Pollutants. *J. Hazard. Mater.* **2014,** *270*, 61–70.

Yao, Z. P.; Jia, F. Z.; Tian, S. J.; Li, C. X.; Jiang, Z. H.; Bai, X. F. Microporous Ni-Doped $TiO_2$ film Photocatalyst by Plasma Electrolytic Oxidation. *Appl. Mater. Interf.* **2010,** *2*, 2617–2622.

Ye, A. H.; Fan, W. Q.; Zhang, Q. H.; Deng, W. P.; Wang, Y. CdS–Graphene and CdS–CNT Nanocomposites as Visible-Light Photocatalysts for Hydrogen Evolution and Organic Dye Degradation. *Catal. Sci. Technol.* **2012,** *2*, 969–978.

Yeh, T. F.; Chan, F. F.; Hsieh, C. T.; Teng, H. Graphite Oxide with Different Oxygenated Levels for Hydrogen and Oxygen Production from Water Under Illumination: The Band Positions of Graphite Oxide, *J. Phys. Chem. C.* **2011,** *115*, 22587–22597.

Yeh, T. F.; Syu, J. M.; Cheng, C.; Chang, T. H.; Teng, H. Graphite Oxide as a Photocatalyst for Hydrogen Production from Water, *Adv. Funct. Mater.* **2010,** *20*, 2255–2262.

Yeh, T.-F.; Chen, S.-J.; Yeh, C.-S.; Teng, H. Tuning the Electronic Structure of Graphite Oxide through Ammonia Treatment for Photocatalytic Generation of $H_2$ and $O_2$ from Water Splitting, *J. Phys. Chem. C* **2013,** *117*, 6516–6524.

Yi, Z.; Ye, J.; Kikugawa, N.; Kako, T.; Ouyang, S.; Williams, S. H.; Yang, H.; Cao, J.; Luo, W.; Li, Z.; Liu,Y.; Withers, L. R. An Orthophosphate Semiconductor with Adation Properties Under Visible-Light Irradiation. *Nat. Mater.* **2010,** *9*, 559–564.

Yin, Z.; Zhu, J.; He, Q.; Cao, X.; Tan, C.; Chen, H.; Yan, Q.; Zhang, H. Graphene-Based Materials for Solar Cell Applications. *Adv. Energy Mater.* **2014,** *4*, 1300574.

You, Y.; Sahajwalla, V.; Yoshimura, M.; Joshi, R. K. Graphene and Graphene Oxide for Desalination. *Nanoscale* **2016,** *8*, 117–119.

Yu , L.; Lim, Y.-S.; Han, J. H.; Kim, K. Y.; Kim, J. Y.; Choi, S.-Y.; Wonwoo, K. S. A Graphene Oxide Oxygen Barrier Film Deposited via a Self-Assembly Coating Method. *Synth Met.* **2012,** 162, 710–714.

Yu, P. et al. Mechanically-Assisted Electrochemical Production of Graphene Oxide. *Chem. Mater.* **2016,** *28*, 8429–8438.

Yu, Q.; Lian, J.; Siriponglert, S. Graphene Segregated on Ni Surfaces and Transferred to Insulators. *Appl. Phys. Lett.* **2008,** *93*, 113103.

Yu, X.; Du, R.; Li, B.; Zhang,Y.; Liu, H.; Qu, J.; An, X. Biomolecule-Assisted Self-Assembly of CdS/$MoS_2$/Graphene Hollow Spheres as High-Efficiency Photocatalysts for Hydrogen Evolution Without Noble Metals. *Appl. Catal. B* **2016,** *182*, 504–512.

Zhang, H.; Fan, X. F.; Quan, X.; Chen, S.; Yu, H. T. Graphene Sheets Grafted Ag@AgCl Hybrid with Enhanced Plasmonic Photocatalytic Activity under Visible Light. *Environ. Sci. Technol.* **2011,** *45*, 5731–5736.

Zhang, H.; Lv, X.; Li, Y.; Wang Y.; Li, J. P25-Graphene Composite as a High Performance Photocatalyst. *ACS Nano* **2010,** *4*, 380–386.

Zhang, J. T.; Xiong, Z. G.; Zhao, X. S. Graphene–Metal–Oxide Composites for the Degradation of Dyes Under Visible Light Irradiation. *J. Mater. Chem.* **2011,** *21*, 3634–3640.

Zhang, J.; Qi, L.; Ran, J.; Yu, J.; Qiao, S. Z. Ternary NiS/$Zn_xCd_{1-x}S$/Reduced Graphene Oxide Nanocomposites for Enhanced Solar Photocatalytic $H_2$-Production Activity. *Adv. Energy Mat.* **2014,** *4*, 1301925–1301931.

Zhang, N.; Yang, M.-Q.; Liu, S.; Sun, Y.; Xu, Y.-J. Waltzing with the Versatile Platform of Graphene to Synthesize Composite Photocatalysts. *Chem. Rev.* **2015,** *115*, 10307–10377.

Zhang, N.; Zhang, Y.; Xu, Y.-J. Recent Progress on Graphene-Based Photocatalysts: Current Status and Future Perspectives. *Nanoscale* **2012,** *4*, 5792–5813.

Zhang, X. F.; Quan, X.; Chen, S.; Yu, H. T. Constructing Graphene/InNbO4 Composite with Excellent Adsorptivity and Charge Separation Performance for Enhanced Visible-Light-Driven Photocatalytic Ability. *Appl. Catal. B* **2011,** *105*, 237–242.

Zhang, Y.; Chen, Z.; Liu, S.; Xu, Y. -J. Size Effect Induced Activity Enhancement and Anti-Photocorrosion of Reduced Graphene Oxide/ZnO Composites for Degradation of Organic Dyes and Reduction of Cr(VI) in Water. *Appl. Catal. B* **2013,** *140–141*, 598–607.

Zhao, Z. H.; Schipper, D. E.; Leitner, A. P.; Thirumalai, H.; Chen, J. H.; Xiee, L.; Qin, F.; Alam, Md K.; Grabow, L. C.; Chene, S. D.; Wang, Z. Ren; Wang, Z.; Whitmire, K. H.; Bao, J. Bifunctional Metal Phosphide FeMnP Films from Single Source Metal Organic Chemical Deposition for Efficient Overall Water Splitting. *Nano Energy* **2017,** *39*, 444–453.

Zhou, F.; Shi, R.; Zhu, Y. F. Significant Enhancement of the Visible Photocatalytic Degradation Performances of γ-Bi2MoO6 Nanoplate by Graphene Hybridization. *J. Mol. Catal. A: Chem.* **2011,** *340*, 77–82.

Zhu, M. S.; Chen, P. L.; Liu, M. H. Graphene Oxide Enwrapped Ag/AgX (X = Br, Cl) Nanocomposite as a Highly Efficient Visible-Light Plasmonic Photocatalyst. *ACS Nano* **2011,** *5*, 4529–4536.

CHAPTER 2

# Carbon Derived from Biowastes for Supercapacitors: Synthesis to Applications

ANIL ARYA[1], A. L. SHARMA[2], VIJAY KUMAR[3], and ANNU SHARMA[1]

[1]*Department of Physics, Kurukshetra University, Kurukshetra, Haryana, India*

[2]*Department of Physics, Central University of Punjab, Bathinda, Punjab, India*

[3]*Department of Physics, Institute of Integrated and Honors Studies (IIHS), Kurukshetra University, Kurukshetra, India*

## ABSTRACT

Continuous consumption of traditional sources of energy (fossil fuels, coal, etc.) and a rapid increase in global warming forced the research community working in this line to develop alternative clean and green renewable sources of energy. The most feasible and attractive source of green energy is the symmetric/asymmetric ultra/supercapacitors (SCs). This clean and green source of energy becomes more attractive/attentive if it could be developed from wastage. In such energy source devices, electrode material plays a crucial role in enhancing the capacity and energy/power density. Most of the electrode is made up of carbon. Hence, carbon derived from the biowastes has grabbed the attention of researchers due to abundance, reducing health issues, environmental and disposal issues, and so on. The key advantages are high electronic conductivity and the large

New Forms of Carbon: Nanocarbons. Aneeya Kumar Samantara & Satyajit Ratha (Eds.)

surface area used to be delivered by this biowaste after proper physical/chemical processing. This results in enhancing the remarkable electrochemical properties. This chapter explores the fundamental characteristics, strategies adaptation to prepare biowaste-based SC electrode material, and characterization techniques involved followed by the latest advancements in the carbon derivation from biowastes.

## 2.1 SUBJECT BACKGROUND

Energy plays an important role in daily human life and is linked to the development of human civilization. Most of the energy demand is fulfilled by traditional energy sources, but due to depletion and increased global warming, it becomes necessary to develop alternate sources of sustainable energy. All over the world, the top priority of energy researchers is to search the environment-friendly and green sources of energy that can fulfill the demand for a broad application range (portable electronics, military/space operations, household supplies, electric vehicles, etc.). The energy sources are wind, nuclear, hydro, and solar contributing to the energy need. This energy needs to be stored somewhere for efficient use as per need. Battery and supercapacitor (SC) are two key candidates and have the potential to provide energy as per demand. The high-energy density of the battery and high-power density of the SC strengthen their candidature as an energy storage device. Out of them, the SC is a very interesting candidate owing to its key properties such as fast charge/discharge, better cyclic stability ($>10^6$ cycles), and low cost. A lot of research efforts have been adopted to increase the energy density of SC along with power density to replace the battery. Two important electrochemical parameters are specific capacitance and energy density. Both of these are linked with the voltage stability window of electrolytes (Choi and Aurbach, 2016; Burke, 2000; Arya et al., 2019; Simon and Gogotsi, 2008; Yao et al., 2015; Saykar et al., 2018; Saykar et al., 2019).

Figure 2.1a displays the Ragone plots and variation of energy density against power density for various electrochemical energy storage systems is shown (Wang et al., 2015; Gao et al., 2017). The SC can fill the vacuum between the battery and the traditional capacitors owing to their high power density. Figure 2.1b shows the roadmap for the development of the future energy sector and highlights the need for the development of renewable sources of energy in place of traditional sources. The most promising is to derive the active electrode materials from biowaste (Centi et al., 2011).

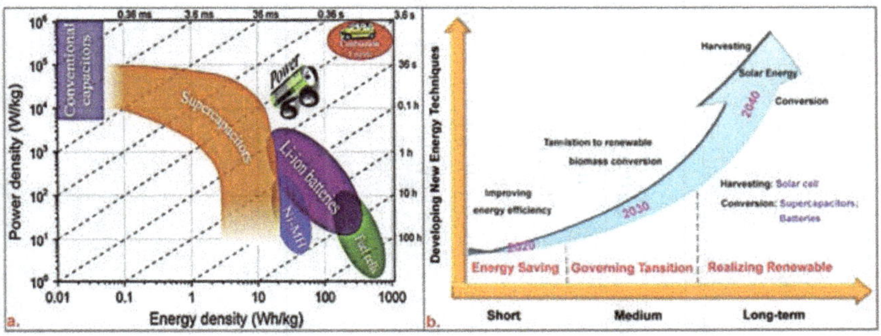


**FIGURE 2.1** (a) Ragone plots for various electrochemical energy storage systems, and (b) indicative roadmap of future energy scenario.

*Source*: Reprinted with permission from Gao et al. 2017. https://creativecommons.org/licenses/by/4.0/

So, the carbon derived from biomass has the potential to fill the gap between the present energy storage material and future needs (Marichi et al., 2017; Senthilkumar et al., 2015; Nanaji et al., 2018). The biomass material advantage is given as follows:

1. Green and cost-effective material.
2. A huge amount of availability
3. High carbon content.
4. Recyclability of waste products is feasible.
5. Various carbon morphologies can be obtained by varying activation/carbonization conditions.
6. Good compatibility with the various range of electrolytes.
7. Doping of hetero-atoms enhance the ion storage properties.

Table 2.1 shows various characteristics of the SC and its comparison with traditional capacitors and batteries. It has been observed that SC has various important characteristics that make it favorable over the battery.

**TABLE 2.1** (a) Characteristics of the supercapacitor, (b) comparison between an electronic capacitor, battery, and supercapacitor.

| (a) | | (b) | | | |
|---|---|---|---|---|---|
| **Features** | **Range** | **Function** | **Electronic capacitor** | **Battery** | **Supercapacitor** |
| Voltage level ($V$) | 50 V–100 V | Charge time | μs–ms | h | ms–min |
| Current ($I$) | 100 A–300 A | Efficiency | ~100 | 70–80 | >80 |
| Pulse duration ($\Delta t$) | 1.0 ms–1 s | Discharge time | μs–ms | 1–900 min | ms–days |
| Capacitance ($C$) | 1–10 F | Energy density | <0.01 W h $L^{-1}$ | 50–300 W h $L^{-1}$ | 0.5–5 W h $L^{-1}$ |
| Power density (kW $L^{-1}$) | 5–180 | Power density | >104 W $L^{-1}$ | <500 W $L^{-1}$ | $10^3$–$3 \times 10^3$ |
| Energy density (kJ $L^{-1}$) | 0.5–0.6 | | | | |
| ESR ($R$) | 20–30 | Cycle life | 106–108 | 200–1000 | $10^6$–$10^8$ |
| Temperature cycle | −20°C to +60°C | | | | |

This chapter presents the fundamental of the SC, key performance parameters, and different charge storage mechanisms. Then, the methods to obtain carbon from biowastes are discussed and followed by characterization techniques. After that, some important studies highlighting the carbon derived from biomass and its use for SC applications are discussed.

## *2.1.1 FUNDAMENTALS OF SUPERCAPACITORS*

SCs are also termed electrochemical capacitors owing to the similar charge storage mechanism and same geometry. However, the classification of SC is done based on the charge storage mechanism. Based on this, two types are (i) electric double-layer capacitor (EDLC): energy stored via ion absorption, and (ii) pseudo capacitor (PC): energy stored by redox reactions (Salanne et al., 2016; Najib et al., 2019; González et al., 2016; Borenstein et al., 2017; Tanwar et al., 2021). Based on their charge storage mechanism, the SCs are classified into the following three as mentioned in Fig. 2.2.

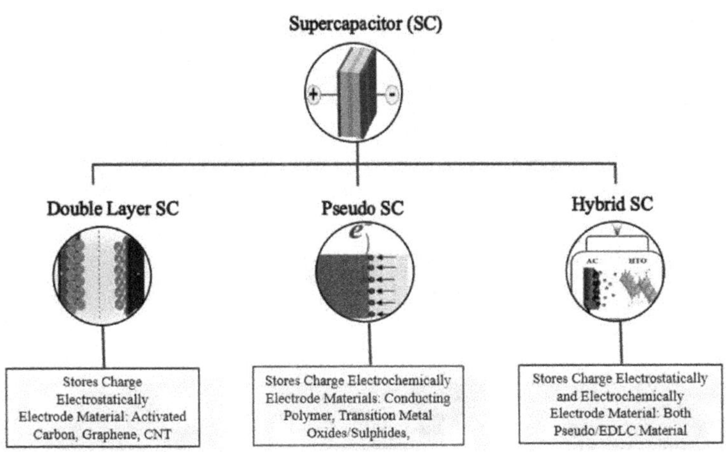


**FIGURE 2.2** Classification of a supercapacitor.

The comparison of EDLC and PC with a traditional capacitor and detailed charge storage mechanism are presented by Zhong's group (2015). In the former one, the ion adsorption/desorption results in charge accumulation at the electrode/electrolyte interfaces and is termed as purely electrostatic charge storage. The built-up of the high charge indicates higher capacitance and is generally observed for carbon-based electrode materials without any redox peaks. The cyclic voltammetry (CV) curve seems to be rectangular. Whereas in PC, the charge storage occurs via reversible redox reactions at electrode/electrolyte interface, and in CV, curve peaks are observed. The electrode material is generally transition metal oxides, hydroxides, and chalcogenides. The specific surface area (SSA), pore size, and morphology of the electrode material affect the charge storage capacity of the SC device. The pseudocapacitive mechanism comprises the EDLC behavior also and indicates the higher capacitance as compared to EDLC only (Zhong et al., 2015).

Based on the architecture, SC is classified as symmetric supercapacitor and an asymmetric supercapacitor (ASSC). The difference between the EDLC and PC is further examined by observing the charging/discharging and CV curves. Figure 2.3 highlights the charge/discharge and the CV curve (Gogotsi and Penner, 2018). The PC electrode material is different from the battery electrodes as evidenced by the nature of the curve in Figure 2.3.

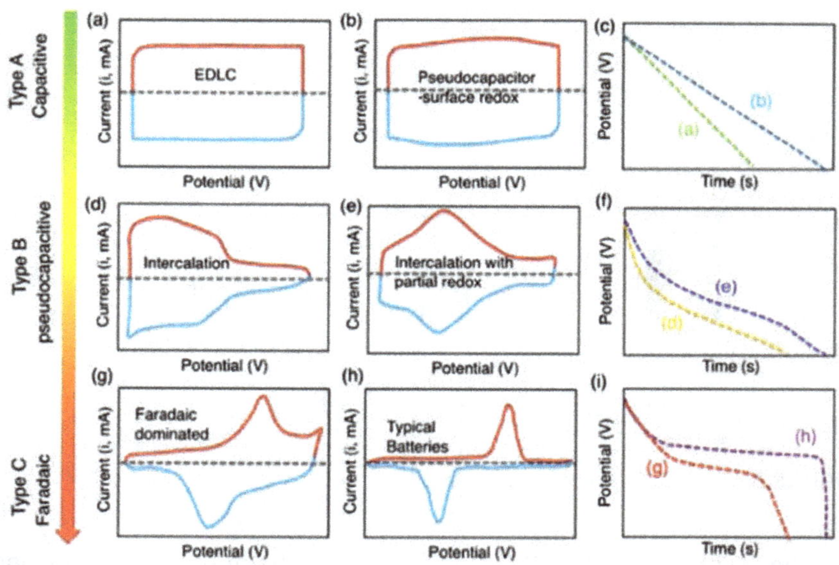


**FIGURE 2.3** The CV curves of (a) EDLC materials, (b, d, e) pseudo-capacitors materials, (g) battery-like materials, and (h) battery materials; and (c, f, i) corresponding galvanostatic discharge curves for various types of energy-storage materials.

*Source*: Reprinted with permission from Gogotsi and Penner, 2018. Copyright © 2018, American Chemical Society

### 2.1.2 CHARACTERISTICS OF SUPERCAPACITOR ELECTRODE MATERIAL

An SC comprises four important components that play every vital role in its charge storage performance. The electrochemical performance is influenced by all internal components. So, electrodes, electrolytes, and separators are chosen after analyzing their characteristic properties, since electrode material plays a very crucial role in the charge storage capacity and must be chosen appropriately for achieving optimum performance. The high SSA and morphology (sheet, rod, wire, flower, wall type, etc.) of the electrode material are important for obtaining enhanced charge storage capacity, energy density, and power density (Simon and Gogotsi, 2008; Saykar et al., 2018). The electrolyte provides the ions in the SC cell, and ions migrate in the system via the electrolyte. It acts as a carpet for ions. So, the low viscosity of the electrolyte, smaller ionic radii, leads to faster ionic mobility and high ionic conductivity. Then, the separator should have high porosity to absorb the electrolyte and should have sufficient thickness. Along with the material selection, the synthesis of electrode material and cell fabrication influences the overall key properties of the cell. So, the preparation method should be easy, cost-effective, and synthesis parameters (reaction temperature, reaction time) need to be optimized for large-scale production. Moreover, good contact of the active material on the current collector is preferred to minimize the interfacial and stability issues by suitable selection of binder. Along with this during cell fabrication, the charge transfer resistance needs to be kept minimum to achieve maximum ion diffusion and hence the higher ion storage capacity in the electrode material. Figure 2.4 depicts the key characteristics of the SC.

### 2.1.3 EXPERIMENTAL METHODOLOGY

#### 2.1.3.1 PREPARATION METHODS

The preparation of electrode material plays a very crucial role in cell performance. The most desirable requirement is that the preparation method should be easy, cost-effective, and environment friendly. The important methods for the preparation of the electrode material are a hydrothermal method, sol–gel method, microwave synthesis, and solid-state reaction method. However, for the preparation of biomass-based electrode materials, carbon is derived from the biowaste by various processes such as carbonization, activation, and hydrothermal carbonization. By varying the activation agent and carbonization conditions, the morphology of the prepared electrode material can be tuned. Figure 2.5 shows the photograph of various biomasses, preparation methods, and the cell fabrication step.

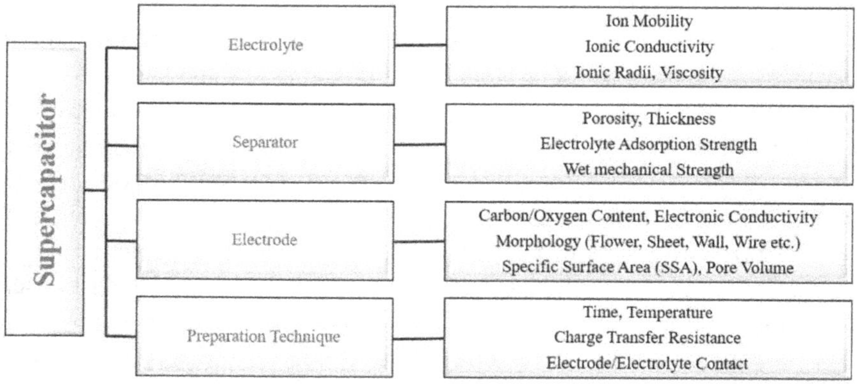


**FIGURE 2.4** Crucial characteristics of different supercapacitor components.

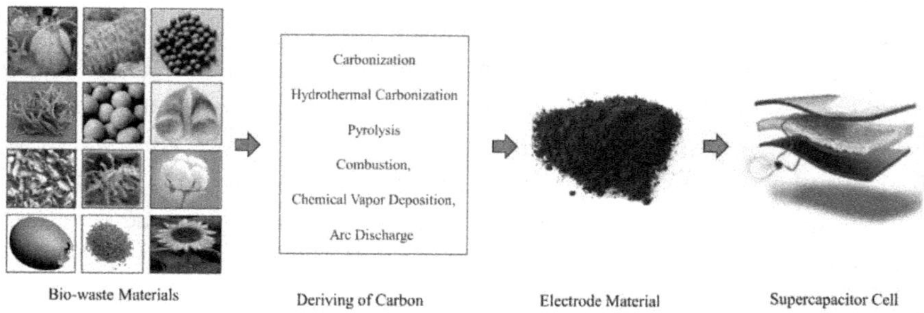


**FIGURE 2.5** Schematic of the cell fabrication from biowaste-derived carbon.

Na et al. in 2018 reported the preparation of electrode material from rice husk. Here, the carbonization and activation technique was adopted to prepare the active material. Han et al. in 2018 derived carbon from plant waste (bamboo shoots shells) via one-step activation. Ma et al. in 2019 derived the hierarchically porous carbons (HPCs) from corn straw. The authors adopted a simple two-step method which is a combination of hydrothermal carbonization and KOH activation (Fig. 2.6).

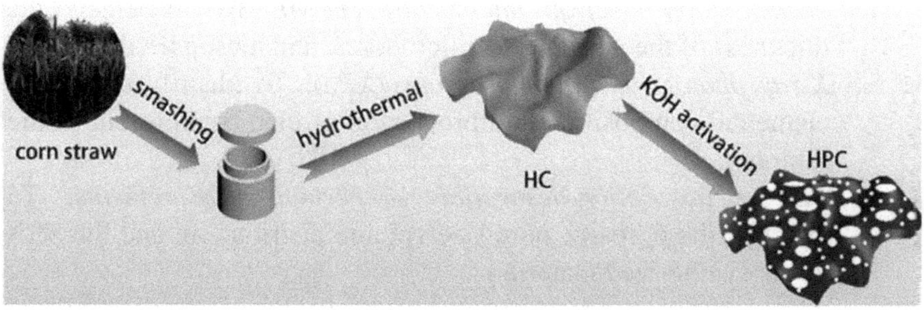


**FIGURE 2.6** Schematic illustration of the synthesis strategy for the HPCs.
*Source*: Reprinted with permission from Ma et al., 2019.

Xu et al. in 2016 prepared the porous active carbon and fluorescent carbon dots from protein-rich biomass precursor—soybean. Here, the two-step carbonization method was adopted. In the first step, low-temperature carbonization is performed to obtain the fluorescent carbon dots. Then, in the second step, porous active carbon is obtained from the high-temperature carbonization. Han et al. in 2018 prepared the nitrogen-doped porous-activated carbon derived from pueraria-based biomass as a precursor. The detailed process is shown in Figure 2.8b.

### *2.1.3.2 CHARACTERIZATION TECHNIQUES*

The structural, morphological, porosity, and elemental composition of the electrode material need to be analyzed before the fabrication of the complete SC cell (Wang et al., 2015; Salanne et al., 2016). So, the important characterization techniques and information obtained from them are given below:

1. *X-ray diffraction (XRD):* To evaluate crystallinity, d-spacing, crystallite size, and form of the carbon.
2. *Raman spectroscopy:* To examine the structural defects and graphitization level of carbon-based materials (by evaluating the ratio of G-band to D-band ($I_G/I_D$).
3. *Scanning electron microscopy (SEM):* To explore the morphology (sheet, wire, rod, and flower) of the electrode material and elemental composition that may be analyzed via energy dispersive spectroscopy (EDS).

4. *Transmission electron microscopy (TEM):* To investigate the thickness of the nanosheets, micropores, and mesopore diameter.
5. *X-ray photoelectron spectroscopy (XPS):* To identify the actual elemental composition and proportion of elements present in the sample.
6. *BET-$N_2$ adsorption-desorption isothermal measurements:* To examine the porosity, pore size/volume distribution, and the SSA of the synthesized material.

Then after checking the material properties and analyzing them properly, the next step is the SC cell fabrication. Cell fabrication can be done in two ways: (i) symmetric and (ii) asymmetric arrangement. After that, the electrochemical properties can be analyzed in the three-electrode and two-electrode systems. The following techniques are adopted to investigate the electrochemical properties.

1. *Complex impedance spectroscopy:* To find out the bulk resistance ($R_b$), charge-transfer resistance ($R_{ct}$), and equivalent series resistance of the SC cell.
2. *CV:* From the CV curve, we can find out the charge storage mechanism hint by using the power law. The power law is the relation between the current and the scan rate: $i = av^b$, here, $i$ is the cathodic current (A), $v$ is the scan rate (mV s$^{-1}$), and $a$ and $b$ are variables (Augustyn et al., 2013). The value of $b$ (designated as $b$-value) is an important factor and defines the charge storage mechanism, (i) $b$ = 1 means charge storage mechanism is a capacitive type, and (ii) $b$ = 0.5 means diffusion-limited charge storage mechanism. Further, the contribution of the capacitive and diffusion capacitance can be separated using the equation; $i = k_1v + k_2v + k_2v^{1/2}$, here $k_1$ and $k_2$ are values (Wang et al., 2007). Here, $k_1v$ is the capacitive contribution and $k_2v^{1/2}$ is the diffusion-limited contribution.
3. *Galvanostatic charge/discharge (GCD)*: The various electrochemical parameters are obtained from the GCD using the formulas given below (Wang et al., 2014).

For the three-electrode system, specific capacitance is obtained from GCD using the formula:

$$C = \frac{I \times \Delta t}{m \times \Delta V}$$

where $I$ is the discharge current, $\Delta t$ is the discharge time, $\Delta V$ is the potential drop during discharge, and $m$ is the mass of active material in the working electrode.

For two-electrode *(symmetric cell configuration)*

*Specific capacitance*

$$C = \frac{2 \times I \times \Delta t}{m \times \Delta V}$$

Here, $I$ is the discharging current, $\Delta t$ is the discharge time, $\Delta V$ is the potential drop during discharge, and $m$ is the mass of active material in the single electrode.

*Energy density and power density*:

$$E = \frac{1}{2 \times 3.6} C(\Delta V)^2$$

$$P = \frac{3600 \times E}{\Delta t}$$

Here, $E$ (W h kg$^{-1}$), $C$, $\Delta V$, $P$ (W kg$^{-1}$), and $\Delta t$ are the specific energy, specific capacitance, potential window, specific power, and discharge time, respectively.

For two-electrode *(asymmetric cell configuration)*

*Specific capacitance*

$$C = \frac{I \times \Delta t}{m \times \Delta V}$$

Here, $I$ is the discharging current, $\Delta t$ is the discharge time, $\Delta V$ is the potential drop during discharge, and $m$ is the total mass of the active electrode materials in both (positive, negative) electrode.

*Energy density and power density*

$$E = \frac{1}{2} C(\Delta V)^2$$

$$P = \frac{3600 \times E}{\Delta t}$$

Here, $E$ (W h kg$^{-1}$), $C$, $\Delta V$, $P$ (W kg$^{-1}$), and $\Delta t$ are the specific energy, specific capacitance, potential window, specific power, and discharge time, respectively.

## 2.2 ADVANCEMENTS IN ELECTRODE MATERIALS DERIVED FROM BIOWASTE

Yu et al. (2018a) reported the preparation of SC cells by collecting the activated carbon from the ground cherry calyces. From this BW, a 3D framework of porous carbon (PC) nanosheets was created via carbonization procedure and used as electrode material. It is denoted as PCNS 7 (activation by 7 M KOH). The SEM analysis confirmed the presence of a 3D framework comprising sheet and hollow structures. The SSA obtained from the BET was very high and is 1612 m$^2$ g$^{-1}$ (pore volume = 1.2 cm$^3$ g$^{-1}$; micropore volume = 0.4 cm$^3$ g$^{-1}$). The higher conductivity of the PCNS7 sample is evidenced by the higher degree of graphitization as obtained from Raman analysis. The CV curve indicates the ideal supercapacitive behavior as evidenced by the rectangular nature. The specific capacitance value was very high and is about 350 F g$^{-1}$ (at 0.1 A g$^{-1}$) with a capacity retention of about 86.3%. Further, the specific capacitance was also checked using the three-electrode system that shows the highest specific capacitance for this system (518 F g$^{-1}$ at 1 A g$^{-1}$). The capacity retention after 10,000th cycle is 94.8%, after 50,000th cycle is 92.2%, and 88.1% after 100,000 cycles. The high specific capacitance was attributed to the high SSA, comprising nanosheets and porous nature, and hierarchical pore-size distribution. The simultaneous presence of the above three allows easier ion migration and enhances the ion storage capacity.

Recently, Goswami et al. in 2019 reported the formation of carbon from the cooking-oven-produced carbon nanoparticles. Then, the carbon NP was incorporated in the polyaniline (PANI; PA) matrix and SC performance has been explored. Here, the polyaniline-waste carbon nanoparticles (PA/WC) are synthesized through in-situ polymerization of PANI with four different polymerization times ($t_1$: 1.5 h, $t_2$: 2 h, $t_3$: 2.5 h, $t_4$: 3 h). The purity and surface morphology of the as-derived samples were analyzed by FESEM, EDS mapping, and XPS. The Raman analysis confirmed that the WCP nanoparticles are well connected with the PANI. The CV plot for the three different composite samples at constant polymerization time is $t_4 = 3.0$ h with control samples and GCE + NF electrode. The CV curve for the PA/WC composite shows the presence of EDLC nature due to carbon and the pseudocapacitive nature due to PANI. The WCP acts as the electron-trapping center and enhances the surface area that is reflected in the arial capacitance. From the GCD, it was concluded that the composite electrodes possess capacity retention of about 80% after 1000 cycles and approximately 90–100% retention of $C$ values with increasing scan rates and current densities. The authors, in brief, claimed that the present composite electrode material can be a promising material for green energy storage devices.

Ma et al. (2015) synthesized hierarchical PC using the sunflower heads collected from sunflower crops. The PC was having abundant oxygen functional groups that make it unique for the SC electrode. SEM and TEM analysis confirmed the micropores, mesopore nature of the BW-derived carbon. The former leads to a high surface area, and the latter leads to the high surface area along with wider transport channels to micropores. Both are important for the SC application. XRD and Raman confirmed the disordered structure with defects. The specific capacitance obtained from GCD is 317 F $g^{-1}$, 345 F $g^{-1}$, and 297 F $g^{-1}$ for SFHC—1:1, SFHC—1:1.5, and SFHC—1:2, respectively (at 1 A $g^{-1}$). Here, 1:1, 1:1.5, and 1:2 are the weight ratios of the precarbonization sample with KOH solid.

The SFHC—1:1.5 shows the highest specific capacitance and is attributed to the (i) abundant oxygen functional groups, (ii) sufficient micropores/mesopores, and (iii) low charge transfer resistance (0.13 Ω). Figure 2.17(ii) shows the SC cell performance for the optimum system (SFHC—1:1.5) in the two-electrode system. The shape of the CV curve and the GCD curve indicates improved stability and faster ion migration. It gives a specific capacitance plot and is 297 F $g^{-1}$ at 0.5 A $g^{-1}$ and 217 F $g^{-1}$ at 10 A $g^{-1}$. This cell shows 83% capacity retention after 5000 cycles and is within a desirable limit. The energy density and owed density for the SC cell are 9.2 W h $kg^{-1}$ and 482 W $kg^{-1}$, respectively. In another work, Zequine et al. (2017) reported SC fabrication by obtaining carbon electrode materials by carbonizing abundantly available and recyclable jute. The jute was chosen over other biomass products due to the fibrous structure and comprises cellulose and lignin. Figure 2.7(i) shows the synthesization of electrodes for SC cells from jute fibers. The XRD diffractograms and Raman spectra confirmed the presence of conducting the graphitic phase of carbon. The surface area obtained from the BET is about 1769 $m^2 g^{-1}$. TEM analysis confirmed that the carbonized jute comprises various graphic thin layers. The CV curve shows the absence of redox peaks and indicates the EDLC nature supported by the porous nature that enhances the charge storage capacity. Also, there was almost no effect of the bending angle on the CV curve, which suggests its use for the flexible SC. The specific capacitance was 408 ± 15 F $g^{-1}$ at 1 mV $s^{-1}$ and 223 ± 6.6 F $g^{-1}$ at 200 mV $s^{-1}$. The energy density and power density for the carbonized jute are 21 W h $kg^{-1}$ and 1.82 kW $kg^{-1}$, respectively. Figure 2.7(ii) shows the cyclic stability obtained from the CV and GCD.

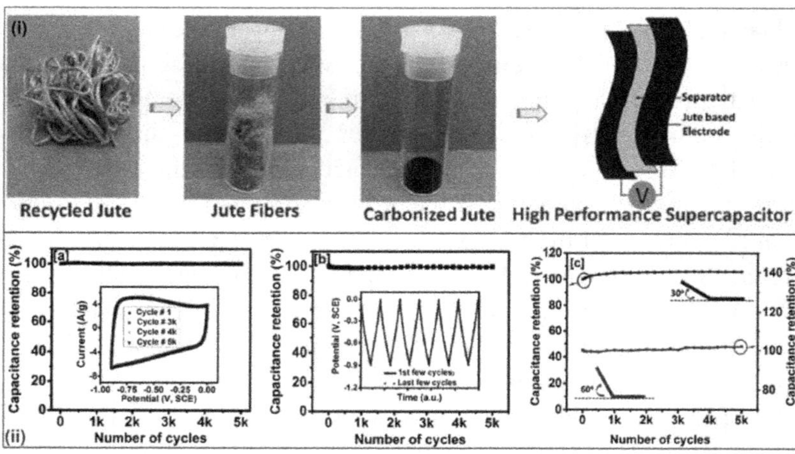


**FIGURE 2.7** (i) Schematic of the device fabrication process. (ii) (a) Capacitance retention verse number of CV cycles, inset figure shows CV curves at various cycles (b) capacitance retention verse number of charge–discharge cycles, inset figure shows potential versus time plot for first and last few cycles of charge–discharge, and (c) capacitance retention verse number of charge–discharge cycles at various bending angles.

*Source*: Reprinted with permission from Zequine et al., 2017. http://creativecommons.org/licenses/by/4.0/

Figure 2.7(ii) a shows good cyclic stability for 5000 cycles and is also supported by the GCD curve as shown in Figure 2.7(ii) b. The inset shows the exact overlap between the first and last cycles indicating better stability for the jute-based electrode. Figure 2.7(ii) c shows the capacity retention plot at different angles and is almost 100% at different banding angles. Further temperature-dependent investigations were performed via CV and

GCD. Both techniques showed the increase of specific capacitance with an increase in temperature (45% higher charge storage at 75°C). Wang et al. in 2018 derived the 3D interconnected HPC from sunflowers stem. The presence of high oxygen content and a 3D interconnected network boosted the SC performance.

HPC-2 designates the HPC with 2 g KOH and the stem of the sunflower shows a 3D interconnected honeycomb-like porous structure. This unique morphology facilitates faster electron transfer and ion diffusion. It was also supported by the TEM analysis which indicates the presence of pores on carbon walls. XRD and Raman spectra evidenced the disordered and defected structure for HPC-2 as compared to the HPC-1 and HPC-3. The highest surface area was exhibited by HPC and is about 1657 $m^2\ g^{-1}$. The specific capacitance obtained from the GCD was 349 F $g^{-1}$ at 1 A $g^{-1}$ (247 F $g^{-1}$ at 50 A $g^{-1}$) and is higher than the commercial activated carbon (YP-50).

The electrochemical stability was checked by CV and shows capacity retention of about 98.6% after 10,000 cycles (at 200 mV $s^{-1}$). Then, an ASSC was fabricated using the HPC-2/$MnO_2$ as positive electrode material and HPC-2 as negative electrode material. The stability of the cell was checked from the CV and shows capacity retention of about 83.1% after 1000 cycles. The fabricated cell shows a high energy density of about 58.8 W h $kg^{-1}$, power density of 210.7 W $kg^{-1}$, and is higher than the reported SC cell. Recently, Elaiyappillai et al. (2019) derived a low-cost carbon from the *Cucumis melo* (Cm) fruit peel and tested the performance of this electrode for SC application. The activated carbon was prepared by KOH activation. The synthesized samples are designated as Cm-600, Cm-700, Cm-800, and Cm-900 depending upon the activation temperature of 600°C, 700°C, 800°C, and 900°C, respectively. The XRD analysis evidenced the micrographic structure of the electrode material and it was also supported by the Raman spectrum. The mesoporous morphology for the Cm-900 and its benefits will be reflected in the electrochemical analysis. Further, TEM evidenced the interconnected porous morphology and is beneficial for a good electrode–electrolyte interface which minimizes the length of ion-conducting pathways. The surface area for the Cm-900 as obtained from BET was 721.91 $m^2\ g^{-1}$. The specific capacitance as obtained from the CV at 5 mV $s^{-1}$ for Cm-900 is 376 F $g^{-1}$ and decreases with the scan rate for all samples. The high capacitance is attributed to the EDLC nature of carbon, pseudocapacitance contribution from the oxygen present in Cm, and high surface area. Further, GCD showed the specific capacitance of about 404 F $g^{-1}$ at 1 A $g^{-1}$ for the Cm-900. The high capacitance was associated with the high porosity and oxygen functionalities in the Cm. Then, the symmetric two-electrode SC cell was fabricated for testing. The GCD of the cell showed the specific capacitance of about 211 F g−1 at 1 A g−1 for Cm-900 and 88% capacity retention was obtained after 4000 cycles (at 2 A $g^{-1}$). The inset shows the fabricated device with a glowing LED. The energy density of the cell is 29.30 W h $kg^{-1}$ and the power density is about 279.78 W $kg^{-1}$.

Yan et al. (2018) derived the carbon from low-cost and environment-friendly waste lotus stems (LC-NCs) and doped it with nitrogen. Here, KOH was used as an activator, and urea as a source of nitrogen. The samples are labeled as LS-NC-X, where *X* represents activation temperature (600°C, 700°C, 800°C, and 900°C). The XRD analysis confirmed the turbostratic graphite structure and Raman analysis shows the disordered carbon structure. The highest area observed from the BET was 2221 $m^2$ $g^{-1}$ for the LS-NCS 800. SEM analysis confirmed the mesoporous structure. The electrochemical performance was examined in the cell configuration as shown in Figure 2.8. The CV analysis suggested the highest specific capacitance for the LS-NC 600. The GCD analysis showed a capacitance of about 303.5 F $g^{-1}$ at 0.5 A $g^{-1}$ and high capacitance is attributed to the presence of pyridine nitrogen (N-6), pyrrolic/pyridinic nitrogen (N-5) on graphene layers as observed from XPS. The nitrogen functional group promotes pseudocapacitive behavior. Figure 2.8b shows the specific capacitance at the different current densities and is highest for LS-NC 600 at all current densities. Figure 2.8c shows cyclic stability in the two-electrode configuration for LS-NC 600. The capacity retention was about 96% after 5000 cycles and is sufficient for SC applications.

Chernysheva et al. (2018) have derived the activated PC from the biomass refinery wastes (humins) by various methods (pyrolysis, physical and chemical activation, or combined approaches). The samples were labeled as per the preparation condition, for example, Chem-900-2 refers to carbon material obtained by chemical activation of humins at 900°C and KOH/humins weight ratio of 2:1. The Chem-900-2-based SC cell shows the highest capacitance of about 370 F $g^{-1}$ at 0.5 A $g^{-1}$. The stability test done by GCD shows capacity retention of about 92% after 10,000 charge/discharge cycles at 10 A $g^{-1}$. The high capacitance was attributed to the micro/mesoporous structure and high porosity of the Chem-900-2.

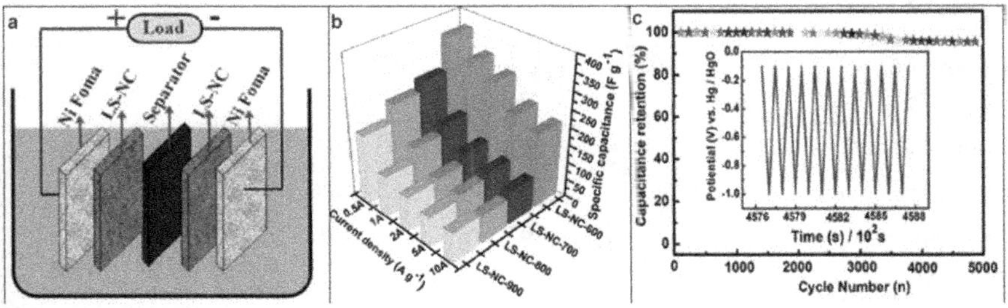


**FIGURE 2.8** (a) Schematic of the LS-NC symmetric supercapacitor, (b) specific capacitance at current densities at different current densities in the range of 0.5–10 A $g^{-1}$, and (c) the cycling performance of the LS-NC-600 as an electrode in three electrodes at the current density of 5 A $g^{-1}$.

*Source*: Reprinted with permission from Yan et al., 2018. https://creativecommons.org/licenses/by/3.0/

The mesoporous nature creates the 3D conducting channels that facilitate ion migration. Further, the cyclic stability was checked by GCD in a two-electrode configuration and shows capacity retention of about 86% even after 10,000 cycles at 0.5 A $g^{-1}$. The potential of the SC cell was demonstrated by powering a calculator and glowing a LED for 5 min. To highlight the superiority of the carbon-derived electrode materials from biowaste, we have made a comparison of various BW-derived carbon material and their performance and electrolyte used. Table 2.2 shows the comparison of various BW-based supercapacitor cells.

**TABLE 2.2** Comparisons of SC Cell Performance Based on Carbon Derived from Different Biowastes.

| Precursor | Electrolyte | Current density | $C_g/(F g^{-1})$ | Reference |
|---|---|---|---|---|
| Broad bean shell | 6 M KOH | 0.5 $Ag^{-1}$ | 202 | Xu et al., 2015 |
| Sugar cane bagasse | 6 M KOH | 1 $Ag^{-1}$ | 323 | Liu et al, 2016 |
| Brussel sprouts | 6 M KOH | 0.5 $Ag^{-1}$ | 255 | Li et al., 2016 |
| Pomelo peel | 6 M KOH | 1 $Ag^{-1}$ | 289 | Liang et al., 2014 |
| Silkworm | 6 M KOH | 1 $Ag^{-1}$ | 304 | Gong et al., 2016 |
| Shiitake mushroom | 6 M KOH | 1 $Ag^{-1}$ | 306 | Cheng et al., 2015 |
| Human hair | 6 M KOH | 1 $Ag^{-1}$ | 340 | Qian et al., 2014 |
| Waste news paper | 6 M KOH | 1 $mAg^{-1}$ | 180 | Adinaveen et al., 2013 |

**TABLE 2.2** *(Continued)*

| Precursor | Electrolyte | Current density | $C_g$/(F g$^{-1}$) | Reference |
|---|---|---|---|---|
| Lotus seedpod shell | 3 M KOH | 0.5 Ag$^{-1}$ | 165 | Sekar et al., 2004 |
| Pomegranate rind | 1M $H_2SO_4$ | 0.1 Ag$^{-1}$ | 268 | Yang et al., 2014 |
| Willow catkin | 3 / 6 M KOH<br>2 / 1 M $Na_2SO_4$ | 0.5 Ag$^{-1}$ | 298 | Li et al, 2016 |
| Almond shells | 3 / 6 M KOH | 1 Ag$^{-1}$ | 272.3 | Wu et al., 2016 |
| Cotton | 3 / 3 M KOH | 0.3 Ag$^{-1}$ | 221.7 | Liu et al., 2016 |
| Melamine | 2 / 1 M $H_2SO_4$ | 0.05 Ag$^{-1}$ | 201 | Wen et al., 2016 |
| Auricularia | 6 M KOH | 1 Ag$^{-1}$ | 347 | Jiang et al., 2016 |
| Chicken eggshell | 1 M KOH | 1 Ag$^{-1}$ | 297 | Li et al., 2012 |
| Rice husk | 1 M $Na_2SO_4$ | 1 Ag$^{-1}$ | 112 | Ganesan et al., 2014 |
| Algae | 6 M KOH | 0.2 Ag$^{-1}$ | 287 | Yu et al., 2018 |
| Ginkgo leaf | 1 M KOH | 0.2 Ag$^{-1}$ | 345 | Wang et al., 2018 |
| Rice straw | 6 M KOH | 1 Ag$^{-1}$ | 337 | Liu et al., 2018 |
| Water chestnut | 6 M KOH | 0.5 Ag$^{-1}$ | 346 | Wei et al., 2018 |
| Puffed rice | 6 M KOH | 0.5 Ag$^{-1}$ | 334 | Hou et al., 2017 |
| Popcorn | 6 M KOH | 0.2 Ag$^{-1}$ | 348 | Hou et al., 2017 |
| Sunfower seed shell | 30 wt % KOH | 250 mAg$^{-1}$ | 311 | Li et al., 2011 |
| Neem dead leaves | 1 M $H_2SO_4$ | 500 mAg$^{-1}$ | 400 | Biswal et al., 2013 |
| Waste coffee beans | 1 M $H_2SO_4$ | 50 mAg$^{-1}$ | 368 | Rufford et al., 2008 |
| Sugar-cane bagasse | 1 M $H_2SO_4$ | 250 mAg$^{-1}$ | 300 | Rufford et al., 2010 |
| Pistachio shell | 0.5 M $H_2SO_4$ | 10 mVs$^{-1}$ | 120 | Wu et al., 2005 |

## 2.3 SUMMARY

To fulfill the room for sustainable, clean, and renewable sources of energy, the scientific community has focused on the development of new electrodes for SC applications. The most demanding is the use of recyclable materials from biowastes. So, biomass has the potential to bridge the gap between environment-friendly and renewable energy devices. The electrode material derived from biowastes has grabbed the attention of researchers to develop SC applications. So, this chapter highlights the development of different PCs derived from various biomass products.

The important characteristics of the electrode material are discussed followed by the preparation and characterization techniques. Then, the development in the carbon derived from biowastes is presented. Various biomass-derived carbon SC cells are compared based on morphology, specific capacity, energy density, and power density. In brief, it may be stated that the carbon derived from biomass has three important benefits that strengthen its adaptability for SC applications: (i) cost-effective and environment-friendly, (ii) waste is used for the energy sector, and (iii) green and sustainable source of energy is developed.

## ACKNOWLEDGMENT

One of the authors (AA) is thankful to the UGC for providing DS Kothari Post Doc Fellowship via ward letter number BSR/PH/20-21/0156.

## KEYWORDS


- **supercapacitor**
- **energy storage**
- **biowastes**
- **carbon**
- **electrode material**


## REFERENCES


Adinaveen, T.; Kennedy, L. J.; Vijaya, J. J.; Sekaran, G. Studies on Structural, Morphological, Electrical and Electrochemical Properties of Activated Carbon Prepared from Sugarcane Bagasse. *J. Ind. Eng. Chem.* **2013,** *19*, 1470.

Arya, A.; Sharma, A. L. Electrolyte for Energy Storage/Conversion ($Li^+$, $Na^+$, $Mg^{2+}$) Devices Based on PVC and Their Associated Polymer: A Comprehensive Review. *J. Solid State Electrochem.* **2019,** *23*, 997.

Augustyn, V.; Come, J.; Lowe, M. A.; Kim, J. W.; Taberna, P. L.; Tolbert, S. H.; Abruna, H. D.; Simon, P.; Dunn, B. High-Rate Electrochemical Energy Storage Through $Li^+$ Intercalation Pseudocapacitance. *Nat. Mater.* **2013,** *12*, 518.

Biswal, M.; Banerjee, A.; Deo, M.; Ogale, S. From Dead Leaves to High Energy Density Supercapacitors. *Energy Environ. Sci.* **2013,** *6*, 1249.

Borenstein, A.; Hanna, O.; Attias, R.; Luski, S.; Brousse, T.; Aurbach, D. Carbon-Based Composite Materials for Supercapacitor Electrodes: A Review. *J. Mater. Chem. A* **2017,** *5*, 12653.

Burke, A. Ultracapacitors: Why, How, and Where Is the Technology. *J. Power Sources* **2000,** *91*, 37.

Centi, G.; Perathoner, S.; Perathoner, S.; Passalacqua, R.; Ampelli, C. *Carbonneutral Fuels and Energy Carriers*; CRC Press: Boca Raton, 2011; Chapter 4, Solar Production of Fuels from water And CO2; p 291.

Cheng, P.; Gao, S.; Zang, P.; Yang, X.; Bai, Y.; Xu, H.; Liu, Z.; Lei, Z. Hierarchically Porous Carbon by Activation of Shiitake Mushroom for Capacitive Energy Storage. *Carbon* **2015,** *93*, 315.

Chernysheva, D. V.; Chus, Y. A.; Klushin, V. A.; Lastovina, T. A.; Pudova, L. S.; Smirnova, N. V.; Kravchenko, O. A.; Chernyshev, V. M.; Ananikov, V. P. Sustainable Utilization of Biomass Refinery Wastes for Accessing Activated Carbons and Supercapacitor Electrode Materials. *Chem. Sus. Chem.* **2018,** *11*, 3599.

Choi, J. W.; Aurbach, D. Promise and Reality of Post-Lithium-Ion Batteries with High Energy Densities. *Nat. Rev Mater.* **2016,** *1*, 16013.

Elaiyappillai, E.; Srinivasan, R.; Johnbosco, Y.; Devakumar, P.; Murugesan, K.; Kesavan, K.; Johnson, P. M. Low Cost Activated Carbon Derived from Cucumis Melo Fruit Peel for Electrochemical Supercapacitor Application. *Appl. Surf. Sci.* **2019,** *486*, 527.

Ganesan, A.; Mukherjee, R.; Raj, J.; Shaijumon, M. M. Nanoporous Rice Husk Derived Carbon for Gas Storage and High Performance Electrochemical Energy Storage. *J. Porous Mater.* **2014,** *21*, 839.

Gao, Z.; Zhang, Y.; Song, N.; Li, X. Biomass-Derived Renewable Carbon Materials for Electrochemical Energy Storage. *Mater. Res. Lett.* **2017,** *5*, 69.

Gong, C.; Wang, X.; Ma, D.; Chen, H.; Zhang, S.; Liao, Z. Microporous Carbon from a Biological Waste-Stiff Silkworm for Capacitive Energy Storage. *Electrochim. Acta* **2016,** *220*, 331.

González, A.; Goikolea, E.; Barrena, J. A.; Mysyk, R. Review on Supercapacitors: Technologies and Materials. *Renew. Sustain. Energy Rev.* **2016,** *58*, 1189.

Goswami, S.; Dillip, G. R.; Nandy, S.; Banerjee, A. N.; Pimentel, A.; Joo, S. W.; Martins, R.; Fortunato, E. Biowaste-Derived Carbon Black Applied to Polyaniline-Based High-Performance Supercapacitor Microelectrodes: Sustainable Materials for Renewable Energy Applications. *Electrochim. Acta* **2019,** *316*, 202.

Han, J.; Li, Q.; Wang, J.; Ye, J.; Fu, G.; Zhai, L.; Zhu, Y. Heteroatoms (O, N)-Doped Porous Carbon Derived from Bamboo Shoots Shells for High Performance Supercapacitors. *J. Mater. Sci.* **2018a,** *29*, 20991.

Han, X.; Jiang, H.; Zhou, Y.; Hong, W.; Zhou, Y.; Gao, P.; Ding, R.; Liu, E. A High Performance Nitrogen-Doped Porous Activated Carbon for Supercapacitor Derived from Pueraria. *J. Alloys Compd.* **2018b,** *744*, 544.

Hou, J. H.; Jiang, K.; Tahirc, M.; Wu, X.; Idreesc, F.; Shen, M.; Cao, C. B. Tunable Porous Structure of Carbon Nanosheets Derived from Puffed Rice for High Energy Density Supercapacitors. *J. Power Sources* **2017,** *371*, 148.

Hou, J. H.; Jiang, K.; Wei, R.; Tahir, M.; Wu, X. G; Shen, M.; Wang, X. Z.; Cao, C. B. Popcorn-Derived Porous Carbon Flakes with an Ultrahigh Specific Surface Area for Superior Performance Supercapacitors. *ACS Appl. Mater Interfaces* **2017,** *9*, 30626.

Jiang, L. L.; Sheng, L. Z.; Chen X.; Wei, T.; Fan, Z. J. Construction of Nitrogen-Doped Porous Carbon Buildings Using Interconnected Ultra-Small Carbon Nanosheets for Ultra-High Rate Supercapacitors. *J. Mater. Chem. A* **2016,** *4*, 11388.

Khan, N.; Mariun, N.; Zaki, M.; Dinesh, L. Transient Analysis of Pulsed Charging in Supercapacitors. In *2000 TENCON Proceedings. Intelligent Systems and Technologies for the New Millennium (Cat. No. 00CH37119)* **2000,** *3*, 193.

Li, J.; Zan, G.; Wu, Q. Nitrogen and Sulfur Self-Doped Porous Carbon from Brussel Sprouts as Electrode Materials for High Stable Supercapacitor. *RSC Adv.* **2016,** *6*, 57464.

Li, X.; Xing, W.; Zhuo, S.; Zhou, J.; Li, F.; Qiao, S. Z.; Lu, G. Q. Preparation of Capacitor's Electrode from Sunflower Seed Shell. *Bioresour. Technol.* **2011,** *102*, 1118.

Li, Y.; Wang, G.; Wei, T.; Fan, Z.; Yan, P. Nitrogen and Sulfur Co-Doped Porous Carbon Nanosheets Derived from Willow Catkin for Supercapacitors. *Nano Energy* **2016,** *19*, 165.

Li, Z.; Zhang, L.; Amirkhiz, B. S.; Tan, X. H.; Xu, Z. W.; Wang, H. L.; Olsen, B. C.; Holt, C. M. B.; Mitlin, D. Carbonized Chicken Eggshell Membranes with 3D Architectures as High-Performance Electrode Materials for Supercapacitors. *Adv. Energy Mater*. **2012,** *2*, 431.

Liang, Q.; Ye, L.; Huang, Z.; Xu, Q.; Bai, Y.; Kang, F.; Yang, Q. A Honeycomb-Like Porous Carbon Derived from Pomelo Peel for Use in High-Performance Supercapacitor. *Nanoscale* **2014,** *6*, 13831.

Liu, J.; Deng, Y.; Li, X.; Wang, L. Promising Nitrogen-Rich Porous Carbons Derived from One-Step Calcium Chloride Activation of Biomass-Based Waste for High Performance Supercapacitor, *ACS Sustain. Chem. Eng.* **2016,** *4*, 177.

Liu, S. B.; Zhao, Y.; Zhang, B. H.; Xia, H.; Zhou, J. F.; Xie, W. K.; Li, H. J. Nano-Micro Carbon Spheres Anchored on Porous Carbon Derived from Dual-Biomass as High Rate Performance Supercapacitor Electrodes. *J Power Sources* **2018,** *381*, 116.

Liu, Y.; Shi, Z.; Gao, Y.; An, W.; Cao, Z.; Liu, J. Biomass-Swelling Assisted Synthesis of Hierarchical Porous Carbon Fibers for Supercapacitor Electrodes. *ACS Appl. Mater. Interfaces* **2016,** *8*, 28283.

Ma, D.; Wu, G.; Wan, J.; Ma, F.; Geng, W.; Song, S. Oxygen-Enriched Hierarchical Porous Carbon Derived from Biowaste Sunflower Heads for High-Performance Supercapacitors. *RSC Adv.* **2015,** *5*, 107785.

Ma, H.; Chen, Z.; Wang, X.; Liu, Z.; Liu, X. A Simple Route for Hierarchically Porous Carbon Derived from Corn Straw for Supercapacitor Application. *J. Renew. Sustain. Energy* **2019,** *11*, 024102.

Marichi, R. B.; Sahu, V.; Sharma, R. K.; Singh, G. Efficient, Sustainable, and Clean Energy Storage in Supercapacitors Using Biomass-Derived Carbon Materials. *Handbook Ecomater*. **2017,** 1–26.

Na, R.; Wang, X.; Lu, N.; Huo, G.; Lin, H.; Wang, G. Novel Egg White Gel Polymer Electrolyte and a Green Solid-State Supercapacitor Derived from the Egg and Rice Waste. *Electrochim. Acta* **2018,** *274*, 316.

Najib, S.; Erdem, E. Current Progress Achieved in Novel Materials for Supercapacitor Electrodes: Mini Review. *Nanoscale Adv*. **2019,** *1*, 2817.

Nanaji, K.; Upadhyayula, V.; Rao, T. N.; Anandan, S. Robust, Environmentally Benign Synthesis of Nanoporous Graphene Sheets from Bio-waste for Ultrafast Supercapacitor Application. *ACS Sustain. Chem. Eng*. **2018,** *7*, 2516.

Qian, W.; Sun, F.; Xu, Y.; Qiu, L.; Liu, C.; Wang, S.; Yan, F. Human Hair-Derived Carbon Flakes for Electrochemical Supercapacitors. *Energy Environ. Sci.* **2014,** *7*, 379.

Rufford, T. E.; Hulicova-Jurcakova, D.; Khosla, K.; Zhu, Z.; Lu, G. Q. Microstructure and Electrochemical Double-Layer Capacitance of Carbon Electrodes Prepared by Zinc Chloride Activation of Sugar Cane Bagasse. *J. Power Sources* **2010,** *195*, 912.

Rufford, T. E.; Hulicova-Jurcakova, D.; Zhu, Z.; Lu, G. Q. Nanoporous Carbon Electrode from Waste Coffee Beans for High Performance Supercapacitors. *Electrochem. Commun.* **2008,** *10*, 1594.

Salanne, M.; Rotenberg, B.; Naoi, K.; Kaneko, K.; Taberna, P. L.; Grey, C. P.; Dunn, B.; Simon, P. Efficient Storage Mechanisms for Building Better Supercapacitors. *Nat. Energy* **2016,** *1*, 16070.

Saykar, N. G.; Phatangare, A. B.; Banerjee, I.; Bhoraskar, V. N.; Ray, A. K.; Mahapatra, S. Electron Beam Induced Synthesis of Ru-rGO and Its Super Capacitive Behavior. *2D Mater*. **2019,** *6*, 045030.

Saykar, N. G.; Pilania, R. K.; Banerjee, I.; Mahapatra, S. K. Synthesis of NiO–$Co_3O_4$ Nanosheet and Its Temperature-Dependent Supercapacitive Behavior. *J. Phys. D* **2018,** *51*, 47**5501,**

Sekar, M.; Sakthi, V.; Rengaraj, S. Kinetics and Equilibrium Adsorption Study of lead (II) Onto Activated Carbon Prepared from Coconut Shell. *J. Colloid Interface Sci.* **2004,** *279*, 307.

Senthilkumar S. T.; Selvan, R. K. Flexible Fiber Supercapacitor Using Bio-Waste Derived Porous Carbon. *Chem. Electro. Chem.* **2015,** *2*, 1111.

Simon, P.; Gogotsi, Y. Materials for Electrochemical Capacitors. *Nanosci. Technol.* **2008,** *7*, 845.

Tanwar, S.; Arya, A; Gaur, A; Sharma, A. L. Transition Metal Dichalcogenide (TMDs) electrodes for Supercapacitors: A Comprehensive Review. *J. Phys.* **2021,** *33*, 303002.

Wang, H.; Yi, H.; Chen, X.; Wang, X. Asymmetric Supercapacitors Based on Nano-Architectured Nickel Oxide/Graphene Foam and Hierarchical Porous Nitrogen-Doped Carbon Nanotubes with Ultrahigh-Rate Performance. *J. Mater. Chem. A* **2014,** *2*, 3223.

Wang, J. G.; Kang, F.; Wei, B. Engineering of $MnO_2$-Based Nanocomposites for High-Performance Supercapacitors. *Progress Mater. Sci.* **2015,** *74*, 51.

Wang, J.; Polleux, J.; Lim, J.; Dunn, B. Pseudocapacitive Contributions to Electrochemical Energy Storage in $TiO_2$ (Anatase) Nanoparticles. *J. Phys. Chem. C* **2007,** *111*, 14925.

Wang, J.; Zhang, P. X.; Liu, L.; Zhang, Y.; Yang, J. F.; Zeng, Z. L.; Deng, S. G. Controllable Synthesis of Bifunctional Porous Carbon for Efficient Gas-Mixture Separation and Highperformance Supercapacitor. *Chem. Eng. J.* **2018,** *348*, 57.

Wang, Y.; Zhao, Z.; Song, W.; Wang, Z.; Wu, X. From Biological Waste to Honeycomb-Like Porous Carbon for High Energy Density Supercapacitor. *J. Mater. Sci.* **2019,** ***54***, 4917.

Wang, Z.; Tian, S.; Xia, G.; Li, K. Discussion on Green Energy Storage Component-Super Capacitor. *Low Voltage Apparatus* **2011,** *7*, 18.

Wei, H. M.; Chen, J.; Fu, N.; Chen, H. J.; Lin, H. L.; Han, S. Biomass-Derived Nitrogen-Doped Porous Carbon with Superior Capacitive Performance and High $CO_2$ Capture Capacity. *Electrochim. Acta* **2018,** *266*, 161.

Wen, Y.; Rufford, T. E.; Hulicova-Jurcakova, D.; Zhu, X.; Wang, L. Structure Control of Nitrogen-Rich Graphene Nanosheets Using Hydrothermal Treatment and Formaldehyde Polymerization for Supercapacitors. *ACS Appl. Mater. Interfaces*. **2016,** *8*, 18051.

Wu, C.; Yang, S.; Cai, J.; Zhang, Q.; Zhu, Y.; Zhang, K. Activated Microporous Carbon Derived from Almond Shells for High Energy Density Asymmetric Supercapacitors. *ACS Appl. Mater. Interfaces* **2016,** *8*, 15288.

Wu, F. C.; Tseng, R. L.; Hu, C. C.; Wang, C. C. Effects of Pore Structure And Electrolyte on the Capacitive Characteristics of Steam-and KOH-Activated Carbons for Supercapacitors. *J. Power Sources* **2005,** *144*, 302.

Xu, G.; Han, J.; Ding, B.; Nie, P.; Pan, J.; Dou, H.; Li H.; Zhang, X. Biomass-Derived Porous Carbon Materials with Sulfur and Nitrogen Dual-Doping for Energy Storage. *Green Chem.* **2015,** *17*, 1668.

Xu, M.; Huang, Q.; Sun, R.; Wang, X. Simultaneously Obtaining Fluorescent Carbon Dots and Porous Active Carbon for Supercapacitors from Biomass. *RSC Adv*. **2016,** *6*, 88674.

Yan, S.; Lin, J.; Liu, P.; Zhao, Z.; Lian, J.; Chang, W.; Yao, L.; Liu, Y.; Lin, H.; Han, S. Preparation of Nitrogen-Doped Porous Carbons for High-Performance Supercapacitor Using Biomass of Waste Lotus Stems. *RSC Adv*. **2018,** *8*, 6806.

Yang, C. S.; Jang, Y. S.; Jeong, H. K. Bamboo-Based Activated Carbon for Supercapacitor Applications. *Curr. Appl. Phys*. **2014,** *14*, 1616.

Yao, F.; Pham, D. T.; Lee, Y. H. Carbon Based Materials for Lithium Ion Batteries, Electrochemical Capacitors, and Their Hybrid Devices. *Chem. Sus. Chem.* **2015,** *8*, 2284.

Yu, D.; Chen, C.; Zhao, G.; Sun, L.; Du, B.; Zhang, H.; Li, Z.; Sun, Ye,; Besenbache, F.; Yu, M. Biowaste Derived Hierarchical Porous Carbon Nanosheets for Ultrahigh Power Density Supercapacitors. *Chem. Sus. Chem.* **2018a,** *11*, 1678.

Yu, S.; Zhu, X. Q.; Lou, G. B.; Wu, Y. T.; Xu, K. T.; Zhang, Y.; Zhang, L. M.; Zhu, E. H.; Chen, H.; Shen, Z. H.; Bao, B. F.; Fu, S. Y. Sustainable Hierarchical Porous Biomass Carbons Enriched with Pyridinic and Pyrrolic Nitrogen for Asymmetric Supercapacitor. *Mater. Des.* **2018b,** *149*, 184.

Zequine, C.; Ranaweera, C. K.; Wang, Z.; Dvornic, P. R.; Kahol, P. K.; Singh, S.; Tripathi, P.; Srivastava, O. N.;Singh, S.; Gupta, B. K.; Gupta, G.; Gupta, R. K. High-Performance Flexible Supercapacitors Obtained Via Recycled Jute: Bio-Waste to Energy Storage Approach. *Sci. Rep*. **2017,** *7*, 1174.

Zhong, C.; Deng, Y.; Hu, W.; Qiao, J.; Zhang, L.; Zhang, J. A Review of Electrolyte Materials and Compositions for Electrochemical Supercapacitors. *Chem. Soc. Rev.* **2015,** *44*, 7484.

CHAPTER 3

# Effect of Doping on the Electronic and Physicochemical Properties of the Atomic Carbon Clusters: A Theoretical Perspective

ANOOP KUMAR KUSHWAHA, SUSHRI SOUMYA JENA, and MIHIR RANJAN SAHOO

*School of Basic Sciences, Indian Institute of Technology Bhubaneswar, Bhubaneswar, India*

## ABSTRACT

The element, carbon, has been widely studied and investigated since many decades at academic and industrial level depending on its size, that is, atomic, molecular, and bulk. The essential properties such as electronic, physical, and chemical are strongly dependent on the particle size and differ significantly for carbon clusters in comparison to their bulk counterparts. In general, atomic clusters are the aggregation of few or few thousand atoms in a specific shape. The carbon clusters exist in various structures such as atomic chain, carbon dust, single/polycyclic rings, 2D sheet graphene, closed-cage fullerene, nanotube, and graphite. Recently, the study of small atomic carbon clusters is gaining attention due to their application in energy storage/harvesting devices, biological processes, and astrophysical investigations. The pristine and doped atomic carbon cluster carries various significant properties, which vary nonmonotonically with size, shape, and composition. Sometimes, properties of these small carbon

New Forms of Carbon: Nanocarbons. Aneeya Kumar Samantara & Satyajit Ratha (Eds.)

clusters change drastically with addition/removal of even a single atom. In this chapter, we have discussed the geometrical structure, electronic properties (e.g., ionization potential, electron affinity, polarizability, binding energy, etc.), physical and chemical properties, of small carbon clusters along with the effect of doping with various atoms, especially boron, on the pristine carbon cluster. Furthermore, we have discussed the application of pristine and doped carbon clusters in the field of energy storage/harvest.

## 3.1 INTRODUCTION

The electronic, physical, chemical, and magnetic properties of the matters usually show drastic variation with the reduction in the size from bulk to nanoscale. The primary reason behind such changes in the material properties is associated with the ratio of surface and volume as well as quantum confinement. With modifications in the shape, size, and dimension of the matters, aforementioned properties show variations, which could be utilized for the engineering and designing of novel materials. In the context of size and dimension of the matters, both nanotechnology and nanoscience have emerged as one of the fastest-growing research fields since the last few decades. For example, by reducing the size of the bulk transition metal to the atomic level, the catalytic activity improved significantly (Jimenez-Izal and Alexandrova, 2018). Furthermore, the reduction in the dimension from 3D graphite to 2D graphene has opened up a plethora of applications from energy storage to biological engineering. With the development of advanced theoretical and experimental techniques, the properties of matters having various sizes have been probed for understanding their characteristics at the atomic level. On the basis of the variation of properties from bulk to atom, the matters could be divided into three distinct regimes (Arenz et al., 2007), a brief discussion on which has been provided in the succeeding section.

The matters having an average diameter of more than 100 nm fall in the first regime, where the properties of the materials do not show direct dependency on the size. In the second regime, matters which are having an average size ranging from 1 to 100 nm are considered. In this regime, the properties of matters (or nanoparticles) show steady variations and follow a set of specific laws unless and until they reached the bulk limit. For example, in this regime, the ionization potential of the metallic clusters could be expressed as a function of the size, that is, the number of atoms ($N$) in the nanoparticle and work function ($W$) of the bulk material, as has been expressed in the following mathematical relation:

$$IP(N) = W + \alpha N^{\frac{-1}{3}}$$

where $\alpha$ is a constant and has a specific value for each individual atom. for example, $\alpha = 2.4$ eV for potassium. This regime is also known as the scalable regime. In the last few decades, several groups have summarized the size-dependent properties (such as electronic, physical, and chemical) of matters employing different characterization techniques (Borman, 1990; Castleman Jr, 2011; Jena, 2013a). In the third regime, matters with smaller sizes are considered and their sizes are generally characterized by the number of atoms present in the structure. The material having a size of up to 100 atoms falls in the third regime, which is otherwise known as the nonscalable regime because of quantum confinement and boundary effects. The quantum size effect controls the matter's properties.

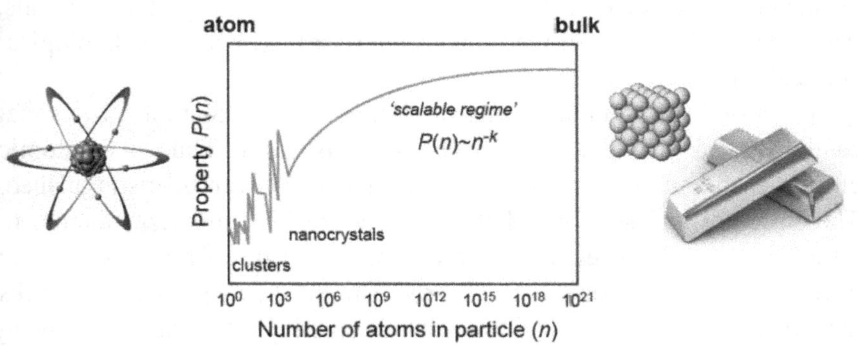


**FIGURE 3.1** Size dependence of a cluster property $P(n)$ on the number ($n$) of the cluster constituents.

### 3.1.1 *ATOMIC CLUSTERS*

The materials, consisting of a few atoms to a few hundred atoms, are called clusters. The atomic clusters provide a bridge between atoms and the bulk. Similar to a molecule, clusters are basically the aggregation of atoms; however, both can be distinguished on the basis of size, composition, and interaction. For example, size and composition might vary in the case of clusters, while they both are fixed in the case of a molecule. Further, the clusters have a tendency to strongly interact with each other and form large clusters, whereas the interaction between two or more molecules has been found to be very weak. In addition, the clusters exhibit a large number of isomers in which the atoms could be arranged in various geometrical configurations. With the variation of the size, the cluster of a nonmagnetic element can show magnetic properties, and a cluster of a semiconducting element might possess metallic behavior and vice versa. Furthermore, through the changes in cluster size, noble metals become reactive. Such behavior of clusters is associated with the unusual structure and the presence of electrons in the molecular orbitals. The molecular orbital analysis provides information regarding the energy gap which is the difference between the highest occupied molecular orbital (HOMO) and the lowest unoccupied molecular orbital (LUMO). Through the study of molecular orbitals including HOMO and LUMO energy levels, one can not only get information about stability but also several associated properties about the clusters. Therefore, the detailed study of the structural parameters and electronic and physicochemical properties of the atomic clusters of various elements provides a new way to design the materials, which could play an important role in energy storage and biotechnological application.

The atomic clusters may consist of single or hetero-atom species that can be studied individually because of mass isolation. Thus, in an atomic cluster, each and every atom as well as an electron can be distinguished. The structure and stability of the clusters are two important factors to afford unique properties. For example, mass spectra of the clusters with different number of atoms show clearly distinguishable pronounced peaks in comparison to others, which exhibit high stability. The clusters, showing conspicuous peaks in the mass spectra, are known as a magic number. The magic numbers differ with the selection of elements. Similar to the stability, the equilibrium geometry of the clusters does not exhibit any resemblance with their respective bulk. Furthermore, the size-dependent evolution of atomic clusters differs from one element to other. The atomic clusters consist of the same number of atoms, though different elements can have different atomic structures. In Figure 3.2, carbon, sodium, magnesium, and argon clusters comprising 20 atoms each have been schematically depicted, which shows that the low-lying geometrical structures differ from element to element. The primary reason behind such differences in the geometrical structure of the cluster is associated with the type of forces between the atoms inside the cluster.

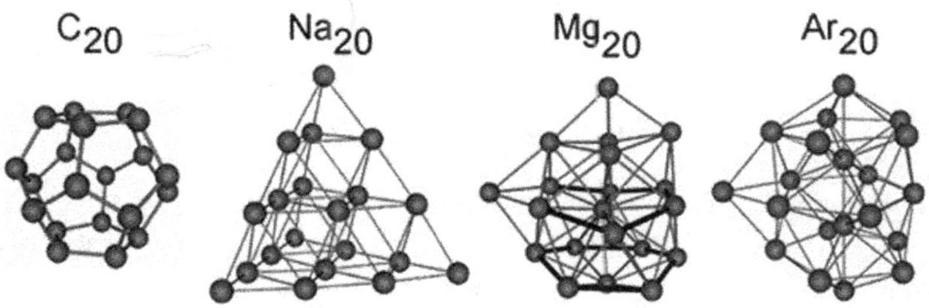


**FIGURE 3.2** Carbon, sodium, magnesium, and argon cluster with size of 20 atoms, which shows that the low-lying geometrical structures differ element by element.

In nature, atomic cluster exists in all forms of matter, that is, solid state, gas, liquid, and plasma. These clusters can be categorized into various types on the basis of availability, for example, metallic cluster, van der Waals cluster, fullerene, semiconductor cluster, and mixed cluster. The most common shape in which the clusters are formed is spherical in nature; however, sometimes the atoms in a cluster might arrange in completely different shapes such as linear, spiral, and arborescent. The atomic clusters are distinguished from the other clusters on the basis of the nature of the forces between the atoms, or the spatial arrangement of the atoms inside the clusters. In the last few decades, with the development of experimental techniques, clusters have been generated by several reported methods. Earlier, the atomic clusters were generated through the pulsed laser ablation technique, where the laser was used to locally heat up the target material for evaporation (Bondybey and English, 1982; De Heer, 1993). Later, spark discharge generation method was introduced, where an externally applied potential difference is used as a cluster generation source (Schwyn et al., 1988). Recently, this method has become one of the effective tools to generate small particles, where the process of cooling down the ion cloud at temperature below the boiling point occurs (Tabrizi et al., 2009). Furthermore, atmospheric pressure spark ablation is another method that has been developed for the generation of small atomic clusters (Maisser et al., 2015). Apart from these experimental techniques, theoretical models based on the ab-initio first principle method and molecular dynamics have been developed for the understanding of the geometrical structure, electronic, and physicochemical properties. The experimental and theoretical investigations have found various types of clusters, which have been discussed in detail including both geometry and stability, in the following section.

### *3.1.1.1 NOBLE GAS CLUSTER*

These types of clusters comprise noble gas atoms which are chemically inert due to their closed electronic shells. The bonding between the atoms is weak, which is generally governed by the van der Waals interaction. In this case, the electronic structure does not play any significant role, and therefore, the geometrical structures are determined by the close-packing model. For some specific-sized clusters, the atomic shells are found to be closed, which leads to high stability in the clusters and the size is represented by the magic numbers. These clusters are assumed to have a typical icosahedral pattern that is calculated by the following equation:

$$N = [\{(10/3)\times K_3\} - (5 \times K_2) + \{(11/3) \times K\} - 1],$$

Here, $K$ is the number of shells. Corresponding to the $K$ value ($K$ = 2, 3, 4, ...), the cluster with sizes 13, 55, 147, ... are found to be the most stable structures as shown in Figure 3.3. Experimentally, the stability for magic number cluster is further confirmed by mass spectra analysis. For example, in Figure 3.3, the mass spectra analysis of xenon cluster shows conspicuous peaks at 13, 55, and 147 which signify a stable cluster.

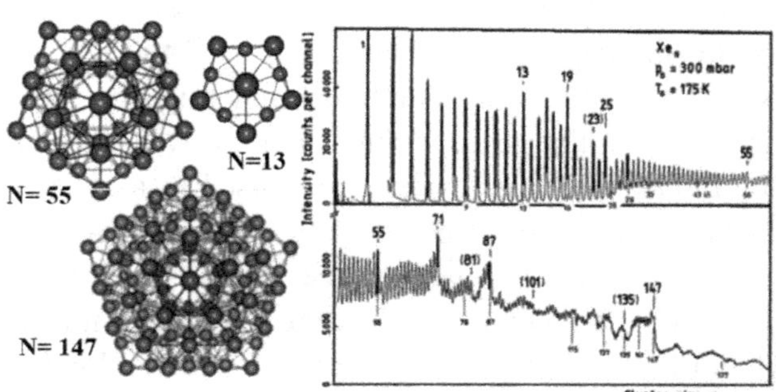


**FIGURE 3.3** (A) The icosahedric structure of noble gas with sizes 13, 55, and 147 atoms. The peaks of mass spectra of Xe clusters (B) show the magic numbers.

*Source*: A- Reprinted with permission from Jena et al., 2018. Copyright © 2018, American Chemical Society. B- Reprinted with permission from Echt et al., 1981.

### 3.1.1.2 SIMPLE METAL CLUSTER

In general, simple metal clusters consist of alkali metal or alkaline earth metal, which has been well investigated by numerous experimental and theoretical techniques in the last few decades. The first principle-based theoretical calculation has been widely applied for the structural, electronic, and vibrational analysis (Chandrakumar et al., 2004; Donoso et al., 2014; Fournier et al., 2003; Guo et al., 2010; Kumar Kushwaha and Kumar Nayak, 2017; Kümmel et al., 2000; Kushwaha et al., 2017; Lin et al., 2006; Manninen, 1986; Wheeler et al., 2004). The shell structure of such metal clusters could be obtained by the number of electrons as well as the available energy state for the electrons. Figure 3.4 shows the geometrical structure of highly stable small Na clusters, including clusters having 55–61 atoms. The numbers of electrons corresponding to close electron shells in the sodium clusters are 8, 20, 40, and 58, which can be confirmed by the mass-spectra analysis. The cluster of aforementioned size, also called magic number cluster, is found to have higher stability in comparison to their neighbor. In the mass spectra of sodium, each peak corresponds to the magic number cluster as shown in Figure 3.4. The formation and bonding of the small metal clusters could be explained by the jellium model. In this model, the cluster is defined by the radius with the homogeneously distributed positive charge on each and every atom.

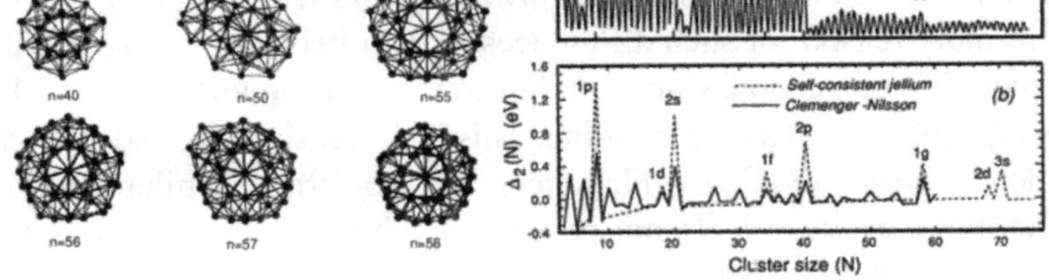


**FIGURE 3.4** Ground-state stable geometry of Na clusters.

*Source*: Reprinted with permission from De Heer., 1993; and Lee and Kanhere, 2007.

#### 3.1.1.3 TRANSITION METAL CLUSTER

The transition metal clusters are composed of transition metal elements with a specific shape and size. In transition metals, the free atom has an incomplete d-shell in the ground state or in an excited state with small energy. The electrons occupied in the d-shell are responsible for all the properties including metallic characteristics at the atomic level as well as at the bulk level. In the case of the transition metal clusters, most of the properties of elements reflect the localized behavior of the d-electrons; the delocalized behavior of the external sp electrons also plays a significant role (Alonso, 2000; Alonso and March, 2012; Cohen et al., 1987; de Heer et al., 1987). The most important characteristic of clusters such as electronic shell formation and closing shell effect for magic number occurs, due to the delocalized character. The electronic shell effects present in the clusters of coinage metals, for example, Cu, Ag, and Au, as d-shell electron form full $d^{10}$ shell in such atoms. Both the jellium model and hard sphere model failed to explain the formation and stability of transition metal clusters, because mass spectra of iron cluster show high-intensity peak for cluster sizes 7, 13, and 15.

#### 3.1.1.4 COVALENTLY BONDED CLUSTER

In the periodic table, C and Si are the two most important elements which form a covalently bonded cluster. The carbon clusters have been discussed in the next section, while the current section is focused only on silicon clusters. C and Si both belong to the same group in the periodic table; however, their bonding patterns are different. Graphite, the bulk form of carbon, always shows metallic behavior while diamond shows semiconducting behavior. On the other hand, silicon always shows semiconducting nature. The primary reason for such differences in their behavior is associated to their bonding. In general, silicon prefers to form sp3 bonding while carbon forms sp, sp2 as well as sp3 bonding. This is the reason why carbon forms a fullerene structure ($C_{60}$) while silicon fails to form a similar structure. Furthermore, Si clusters do not possess the chain or ring structure as in the case of C clusters. The dissociation experiments provide the existence of the silicon magic cluster with sizes $Si_6$ and $Si_{10}$. The spectra of small/medium silicon clusters have been represented in Figure 3.5, peaks of which show higher stability in comparison to the neighbor (Bloomfield et al., 1985).

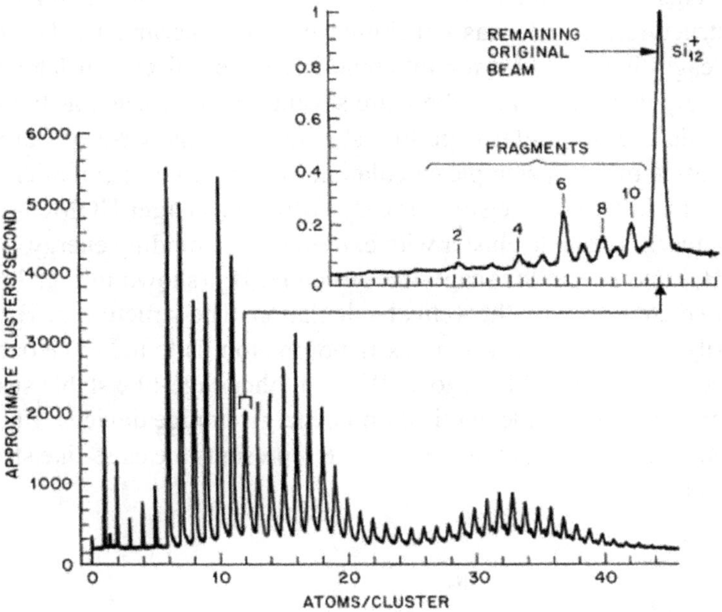


**FIGURE 3.5** Spectra of small and medium size silicon clusters.
*Source*: Reprinted with permission from Bloomfield et al., 1985.

### *3.1.1.5 IONICALLY BONDED CLUSTER*

In the ionically bonded clusters, the atoms are bonded by ionic interactions. Such kind of interaction is found in the alkali halide and metal nitride clusters. The major difference between other clusters and ionically bonded clusters is associated with the behavior of elements at the atomic level and bulk level. In general, the behavior of the small atomic clusters does not show a resemblance with their respective bulk; however, in case of the ionically bonded clusters, the properties exhibit a great resemblance with their bulk part. For example, the dimer of tungsten oxide $(WO_3)_2$ possesses properties similar to their bulk (Sun et al., 2004). Recently, the "baby" crystal of PbS has been found and its 32 monomers show resemble with the bulk crystalline form (Kiran et al., 2012).

#### *3.1.1.6 CLUSTER WITH CAGE STRUCTURE*

The cage-like structure was originated due to the unusual stability of the atomic cluster in such structures. $C_{60}$ fullerene is one of the first hollow-cage structures, which was confirmed by the experiments. The details of the cage-like structure for fullerene have been discussed later in this chapter. Apart from carbon, there are several elements such as boron and gold, which form similar cage-like structures. In the case of boron, the combination of first-principle calculations and photo-electron microscopy identified a fullerene-like structure, $B_{40}$, which has larger HOMO–LUMO gap, that is, it's a stable cluster with extremely low binding energy. Similar to the $B_{40}$, another hollow-cage-like structure, $B_{80}$ (shown in Fig. 3.6), has been predicted through theoretical calculation. The structure of $B_{80}$ has a similarity with $C_{60}$; however, an extra boron atom is found to be occupied at the center of each 20 hexagons. $B_{12}^{2}$ is another cage-like stable structure of boron, which has wide application in energy storage devices, when it is functionalized with hydrogen. Figure 3.6 represents the cage-like structure of $B_{12}^{-2}$ and $(B_{12}H_{12})^{2-}$.

**FIGURE 3.6** Equilibrium and optimized structure of $B_{80}$, $B_{12}$, and $(B_{12}H_{12})^{2-}$.

*Source*: Reprinted with permission from Prasad and Jemmis. 2008; and Szwacki et al. 2007.

### *3.1.2 GENERAL CHARACTERISTICS OF ATOMIC CLUSTERS*

As discussed in the previous section about the classification of small atomic clusters, this section leads toward the general characteristics such as morphology, electronic properties, and magnetic properties.

### 3.1.2.1 GEOMETRICAL STRUCTURE

For the understanding of the electronic and physicochemical properties of atomic clusters, it is essential to achieve an accurate geometrical structure. The atomic structure plays an important role in case of clusters, as a specific size of the cluster can have various isomers with very close ground-state energy values. Finding the accurate geometrical structures for the cluster is always a crucial problem for the experimentalist as well as the theorist. In the current scenario, no such experimental technique is available which could independently determine the geometrical structure of the cluster including structural parameters such as bond length and bond angles. The available experimental techniques such as ion mobility, electron diffractions, Raman spectroscopy, and infrared spectroscopy provide essential information about clusters' structure. The theoretical technique based on the first principles calculations provides a way for the determination of the structure. With an initial guess about the structure, the cluster is optimized by a theoretical algorithm such as the Berny algorithm and generic algorithm for various isomers. The isomer having the lowest ground-state energy is considered as the correct geometrical structure. Recently, the combination of IR spectroscopy and first principle calculations has provided direct information about the geometrical structure for both monoatomic and heteroatomic clusters. In the following section, the geometrical structure of the carbon clusters with different sizes has been discussed in detail.

### 3.1.2.2 STABILITY

Stability is another important factor in small atomic clusters. In general, the stability of the cluster is governed by two factors, that is, electron shell closure and atomic packing. These factors are responsible for the formation of magic numbered clusters, which are highly stable in comparison to the neighbor. The stability of the cluster in terms of the magic number has been already discussed for various types of clusters in the previous section. Further, the cluster stability can be investigated in terms of binding energy, second-order binding energy, and HOMO–LUMO gap. The clusters, where the atomic interaction is more, have higher binding energy, leading to high stability. Similarly, the clusters with wider HOMO–LUMO gaps generally have high chemical stability. The second difference of the binding energy provides information about the relative stability of the specific size of clusters in comparison to the neighbors. Furthermore, the fragmentation energy is another quantity that provides qualitative information about the stability. The higher value of fragmentation energy and higher negative value of the second difference of binding energy lead to higher stability in the cluster. The normalized binding energy ($E_b$), second difference of binding energy ($\Delta^2 E_b$), and fragmentation energy ($D(n, n-1)$) for the cluster ($M$) of size $n$ have been expressed with the following relations:

$$E_b(M_n) = \frac{nE(M) - E(M_n)}{n}$$

$$E_b(M_n^+) = \frac{(n-1)E(M) - E(M^+) - E(M_n^+)}{n}$$

$$D(n, n-1) = E(M_{n-1}) + E(M) - E(M_n)$$

$$\Delta^2 E_b(n) = 2E_b(n) - E_b(n-1) - E_b(n+1)$$

#### *3.1.2.3 ELECTRONIC AND CHEMICAL PROPERTIES*

On the basis of electronic structures, materials could be classified into three classes. In the first kind of materials, the electronic structures show very small overlapping between the atomic orbital as in the case of noble atom clusters, where the clusters are weakly bound systems. In the second case, the valence electron of the atoms becomes free from their ions; therefore, the energy gap vanishes at the Fermi level. In the last case, that is, in the semiconductor and insulators, the bands (valence and conduction bands) are separated by a significant band gap. In the case of the clusters, the atomic energy levels are represented by the molecular energy levels rather than the energy band. With increasing cluster size, the atomic energy levels exhibit broadening as shown in Figure 3.7, where for Mg clusters, the HOMO–LUMO gap changes with size and shows bulk behavior for critical size clusters. Other electronic properties such as ionization potential, electron affinity, show odd–even variation with cluster size, especially in the case of lithium clusters (Dugourd et al., 1992; Kushwaha et al.,

2017). Similar to the variation of the ionization potential of the lithium cluster, both free energy and redox potential show odd–even oscillation.

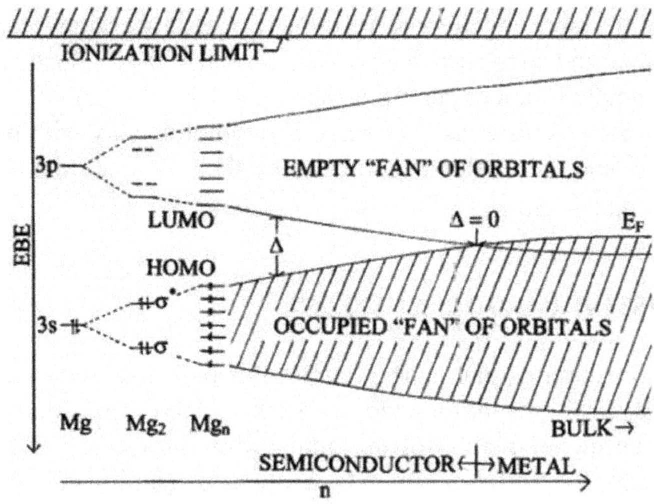


**FIGURE 3.7** Evolution of electronic structure of magnesium represented with a schematic diagram.

*Source*: Reprinted with permission from Thomas et al., 2002.

### 3.1.2.4 MAGNETIC PROPERTIES

The magnetic properties of the clusters are generally applicable for the development of high-density magnetic storage materials as well as nano-electronic devices. The total magnetic moment of clusters is composed of two major components, that is, spin magnetic moment and orbital magnetic moment of the electron. In general, the spin-magnetic moment of electrons is dominant in the total magnetic moment of clusters. The primary reason behind such domination is associated with the smaller value of the orbital magnetic moment in comparison to the spin magnetic moment, which leads to domination in the total magnetic moment. The magnetic moments are affected by the interatomic distances between the atom as well as coordination and dimension. With the increase in the cluster size, the overlap between electrons at neighboring sites increases, which leads to a reduction in the magnetic moment. For example, in case of the Ag clusters, the total magnetic moment shows odd–even oscillation, while with vanadium doping, the total magnetic moment shows a reduction with increasing cluster size (Xiong et al., 2017). Furthermore, the atoms in the linear chain are more magnetic than the surface or bulk atom. As most of the atoms in clusters are at the surface, lower dimension, lower coordination, and increased interatomic distance cause clusters to have higher magnetic moment per atom than bulk. The magnetic moment per atom of clusters is observed to change nonmonotonically with increasing cluster size, and their value lies in between those of individual atom and atoms in a bulk system.

## 3.2 PRISTINE CARBON CLUSTERS

With the detection of pure carbon molecules before a century ago from the astrophysical source, the study of such molecules has been of great interest. The detailed study of the pure carbon molecule has been able to provide the scientific community plethora of significant information in order to investigate the chemical characteristics and compositions of carbon stars, comets, and interstellar clouds. Apart from the carbon molecule, another form of carbon, that is, "carbon cluster," has been detected in the hydrocarbon flames and soot-forming flames. The flexibility in the formation of bonding, that is, single, double, and triple bonds, the carbon clusters appear in a wide range of structural forms that are spontaneously being synthesized in the hot carbon plasma. The evolution of carbon cluster structure, from linear chain to ring and from close-spherical cage to nanotube, creates a challenge and requires the development of both experimental and theoretical techniques. After the discovery of fullerenes in 1985, several stable sp-allotropes of carbon were proposed, one of which consisted of long acetylenic chains of sp-hybridized carbon atoms, that is, polyynes. This carbon chain is observed in the interstellar medium and the larger carbon chain species were developed in the laboratory environment. Therefore, the detailed knowledge of the physical and chemical properties of carbon clusters is important for understanding the properties as well as for the implementation of these clusters in various kinds of applications.

The sp-hybridized pristine carbon molecules are found in several structures such as chain, monocyclic, and polycyclic. These molecules are transient species that are highly reactive; therefore, it is very difficult to produce and study their physicochemical properties. Nevertheless, with the development of efficient theoretical method and experimental technique, great progress in the study of these systems has been reported in the last two decades. The structural analysis of the anionic carbon clusters ($C_n^-$) with the photoelectron spectroscopy shows a linear structure, for $n$ = 2–9, and monolayer rings for the $n$ = 10–29 atoms (Yang et al., 1988). The structural analysis with the ion-chromatograph shows a significant difference in the structure of the anion and cation of the carbon clusters (Gotts et al., 1995). For example, linear structures are found for $C_{30}^-$ anion cluster while cationic clusters are not found beyond $C^+_0$ cluster (Gotts et al., 1995). The annealing investigation of anion isomers for $n$ > 30 has resulted in the discovery of monocyclic rings, bicyclic rings, tricyclic rings, graphite, and fullerene structures. In this section, we have discussed the structural parameters, electronic structure, and physicochemical properties of the pristine carbon clusters including their cationic and anionic parts.

### 3.2.1 GEOMETRICAL STRUCTURE

For the investigation of any clusters, it is important to obtain the stable and optimized isomer of the specific-sized clusters. In this section, we have discussed about the geometrical structure of the carbon cluster ($C_n$; $n$ = 2–10), which is depicted in Figures 3.9 and 3.10. In the case of the carbon dimer ($C_2$), the only possible configuration is a linear arrangement with different bond lengths. Earlier theoretical investigation has found the bond length of C–C between the range of 1.24–1.41 Å by employing different methods (Fura et al., 2002; Raghavachari and Binkley, 1987); however, the experimental study shows a bond length of 1.24 Å (Huber and Herzberg, 1979). Few reports exhibit the bond length to be 1.325 Å, which corresponds to the typical C–C double bond (Kosimov et al., 2008). There are two possible configurations, that is, linear and triangular, which have been found in the case of the trimer carbon cluster ($C_3$). Here, the linear configuration shows minimum energy, hence higher stability in comparison to the triangular configuration. In the linear configuration, the average bond length between the carbon atoms is 1.310 Å, which has been found from experimental as well as theoretical investigations (Martin et al., 1995; Zhang et al., 2002). Triangular structures are the second-most stable structures, which have a bond angle of 60° between the C–C–C bonds as shown in Figure 3.9(A) (Kosimov et al., 2008). Other irregular triangular structures are found to have much higher energies, that is, they have less stability; therefore, such structures could not be a part of the investigation.

The carbon clusters with size 4 along with various isomers have been shown in Figure 3.8(A). In the case of the linear structure 4(1) (in Figure 3.8(A)), the inner and outer bond lengths have been observed to be close values, that is, 1.331 and 1.330 Å, respectively (Kosimov et al., 2008). However, few reports suggest that the outer bond length is longer than the inner one (Jena, 2013b; Martin et al., 1995). Apart from the linear configuration of $C_4$, other isomers have also been shown in structure 4(2–5) (in Figure 3.8(A)). The threefold symmetry structure 4(2) with a bond length 1.364 Å shows the next stable structure after the linear structure. Furthermore, branched structure, 4(3), consists of single and double bond for its triangular and outer atom part, respectively. Other configurations of $C_4$ clusters, that is, square, rhombic, and tetrahedral structures, are found with comparatively lower stabilities; however, few reports have found that rhombic structure is more stable than the linear form (Raghavachari and Binkley, 1987; Whiteside et al., 1981). For the cluster size 5 (C5), there are 11 structures which are depicted as 5(1–11) in Figure 3.9(A). In the case of linear configuration, 5(1), the inner bond length is found longer than the outer bond length, similar to the earlier cases. The regular cyclic pentagonal structure with the bond length 1.437 Å and internal bond angle 108° has been shown in Figure 3.8(A). The irregular cyclic structures of stable form (5(4, 6, 8, 10) and highly unstable form (5(5, 7)) with lesser symmetry have been shown in Figure 3.8(A).

The stable geometries of $C_6$ have been depicted in Figure 3.8(B), where the cyclic structure (benzene ring) has been found to be the most stable structure than other structures including linear configuration. The benzene ring has been recognized as the ground state for the $C_6$ clusters, which has a slightly higher total energy than that of linear configuration. The structures, 6(3–11), are planar, while the structures, 6(12–17), are nonplanar in nature. Pentagonal pyramidal 6(14), triangular prism 6(15), fish-like structure 6(16), and square di-pyramidal 6(17) show lower stability in comparison to cyclic and linear configurations. Figure 3.8 (C and D) shows the stable geometries of $C_7$ and $C_8$, respectively. Their ring-like structure has been found as a ground-state structure. Apart from the ring-like structure for C7, there are other 18 (7(2–19)) structures, both planar and nonplanar has been investigated, which are found to have comparatively lower stability. In the case of $C_8$, the planar and nonplanar structures along with their combinations show about 29 stable structures. 8(1–29), as depicted in Figure 3.8(D). Similar to $C_7$, the ring structure, 8(1), of cluster $C_8$ has been found to have comparatively higher energy and is therefore considered as the ground state. It is noticed that seeds for the planar graphene structure are found for $C_8$ as double-ring structure consists of two pentagons 8(8).

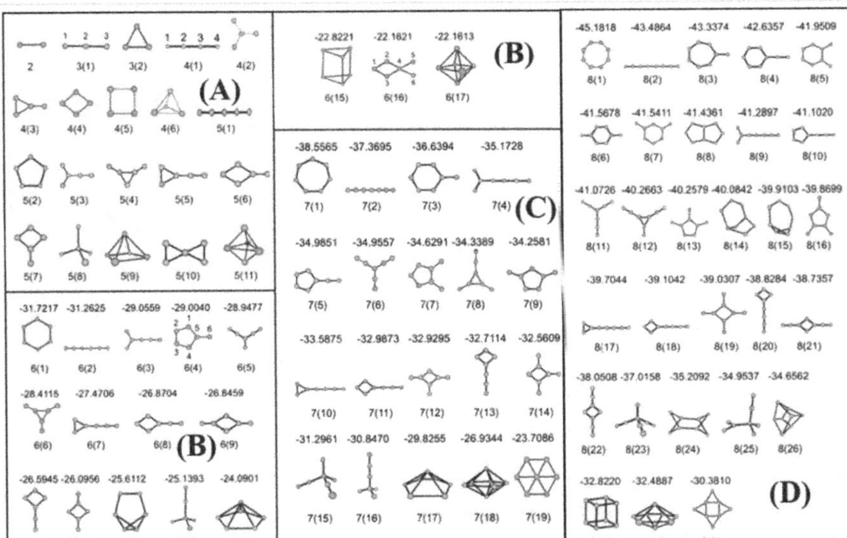


**FIGURE 3.8** Different isomers of carbon clusters from 2 to 8.

*Source*: Reprinted with permission from Kosimov et al., 2008.

The stable structures for the clusters with atomic sizes 9 and 10 (C9 and C10) have been depicted in Figure 3.9(A and B), respectively. The modification in the cyclic configuration of the C9 and C10 structures creates other structures which lie between the most stable ring and their linear structure. In both cases, the cyclic structures are the most stable structures, which can be considered as the ground state. The comparative analyses of the bond length (C–C) for both cases, that is, linear and cyclic ring with increasing cluster size, have been depicted in Figure 3.10. In the case of cyclic structures, the bond length decreases initially and attains saturation, while the bond length in the linear configurations does not show any change after $C_4$. Furthermore, the bond length for the cyclic case is comparatively longer in comparison to the linear configuration. The ground-state configuration of the higher numbered clusters $C_n$ ($n = 11–55$) has been depicted in Figure 3.9(C). The ground state has been determined using the energy minimization technique with Brenner potential function (Kosimov et al., 2010).

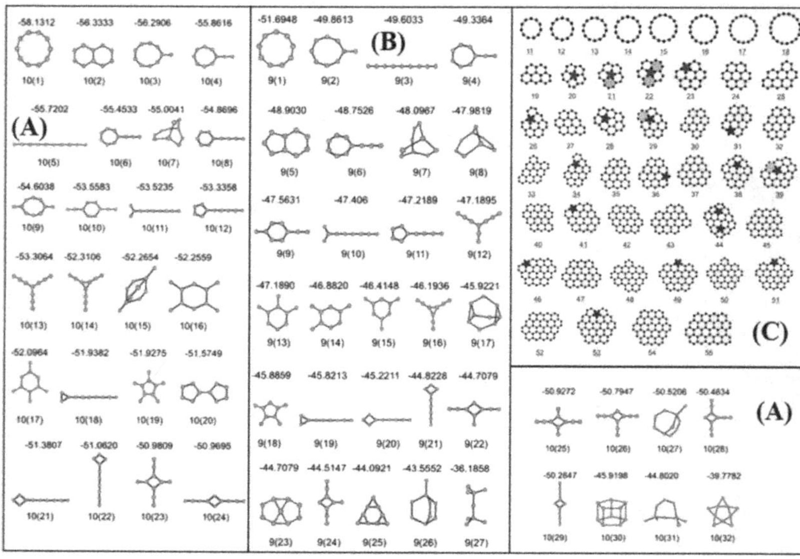


**FIGURE 3.9** (A) and (B) represent the different isomers of carbon clusters from 9 to 10 and (C) is the most stable isomer of carbon clusters with a range of 11–55.

*Source*: Reprinted with permission from Kosimov et al., 2008.

### 3.2.2 *ELECTRONIC PROPERTIES*

The complexity and diversity of the bonding between the carbon atoms within the clusters are of great interest due to their application as potential components. To explore their wide range of applications, it is essential to study electronic properties, which could provide initial information about the nature of interaction between the atoms. Furthermore, with the doping of the other atoms in the carbon clusters, the electronic properties could be tuned to suit other applications. The normalized binding energies of the most stable cyclic and linear structures of carbon cluster are depicted in Figure 3.10(B) and compared with the earlier results calculated with Hartree–Fock (HF) method (Kosimov et al., 2008; Raghavachari and Binkley, 1987). In both cases, the normalized binding energy initially increases steeply and later increases gradually with an increase in the cluster size. It has been found that linear structures are highly favorable for small-sized clusters while cyclic structures are stable for clusters with atomic size equal to or greater than 6. Tomanek et al. have investigated the binding energy of the linear carbon clusters through the combination of the adaptive simulation and tight-bonding method (Tománek and Schluter, 1991). Their study reveals the odd–even variation of normalized binding energy with increasing cluster size and higher energy values for odd clusters in comparison to their neighboring even clusters.

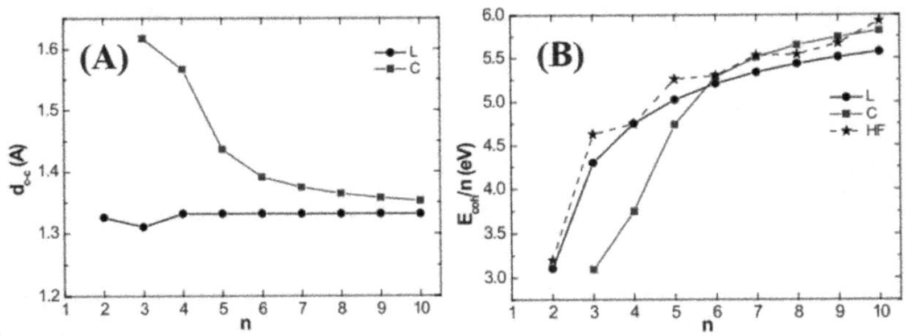


**FIGURE 3.10** Figure (A) represents the changing in the bond length for linear and cyclic carbon cluster. Figure (B) shows the variation of normalized binding energy with linear and cyclic carbon clusters.

*Source*: Reprinted with permission from Kosimov et al., 2008.

Ionization potential and electron affinity are two fundamental properties of the carbon clusters, which are highly influenced by the geometrical structures. The determination of ionization potential with high accuracy is still a challenge in the case of carbon clusters; however, in recent years, first principle-based theoretical technique has been applied for the prediction of the ionization potential of clusters which is depicted in Figure 3.11(A). The first and second ionization potential values exhibit drastic changes with increasing cluster size, as shown in Figure 3.11(A). The first ionization potential decreases slowly with cluster size nonmonotonically, as some oscillations are present with the increasing size. The second ionization potential also decreases with increasing cluster size; however, it shows a stable value for cluster size $C_8$–$C_{10}$. The first ionization potential is smaller than the second ionization potential and their difference is larger for small-sized clusters. The primary reason behind such behavior could be attributed to the accommodation of multiple charges for larger cluster in comparison to small clusters. Similar to the ionization potential, electron affinity is also a very sensitive quantity and drastically changes with cluster size. The electron affinities including vertical electron detachment energy, adiabatic electron affinity (AEA), and their comparison with experimental results have been depicted in Figure 3.11(B), for increasing carbon chain size. The theoretically calculated vertical and AEA show odd–even alternation with increasing size of carbon clusters, which is in good agreement with the experimental results (Lépine et al., 2002). The electron affinity for the even-numbered carbon clusters seems to have larger value in comparison to odd-numbered clusters due to the partial occupying nature of the π-orbital of linear even-numbered clusters.

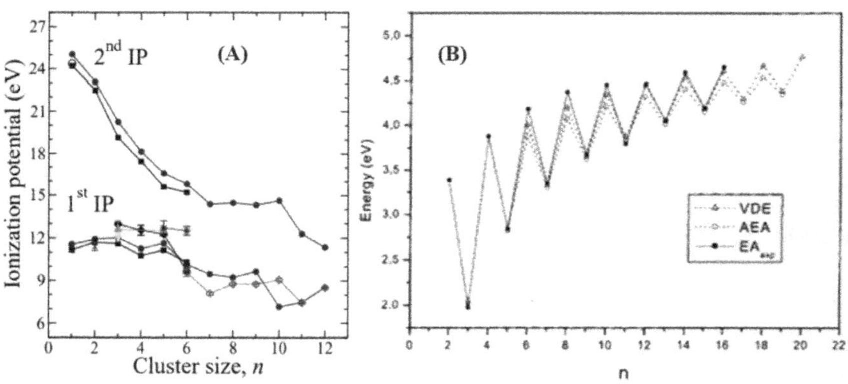


**FIGURE 3.11** (A) First and second ionization potential of carbon clusters with increasing cluster size. (B) Electron affinities of linear carbon chains as a function size.

*Source*: (A) Reprinted with permission from Diaz-Tendero et al., 2006 and (B) Lepine et al. 2002, Copyright © 2002, American Chemical Society

The discussion of the electronic properties of the carbon clusters is not complete without mentioning the energy band gap and density of states (DOS). The energy band gap is represented by the nonavailability of state at a particular energy level. Those states which are comparatively lower than the band gap are defined as the occupied states and the highest occupied orbital is known as HOMO. Similarly, those states which have higher energies than band gap are defined as unoccupied orbital and the state with the lowest possible occupancy is known as LUMO. The difference of the energy levels, that is, HOMO–LUMO energy gap, provides vital information about the stability, reactivity, and other physicochemical properties of molecules and clusters. The variation of the HOMO–LUMO energy gap occurs with increasing carbon cluster size (specific structures are taken from Qiang et al. (2015)). The variation shows a nonmonotonic nature with an increase in the number of atoms in the clusters. The HOMO–LUMO energy gap of carbon clusters has been found to fall in the range of 1.19–2.52 eV (except dimer), which shows their semiconductor-like behavior. The DOS of a system describes the number of states per interval of energy at each energy level that are available to be occupied by the electrons. A high value of DOS at a particular energy interval suggests that multiple states are available for the occupation of the electron, while a zero DOS indicates that no states are available for the electron to be occupied at the particular energy interval. The partial density of states for the carbon cluster $C_4$ shows the availability of s and p states at the Fermi level. Furthermore, it has been found that with an increase in cluster size, the energy of s-state falls while that of p-state rises (Qiang et al., 2015).

## 3.3 DOPED CARBON CLUSTERS

As discussed in the previous section, the electronic properties of the carbon clusters show drastic changes with increasing cluster size. These size-dependent cluster properties of the carbon clusters are altered by the introduction of dopant atoms. Due to the doping of an atom in the small atomic clusters, the stability, electronic properties, reactivity, optical, and magnetic properties are affected. In this section, we have discussed the effect of doping of various atoms such as boron, silicon, and gold in the carbon clusters.

### *3.3.1 BORON-DOPED CARBON CLUSTER*

The ability of formation of the atomic clusters of boron resembles carbon; however, their bulk behaviors are different. By doping with the boron atom, various properties of carbon clusters are found to improve and show superior performance in comparison to the parent clusters. For example, the replacement with the boron atom in the trigonal site of graphite lattice leads to an improved resistance toward the oxidation process. Furthermore, their application in hydrogen storage devices is reported on the basis of adsorption of hydrogen atom on boron-doped carbon moieties. The study of the interaction between boron and carbon for the smaller clusters provides understanding of the process for the formation of larger clusters. The boron-doped small carbon clusters are well investigated through experimental tools and technique, such as electronic emission and absorption spectroscopy, mass spectroscopy, and IR spectroscopy. However, critical information about the bonding and electronic properties of such clusters is highly essential. In this context, advanced theoretical techniques are found to be the appropriate tools which could provide detailed and consistent information of boron-doped carbon clusters.

Saloni et al. (2011) have investigated the boron-doped carbon clusters with molecular orbital formulism-based first principle calculations and reported their structural, vibrational, and thermodynamic properties. In this section, we have discussed the characteristics of boron-doped clusters briefly on basis of recent reports (Saloni et al., 2011). The molecule, BC, is the smallest unit of clusters, which is widely investigated using both theoretical and experimental methods. BC exhibits a bond length of 1.48 Å, atomization energy of 98 kcal/mol, ground state $^4\Sigma^-$ ($\ldots 5\sigma^1 1\pi^2$), and ground-state frequency of 1197 cm$^{-1}$ (Saloni et al., 2011). Further, $BC_2$ clusters are found in both cyclic and linear forms; however, the cyclic form is highly stable. In the case of $BC_3$, the cyclic kite-like structure with the shortest C–C diagonal bond is found to be the most stable structure in comparison to other isomers. The vibrational analysis suggests that the vibrational spectrum of $BC_3$ is similar to that of $C_4$. For the cluster, $BC_4$, both cyclic and linear structures have been studied with first principle-based model which suggests four stable isomers. Among these isomers, the linear structure (BCCCC) has been found to have the lowest energy. In the case of the clusters, $BC_5$ and $BC_6$, the cyclic structures are found to be the most stable structures. Apart from the cyclic structure, the linear configuration has been found to be the second most stable structure (Saloni et al., 2011).

The relative variation of the boron-doped carbon structure with increasing cluster size could be categorized into two parts, that is, cyclic and linear structures. In the linear structure, one carbon atom is replaced by the boron atom inside the chain as well as at the terminal. The small linear clusters with terminal boron atom show higher stability in comparison with the boron atom present inside the chain. However, with the increase in cluster size, the linear structures lose stability. For odd and even cluster size of boron-doped carbon clusters, the ground states are represented by quartet and doublet, respectively. Furthermore, the HOMO possesses the $\pi$-character which is delocalized over the whole linear cluster. In the case of the boron present inside the linear chain, the stable linear structure is the doublet. Further, boron is involved in the conjugate bonding system and radical electrons are located on the terminal carbon atom. Their stability is gained by the stabilization energy with $\pi$-conjugations. On the other hand, larger size of boron-doped carbon clusters shows higher stability for the cyclic structures. In this case, the boron atom is a part of the conjugated system with a single electron delocalized mostly with the $\pi$-orbitals.

### *3.3.2 SILICON-DOPED CARBON CLUSTER*

With the doping of the silicon in the carbon clusters, various properties including electronic, physical, and chemical can be engineered for the application in field of micro-electronics, energy storage, and biomaterials. Furthermore, silicon-doped carbon materials have also gained remarkable attention in the understanding of the interstellar atmospheres (especially CSi, $C_2Si$, $C_3Si$, and $C_4Si$) of carbon stars and of the circumstellar chemical environment. Silicon carbide (SiC) is the smallest unit of the silicon-doped clusters, which has exceptional physicochemical properties such as high thermal conductivity, high strength, low thermal expansion, wide tunable band gap, high refractive index, and chemical inertness. Most of the micro-electronic devices are developed with the growth of films on surfaces. In such cases, SiC clusters can be considered as the building block for the synthesis of ideal materials. With the understanding of the electronic properties, various sizes of silicon-doped carbon materials can be developed. From the earlier research, it has been investigated that there is a strong resemblance between the pure carbon cluster and the silicon-doped carbon cluster which stipulates that silicon is a suitable substitutional dopant in the cluster for the enhancement of cluster's properties. Therefore, this section deals with the structural parameters, electronic properties such as binding energy, and stability of silicon-doped carbon clusters.

There are several theoretical reports which provide the details of geometrical structure and electronic properties of silicon-doped carbon clusters (Hou and Song, 2008; Pascoli and Lavendy, 1998; Pradhan and Ray, 2006; Song et al., 2010). For SiC, the bond length, binding energy, and vibrational frequency have been found to be 1.72 Å, 4.736 eV, and 954.4 $cm^{-1}$, respectively. Furthermore, the $^3\pi$ found as the ground state of the SiC (Hou and Song, 2008). The neutral $SiC_2$ has both linear and cyclic structures, whereas $SiC_3$ and $SiC_4$ have cyclic and linear structures, respectively. In case of cationic silicon carbide ($SiC_n^+$), the $SiC^+$ dimer exhibits a bond length of 1.826 Å, dipole moment of 1.254 D, and $^4\Sigma$ as the ground state. For the trimer, $SiC_2^+$, the HF and configuration-interaction singles and doubles (CISD) level of the theory suggest higher stability of linear structure (Si atom is present on the terminal of the chain), while fourth-order many-body perturbation theory (MP4) shows higher stability for cyclic structure with $C_{2v}$ symmetry. The energy difference of both the structures, that is, cyclic and linear, is very small (~0.116 eV); therefore, both structures can coexist for $SiC_2^+$ cluster (Lavendy et al., 1997; Pascoli and Lavendy, 1998; Pradhan and Ray, 2006). This coexistence of both linear and cyclic structure is also beneficial for the explanation of photoelectron spectroscopy.

In the case of $SiC_3^+$ cluster, various levels of theory including 6-311G* basis set suggest higher stability of rhombic structure with $C_{2v}$ symmetry and $^2A_1$ electronic state. The submolecule $C_3$ exhibits a bond length of 1.351 Å for each C–C bond, while a bond length of 1.930 Å has been found for Si–C bond. The silicon-terminated linear structure is the second low-lying isomer (0.637 eV higher than rhombic) of $SiC_3^+$ clusters (Pradhan and Ray, 2006). In the case of $SiC_4^+$ cluster, the silicon-terminated linear carbon chain with symmetry, $C_{\infty v}$, and electronic state, $^2\Sigma^+$, has been found to be the most stable structure by Pascoli et al., at the B3LYP/6-31G* level of theory (Pascoli and Lavendy, 1998). However, the structural investigation of Pradhan et al. shows higher stability for the fan-shaped planar structure with bond length, 1.912 Å, for Si–C, and 1.386 Å and 1.264 Å for C–C bond (Pradhan and Ray, 2006). For $SiC_5^+$ clusters, the silicon-terminated linear chain with doublet state ($^2\pi$) is energetically favored. The bond length has been found to be 1.78 Å between Si–C. In case of $SiC_6^+$, $SiC_7^+$, and $SiC_8^+$ clusters, the silicon-terminated carbon shows higher stability with corresponding electronic states, $^2\Sigma^+$, $^2\pi$, and $^2\Sigma^+$. The distance between Si and C is found to be ~1.75 Å for aforementioned clusters with single bond (Pascoli and Lavendy, 1998). In the case of silicon-doped carbon clusters with size more than 10 atoms, Si-capped pure monocyclic ring has been found to be significantly more stable than the linear configuration (Pascoli and Lavendy, 1998).

In the case of silicon-doped carbon clusters, the binding energy is strongly dependent on the size. For small cationic $SiC_n^+$ ($n \leq 4$) clusters, the binding energy corresponding to the stable structures (as discussed earlier) is 2.279 eV, 4.568 eV, 5.338 eV, and 5.617 eV for $SiC^+$, $SiC_2^+$, $SiC_3^+$, and $SiC_4^+$ cluster, respectively (Pradhan and Ray, 2006). The relative stabilities of the small clusters with function of size can be investigated with the concept of incremental binding energies. The incremental binding energy is determined from the difference between consecutive binding energies of the adjacent clusters. The incremental binding energy of $C_nSi^+$ ($4 \leq n \leq 15$) has been discussed in detail by Pascoli et al., where they found that the even linear carbon clusters are more stable than their neighbor odd clusters for size less than seven (Pascoli and Lavendy, 1998). In addition, clusters, $C_nSi^+$, with size ($4 \leq n \leq 9$), show higher stability for linear isomers, whereas the size ($10 \leq n \leq 15$) exhibits higher stability for Si-capped pure carbon monocycle structure. In the case of the neutral linear $SiC_n$ clusters, the lowest energy states can be explained with the available number of valence electron in the clusters (Li and Tang, 2003). Except for SiC dimer, the $2n + 4$ valence σ-electrons and $2n$ valence π-electrons are contained by the linear $SiC_n$ clusters. In the case of even clusters ($n$ is even in $SiC_n$ cluster), the $2n$ π-electron constituted in the form of a closed shell with $^1\Sigma$ electronic state, whereas in case of the odd clusters, two of the 2π-electrons are distributed over a pair of degenerate π-orbitals, which leads to half-filled π-orbitals and results in the lowest energy state corresponding to the open shell configuration, $^1\Sigma$ (Li and Tang, 2003).

### *3.3.3 GOLD-DOPED CARBON CLUSTER*

The transition metal atom-doped carbon clusters are getting attention due to their increasing application in the field of astrophysics and astrochemistry, catalysis, and surface sciences. Such atomic clusters have been well investigated with experimental methods as well as theoretical techniques since the few decades. For example, the early transition metals (Sc, Ti, V, Y, La, and Nb)–doped carbon clusters have been studied and a higher stability has been recorded for cyclic and fan isomers as compared to their respective linear structures (Largo et al., 2006; Redondo et al., 2006, 2005;

Roszak and Balasubramanian, 1997). The electronic structure investigation of the carbon clusters doped with late transition metals (Ni, Pd, Pt, and Zn) shows higher stability of linear isomers for small size of clusters; however, their stability for ground state is dominated by cyclic and fan-like structure for larger clusters (Barrientos et al., 2008; Miller and Hall, 1999; Strout et al., 1998). Carbon clusters doped with the noble transition metals (Cu, Ag, and Au) were also investigated; however, their negligible reactivity nature could not provide the details of the electronic structures. To understand the noble-metal-doped carbon clusters, it is important to carry out theoretical investigations. This section deals with the different geometries and electronic structures of gold-doped carbon clusters ($AuC_n$; $n = 1–11$), especially for finding their ground states.

From the previous theoretical investigation, it has been predicted that Au-doped carbon clusters ($AuC_n$) have two types of isomers; one is open-chain structure and the other is the cyclic structure (Sun et al., 2013). In open-chain structures, the gold atom is located at one end of the carbon chain, while in cyclic structure, the gold atom is bonded fundamentally to two carbon atoms of the $C_n$ unit. We can see that all the $n$-even species of the clusters are linear in structure except $AuC_2$ cluster, but in cyclic terms, all the $AuC_n$ species have doublet lowest-lying state except $AuC_6$ and $AuC_8$. Now moving toward the bonding part, it has been reported that all the C–Au bond in the open-chain $Au_{2–11}$ lies in between 1.907 and 1.924 Å in which 1.907 Å is the shortest bond length ($AuC_4$) and 1.924 Å is the longest one ($AuC_3$). On the other hand, the bond length of Au–C in the cyclic species lies in the range of 1.968–2.221 Å, which appears longer than the open-chain isomers. For the measurement of the bond strength of both the structures, that is, the open-chain and the cyclic structure, Sun et al. used Wiberg bond indices which show that in the open-chain structure, except for the dimer, AuC (1.4166), all other clusters lie in the range of 0.7523–0.9595 revealing the single bond characteristics of Au–C (Sun et al., 2013). Furthermore, the cyclic bond strength lies in between 0.7523 and 0.6420, which gives an affirmation that Au–C bond present in the cyclic structure is weaker than the open-chain structure as both bond length and strength in the case of the former are comparatively lower.

The atomic interaction between the carbon and gold exhibits ionic characteristics, where Au atom behaves as the donor and transfers the electron to C atom in the case of all open linear-chain structures. On comparing the dipole moments of both the structure, it has been found that the dipole moment of the open chain goes on increasing with the rise in the number of carbon atoms for both *n*-even and *n*-odd clusters, and predominantly, *n*-even clusters have higher dipole moments. But it is noticeable that the dipole moments of the cyclic isomers are lower than that of open-chain structures. The relative stabilities of the gold-doped small carbon clusters can be investigated through the incremental binding energies. Similar to the previous cases, the incremental binding energy can be determined from the consecutive binding energy difference between the adjacent $AuC_n$ and $AuC_{n-1}$ cluster with the lowest-lying states. The variation of incremental binding energy of gold-doped carbon clusters with increasing size shows odd–even oscillation for the open-chain structure and undefined variation for cyclic structures. The higher value of the incremented binding energy shows higher stability; therefore, the *n*-even clusters show comparatively more stability than the *n*-odd clusters. The embedded binding energy also shows the odd–even alternation and confirms higher stability of even clusters for linear structure. For the small-sized clusters ($n \leq 9$), open-chain structures are comparatively stable, while for large-sized clusters ($n = 10$ and 11), the cyclic structures are preferable to the linear ones (Sun et al., 2013).

### 3.3.4 IRON-DOPED CARBON CLUSTER

Similar to the Au-doped carbon clusters, the study of Fe-doped clusters is also an emerging topic of research. It is interesting to note that through the incorporation of even a single iron atom, the electronic, physical, and chemical properties of carbon clusters can be engineered according to our needs. In general, for neutral $FeC_n$ ($n = 1$–8) clusters, there are three low-lying structures, namely, linear, fan, and cyclic. A linear isomer with the iron atom in an intermediate position or a cyclic isomer with an exocyclic iron atom shows relatively high energy. In the Fe-doped carbon cluster, the bond distance of Fe–C in a cyclic system is found to be somewhere in between the linear and fan-like isomers. In case of a linear isomer, Fe is only bonded to a terminal C atom of the cluster due to the stronger bonding between C–C in comparison to Fe–C bond. However, in the cyclic structure, the Fe atom is bonded with the two terminal C atoms, while in the fan-like structure, the Fe atom is side bonded to the carbon chain. In the fan-like structure, the π-type interaction is dominant between the iron atom and the entire carbon chain.

Similar to the earlier cases, the relative stabilities of the $FeC_n$ clusters with different sizes can be investigated with the concept of incremental binding energies. In this case, this energy is determined from the consecutive binding energy difference between the adjacent $FeC_n$ and $FeC_{n-1}$ cluster of the lowest-lying state. It has been observed that the stability of structures (i.e., linear, fan-like, and cyclic structure) has an odd–even oscillation with increasing cluster size for both linear and fan-like $FeC_n$ clusters. Moreover, the *n*-even members of the doped cluster are supposed to be more stable than the corresponding *n*-odd ones. However, bigger size clusters, especially the linear structures, show small attenuation. Further, in the case of cyclic isomers, the incremental binding energy can only be calculated for the last members, which shows the decrement along the series, though other reports suggest the highest stabilization for $FeC_5$ and $FeC_6$ cyclic clusters. Furthermore, the relative energies of the cyclic and the fan-like structure have negative values with reference to the linear ones, which depict the extra stability for both fan and cyclic structures as compared to the linear one. The electron affinities and the ionization potentials for iron-doped clusters are the most sensitive electronic properties that vary with the cluster size. The analysis of ionization potential shows odd–even oscillation with increasing size of clusters and higher values for *n*-even clusters in comparison to odd ones. Further, the electron affinity shows increment with increasing cluster size from FeC to $FeC_4$ but afterward show steady variation from $FeC_5$ to $FeC_8$ cluster.

### 3.3.5 PHOSPHOROUS-DOPED CARBON CLUSTER

Similar to the earlier cases, the introduction of phosphorous atom in the carbon clusters shows significant changes in the electronic structure (Pascoli and Lavendy, 1999a). The cationic and anionic phosphorus-doped carbon clusters ($C_nP^+$ and $C_nP^-$) are isoelectronic with the neutral $C_nSi$ and $C_nS$ clusters, respectively, due to which many researchers have tried to investigate and understand the origin of such clusters. Earlier, Pascoli and Lavendy (1999b) investigated the geometrical structure, especially linear chain of neutral $PC_n$ ($n$ = 1–7) clusters, for finding the low-lying isomers. The P atom terminated carbon chain shows the electronic ground state $^2\Sigma^+$ and $^2\Pi$ for CP dimer and rest of the other ($n$ = 2–7) clusters, respectively. Similar to the neutral clusters, the cationic $C_nP^+$ clusters show the lowest-lying isomers with electronic state $^3\Pi$ for dimer $CP^+$ and $^3\Sigma^+$ and $^1\Sigma^+$ electronic ground states for respective $n$-odd and $n$-even clusters. In the case of anion $C_nP^-$ clusters, the linear structures are stable only for $C_2P^-$ and $C_3P^-$ clusters, and with increasing cluster size, the structure of higher $n$-odd $C_nP^-$ anion clusters shows slightly bent arrangement for the ground state with a floppy bending motion. The structural analysis suggests that C–P bond length in the CP dimer is approximately same in the case of both cation and anion, that is, ~1.62 Å, which is slightly longer than their neutral (~1.56 Å) counterpart. In the neutral doped clusters, the C–P bond length slightly decreases with increase in the cluster size except $C_2P$ clusters. In the anionic case, the C–P bond length for $n$-odd clusters is slightly shorter than $n$-even clusters (Pascoli and Lavendy, 1999b).

Again, the relative stabilities of the phosphorus-doped carbon clusters with different sizes can be investigated through the concept of the incremental binding energies. In this case, the incremental binding energy is defined as the change in energy accompanying the process, $C_nP \rightarrow C_{n-1}P + C$, and it is determined as the consecutive binding energy difference between the adjacent $C_nP$ and $C_{n-1}P$ clusters with the lowest-lying state in each case. The incremental binding energy of neutral, cationic, and anionic cluster, and ionization potential and electron affinity of neutral phosphorous-doped carbon has been depicted in Figure 3.12. The odd–even oscillation in the incremental binding energy with increasing cluster has been found in all the clusters (Liu et al., 1998). The incremental binding energy of cation infers that the $n$-even clusters are comparatively stable than that of $n$-odd ones. In the case of the neutral and anionic clusters, this variation shows contrasting behavior, where $n$-odd clusters show more stability than $n$-even clusters. The primary reason for such behavior is associated with the bonding and nonbonding nature of π-orbitals for neutral and cationic clusters, respectively (Pascoli and Lavendy, 1999b). In the case of anionic $C_nP^-$ clusters, $n$-odd clusters have fully filled π-orbital which imparts more stability than that of $n$-even one which is half-filled π-orbital. The ionization potentials of the phosphorus-doped small carbon cluster are calculated with the total energy difference between the neutral and cationic clusters, while electron affinities are calculated with the total energy differences between the neutral and the anionic clusters. Figure 3.12 shows the small odd–even oscillation or a parity effect in both the ionization potential and the electron affinity variation, with increasing cluster size. Both energies have higher values for *n*-odd clusters in comparison to *n*-even clusters.

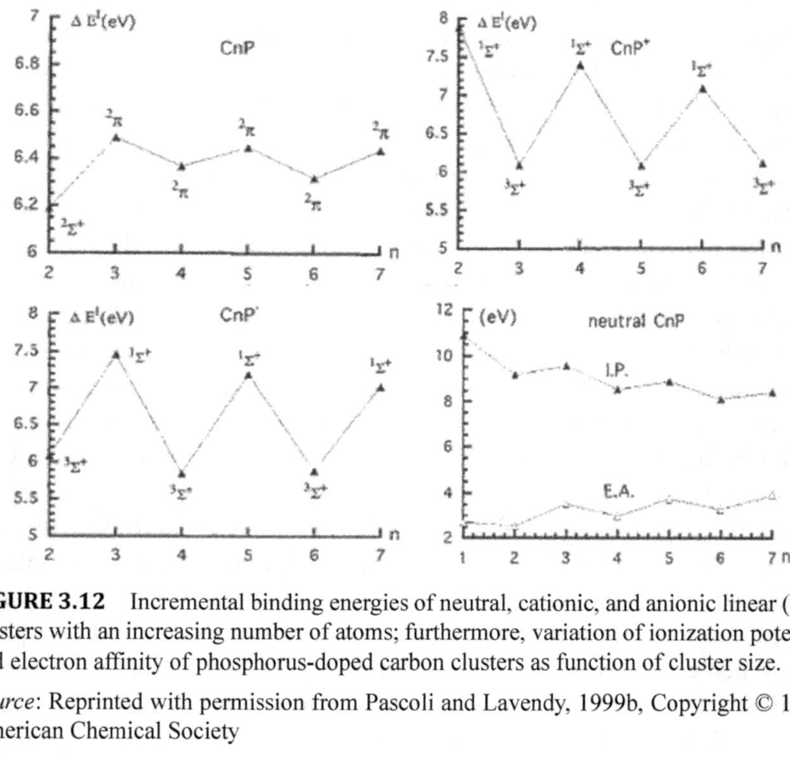


**FIGURE 3.12** Incremental binding energies of neutral, cationic, and anionic linear ($C_nP$) clusters with an increasing number of atoms; furthermore, variation of ionization potential and electron affinity of phosphorus-doped carbon clusters as function of cluster size.

*Source*: Reprinted with permission from Pascoli and Lavendy, 1999b, Copyright © 1999, American Chemical Society

## 3.4 SUMMARY AND FUTURE SCOPE

In this chapter, we have discussed about the geometrical and electronic structures of small atomic carbon clusters along with changes in their properties due to doping of various elements. In the introduction section, the basics of the atomic clusters have been briefly introduced and further these small atomic clusters were classified on the basis of forces acting between the atoms inside the clusters. With the experimental and theoretical investigations, various types of atomic clusters have been discussed in details, especially to understand the geometrical arrangement and unusual stability in the case of different atomic clusters. In general, the atomic clusters are investigated with the calculation of electronic, physical, chemical, and magnetic properties of clusters, which have been discussed along with the respective calculation method. In the case of pristine atomic carbon clusters, up to 10 atoms, the geometrical structure and electronic properties have been extensively discussed. Furthermore, with the doping of other elements like Fe, Au, Si, B, and P, the change in the geometry and electronic properties, that is, incremental binding energy, ionization potential, and electron affinity have been discussed in details. Finally, our discussion concluded that with the doping of the different elements in the small atomic carbon clusters, various properties including electronic, physical, and chemical properties can be engineered for suitable application in the field of micro-electronics, energy storage, and biomaterials.

## KEYWORDS

- **carbon clusters**
- **electronic properties**
- **first principles**
- **energy storage/harvesting devices**

## REFERENCES

Alonso, J. A. Electronic and Atomic Structure, and Magnetism of Transition-Metal Clusters. *Chem. Rev.* **2000,** *100*, 637–678.

Alonso, J. A.; March, N. H. Electrons in Metals and alloys; Academic Press, 1989; p 603. ISBN: 978-0-12-053620-7.

Arenz, M.; Gilb, S.; Heiz, U. Size Effects in the Chemistry of Small Clusters. *Chem. Phys. Solid Surfaces* **2007,** *12*, 1–51.

Barrientos, C.; Redondo, P.; Largo, A. Structural and Electronic Properties of ZnCn+/– clusters. *Int. J. Mass Spectrom.* **2008,** *273*, 87–94.

Bloomfield, L. A.; Freeman, R. R.; Brown, W. L. Photofragmentation of Mass-Resolved Si 2– 1 2+ Clusters. *Phys. Rev. Lett.* **1985,** *54*, 2246.

Bondybey, V. E.; English, J. H. Laser Excitation Spectra and Lifetimes of Pb2 and Sn2 Produced by YAG Laser Vaporization. *J. Chem. Phys.* **1982,** *76*, 2165–2170.

Borman, S. A. Three-Dimensional Periodic Table Of Elements Proposed. *Chem. Eng. News Arch.* **1990,** *68*, 18–21.

Castleman Jr, A.W. From Elements to Clusters: The Periodic Table Revisited. *J. Phys. Chem. Lett.* **2011,** *2*, 1062–1069.

Chandrakumar, K. R. S.; Ghanty, T. K.; Ghosh, S. K. Relationship Between Ionization Potential, Polarizability, and Softness: A Case Study of Lithium and Sodium Metal Clusters. *J. Phys. Chem. A* **2004,** *108*, 6661–6666.

Cohen, M. L.; Chou, M. Y.; Knight, W. D.; De Heer, W. A. Physics of Metal Clusters. *J. Phys. Chem.* **1987,** *91*, 3141–3149.

De Heer, W. A. The Physics of Simple Metal Clusters: Experimental Aspects and Simple Models. *Rev. Mod. Phys.* **1993,** *65*, 611.

de Heer, W. A.; Knight, W. D.; Chou, M. Y.; Cohen, M. L. Electronic Shell Structure and Metal Clusters. In *Solid State Physics*; Elsevier, 1987; pp 93–181.

Donoso, R.; Cardenas, C.; Fuentealba, P. Ab Initio Molecular Dynamics Study of Small Alkali Metal Clusters. *J. Phys. Chem. A* **2014,** *118*, 1077–1083.

Dugourd, P.; Rayane, D.; Labastie, P.; Vezin, B.; Chevaleyre, J.; Broyer, M. Measurements of Lithium Cluster Ionization Potentials. *Chem. Phys. Lett.* **1992,** *197*, 433–437.

Fournier, R.; Bo, Yi Cheng, J.; Wong, A. Theoretical Study of the Structure of Lithium Clusters. *J. Chem. Phys.* **2003,** *119*, 9444–9454.

Fura, A.; Tureček, F.; McLafferty, F. W. Small Carbon Clusters (Cn0, Cn+, Cn−) from Acyclic and Cyclic Precursors: Neutralization–Reionization and Theory. *Int. J. Mass Spectrom.* **2002,** *217*, 81–96.

Gotts, N. G.; von Helden, G.; Bowers, M. T. Carbon Cluster Anions: Structure and Growth from C5− to C62−. *Int. J. Mass Spectrom. Ion Process.* **1995,** *149*, 217–229.

Guo, Z.; Lu, B.; Jiang, X.; Zhao, J.; Xie, R. H. Structural, Electronic, and Optical Properties of Medium-Sized Lin Clusters ($n$ = 20, *30*, *40*, 50) by Density Functional Theory. *Phys. E Low-Dimensional Syst. Nanostruct.* **2010,** *42*, 1755–1762.

Hou, J.; Song, B. Density-Functional Study of Structural and Electronic Properties of Si n C n ($n$ = 1–10) Clusters. *J. Chem. Phys.* **2008,** *128*, 154304.

Huber, K. P.; Herzberg, G. *Molecular Spectra and Molecular Structure*, Vol. 4; Constants Diatomic Mol. Nostrand Reinhold: New York, 1979; p 534.

Jena, P. Beyond the Periodic Table of Elements: The Role of Superatoms. *J. Phys. Chem. Lett.* **2013a,** *4*, 1432–1442.

Jena, P. *Physics and Chemistry of Small Clusters*; Springer Science & Business Media, 2013b.

Jimenez-Izal, E.; Alexandrova, A. N. Computational Design of Clusters for Catalysis. *Annu. Rev. Phys. Chem.* **2018,** *69*, 377–400.

Kiran, B.; Kandalam, A. K.; Rallabandi, R.; Koirala, P.; Li, X.; Tang, X.; Wang, Y.; Fairbrother, H.; Gantefoer, G.; Bowen, K. *(PbS) 32: A Baby Crystal. J. Chem. Phys.* **2012,** *136*, 24317.

Kosimov, D. P.; Dzhurakhalov, A. A.; Peeters, F. M. Theoretical Study of the Stable States of Small Carbon Clusters C n ($n$ = 2–10). *Phys. Rev. B* **2008,** *78*, 235433.

Kosimov, D. P.; Dzhurakhalov, A. A.; Peeters, F. M. Carbon Clusters: From Ring Structures to Nanographene. *Phys. Rev. B* **2010,** *81*, 195414.

Kumar Kushwaha, A.; Kumar Nayak, S. Wobbled Electronic Properties of Lithium Clusters: Deterministic Approach Through First Principles, *Physica E: Low-dimensional Syst. Nanostruct.* **2017,** *97*, 368-374.

Kümmel, S.; Brack, M.; Reinhard, P. G. Ionic and Electronic Structure of Sodium Clusters up to N= 59. *Phys. Rev. B* **2000,** *62*, 7602.

Kushwaha, A. K.; Sahoo, M. R.; Nanda, J.; Nayak, S. K. Engineering Redox Potential of Lithium Clusters for Electrode Material in Lithium-Ion Batteries. *J. Clust. Sci.* **2017,** *28*, 2779–2793.

Largo, L.; Cimas, A.; Redondo, P.; Rayon, V. M.; Barrientos, C. Structure of Small TiCn Clusters: A Theoretical Study. *Chem. Phys.* **2006,** *330*, 431–440.

Lavendy, H.; Robbe, J. M.; Flament, J. P.; Pascoli, G. Density Functional Study of Silicon Carbide Cluster Cations C4+, C3Si+, C2Si2+, CSi3+ and Si4+. *J. Chim. Phys.* **1997,** *94*, 1779–1793.

Lépine, F.; Allouche, A. R.; Baguenard, B.; Bordas, C.; Aubert-Frécon, M. Computed Electron Affinity of Carbon Clusters C n up to n= 20 and Fragmentation Energy of Anions. *J. Phys. Chem. A* **2002,** *106*, 7177–7183.

Li, G.; Tang, Z. Parity Alternation Effects in the Stabilities of the Second-Row-Atom-Doped Linear Carbon Clusters C n X/C n X+/C n X-(n= 1– 10; X= Na, Mg, Al, Si, P, S or Cl). A Comparative Study. *J. Phys. Chem.* **2003,** A *107*, 5317–5326.

Lin, Y. C.; Sundholm, D.; Juselius, J.; Cui, L. F.; Li, X.; Zhai, H. J.; Wang, L. S. Experimental and Computational Studies of Alkali-Metal Coinage-Metal Clusters. *J. Phys. Chem. A* **2006,** *110*, 4244–4250.

Liu, Z.; Huang, R.; Tang, Z.; Zheng, L. Linear Aluminum-and Phosphorus-Doped Carbon Cluster Anions: Mass Distribution and Ab Initio Calculations. *Chem. Phys.* **1998,** *229*, 335–341.

Maisser, A.; Barmpounis, K.; Attoui, M. B.; Biskos, G.; Schmidt-Ott, A. Atomic Cluster Generation with an Atmospheric Pressure Spark Discharge Generator. *Aerosol Sci. Technol.* **2015,** *49*, 886–894.

Manninen, M. Structures of Small Alkali-Metal Clusters. *Phys. Rev. B* **1986,** *34*, 6886.

Martin, J. M. L.; El-Yazal, J.; François, J. P. Structure and Vibrational Spectra of Carbon Clusters Cn (n= 2–10, *12*, *14*, *16*, 18) Using Density Functional Theory Including Exact Exchange Contributions. *Chem. Phys. Lett.* **1995,** *242*, 570–579.

Miller, T. F.; Hall, M. B. Structural and Bonding Trends in Platinum– Carbon Clusters. *J. Am. Chem. Soc.* **1999,** *121*, 7389–7396.

Pascoli, G.; Lavendy, H. Structures and Energies of CnSi+ (4≤ n≤ 15) Silicon Carbide Clusters. *Int. J. Mass Spectrom. Ion Process.* **1998,** *173*, 41–54.

Pascoli, G.; Lavendy, H. Geometrical Structures of the Phosphorus-Doped Carbon Cluster Cations CnP+ (n= 1–20). *Int. J. Mass Spectrom.* 1999a, *189*, 125–132.

Pascoli, G.; Lavendy, H. Theoretical Study of CnP, CnP+, CnP- (n = 1–7) Clusters. *J. Phys. Chem. A* 1999b, *103*, 3518–3524.

Pradhan, P.; Ray, A. K. An Ab Initio Study of the Electronic and Geometric Structures of Si m C n+ Cationic Nanoclusters. *Eur. Phys. J. D-Atomic, Mol. Opt. Plasma Phys.* **2006,** *37*, 393–407.

Qiang, L.; Ling-Ling, S.; Qiang, Z.; Bin, T.; Zhang-Hai, W.; Zeng-Ling, R. First-Principles Study of Structural and Electronic Properties of Carbon Clusters. *Mater. Res. Innov.* **2015,** *19*, S8-14–S8-16.

Raghavachari, K.; Binkley, J. S. Structure, Stability, and Fragmentation of Small Carbon Clusters. *J. Chem. Phys.* **1987,** *87*, 2191–2197.

Redondo, P.; Barrientos, C.; Largo, A. Small Carbon Clusters Doped with Early Transition Metals: A Theoretical Study of ScC n, ScC n+, and ScC n-(n= 1– 8) Open-Chain Clusters. *J. Phys. Chem. A* **2005,** *109*, 8594–8603.

Redondo, P.; Barrientos, C.; Largo, A. Small Carbon Clusters Doped with Vanadium Metal: A Density Functional Study of VC n (n= 1− 8). *J. Chem. Theory Comput.* **2006,** *2*, 885–893.

Roszak, S.; Balasubramanian, K. Theoretical Investigation of Structural and Thermodynamic Properties of Lanthanum Carbides LaC n (n= 2–6). *J. Chem. Phys.* **1997,** *106*, 158–164.

Saloni, J.; Kadłubański, P.; Roszak, S.; Majumdar, D.; Hill Jr, G.; Leszczynski, J. The Evolution of Bonding and Thermodynamic Properties of Boron-Doped Small Carbon Clusters: An Ab Initio Study. *Chem Phys Chem* **2011,** *12*, 1358–1366.

Schwyn, S.; Garwin, E.; Schmidt-Ott, A. Aerosol Generation by Spark Discharge. *J. Aerosol Sci.* **1988,** *19*, 639–642.

Song, B.; Yong, Y.; Hou, J.; He, P. Density-Functional Study of Si n C n (n= 10–15) Clusters. *Eur. Phys. J. D* **2010,** *59*, 399–406.

Strout, D. L.; Miller, T. F.; Hall, M. B. Structure and Stability of Palladium− Carbon Cations. *J. Phys. Chem. A* **1998,** *102*, 6307–6310.

Sun, Q.; Rao, B. K.; Jena, P.; Stolcic, D.; Kim, Y. D.; Gantefor, G.; Castleman Jr, A. W. Appearance of Bulk Properties in Small Tungsten Oxide Clusters. *J. Chem. Phys.* **2004,** *121*, 9417–9422.

Sun, X.; Du, J.; Jiang, G. Au-Doped Carbon Clusters AuC n with n= 1–11: A Theoretical Investigation. *Struct. Chem.* **2013,** *24*, 1289–1295.

Tabrizi, N. S.; Ullmann, M.; Vons, V. A.; Lafont, U.; Schmidt-Ott, A. Generation of Nanoparticles by Spark Discharge. *J. Nanoparticle Res.* **2009,** *11*, 315.

Tománek, D.; Schluter, M. A. Growth Regimes of Carbon Clusters. *Phys. Rev. Lett.* **1991,** *67*, 2331.

Wheeler, S. E.; Sattelmeyer, K. W.; Schleyer, P. v R.; Schaefer III, H. F. Binding Energies of Small Lithium Clusters (Li n) and Hydrogenated Lithium Clusters (Li n H). *J. Chem. Phys.* **2004,** *120*, 4683–4689.

Whiteside, R. A.; Krishnan, R.; Defrees, D. J.; Pople, J. A.; Von R, P. Structures of C4. *Chem. Phys. Lett.* **1981,** *78*, 538–540.

Xiong, R.; Die, D.; Xiao, L.; Xu, Y. G.; Shen, X. Y. Probing the Structural, Electronic, and Magnetic Properties of Ag n V (n= 1–12) Clusters. *Nanoscale Res. Lett.* **2017,** *12*, 625.

Yang, S.; Taylor, K. J.; Craycraft, M. J.; Conceicao, J.; Pettiette, C. L.; Cheshnovsky, O.; Smalley, R. E. UPS of 2–30-Atom Carbon Clusters: Chains and Rings. *Chem. Phys. Lett.* **1988,** *144*, 431–436.

Zhang, C.; Xu, X.; Wu, H.; Zhang, Q. Geometry Optimization of Cn (n= 2–30) with Genetic Algorithm. *Chem. Phys. Lett.* **2002,** *364*, 213–219.

## CHAPTER 4

# Fullerenes: Synthesis and Applications

JAGANNATH PANDA[1], TANASWINI PATRA[2], PRASANNA KUMAR PANDA[1], ROJALIN SAHU[2], BANKIM CHANDRA TRIPATHY[1], and AVIJIT BISWAL[3]

[1]*CSIR-IMMT, Bhubaneswar, Odisha, India*

[2]*School of Applied Sciences, KIIT Deemed to be University, Bhubaneswar, Odisha, India*

[3]*College of Engineering and Technology (Autonomous) Bhubaneswar, Odisha, India*

### ABSTRACT

Day by day with the advancement of science and technology, multidisciplinary applications are possible with the allotropes of carbon. This chapter showcases an ephemeral discussion on one of the unique allotropes of carbon, that is, Fullerene from its history to its multidisciplinary applications in various fields of science and technology. The characteristic features of various types of fullerenes are highlighted. A comprised data is included which discussed the methods of synthesis of fullerenes. The functionalized derivatives of fullerene also find numerous applications in fast-growing areas, like photovoltaics, medicines, gas storage/adsorption, pharmaceuticals, and water treatment. The highly conductive nature and exceptional electronic properties, makes it suitable for application in adsorption–separation, electroactive materials, hydrogen storage and solar energy storage material, and energy accumulation, etc. However, more

New Forms of Carbon: Nanocarbons. Aneeya Kumar Samantara & Satyajit Ratha (Eds.)

exploration is essential to open new avenues for this unique material to fulfill the needs of market in the near future.

## 4.1 INTRODUCTION

One of the most important elements in the periodic table is carbon because it is an important constituent of all living beings. It has also a unique ability to form four covalent bonds with different atoms. Moreover, due to its self-linking property, catenation, carbon forms covalent bonds with other carbon atoms and forms millions of long and short chain molecules. Carbon when linked to carbon exhibit different hybridizations, that is, sp, $sp^2$, and $sp^3$. This is the fact, that carbon exists in different allotropic forms such as diamond, graphite, lonsdaleite, carbon nanotubes, fullerenes, nano-onions, nanodiamond, amorphous carbon, nanoballs, nanospheres, etc. (Delgado et al., 2008). Diamond crystals are 3-D super hard crystals consisting of $sp^3$ hybridized carbon atoms arranged in a tetrahedral manner. On the other hand, Graphite is another form of carbon having $sp^2$ hybridization. It is constructed from 2-D graphene sheets that are connected by weak van der Waals force of attraction. Thirdly, carbon nanotubes and carbon nanobuds are 1-D structures that possess many properties. The zero-dimensional forms of carbon are fullerene, nano-onions, nanodiamonds, etc. Nowadays, more emphasis is given on different properties, structures, and applications of these zero-dimensional forms of carbon. Among these, fullerene is the most emerging field of study (Katsnelson, 2007).

Fullerene is an interesting allotrope of carbon made up of carbon atoms having varieties of shapes such as sphere, tube, ellipsoid or some other shapes (Peter et al., 2015). Spherical fullerenes are known as Buckminsterfullerene or Buckyball. One of the topics of intense research is buckyballs and buckytubes. It may be because of their unique properties and multimodal applications in advance field of science, mainly in nanotechnology, material science, electronics, etc. Structurally fullerenes consist of asymmetrical loaded graphene sheets which look same as that of graphite. They form rings which are hexagonal and pentagonal in shape. Fullerenes are having truncated icosahedral shape in a closed cage with $I_h$ symmetry. In fullerene molecule, each carbon atom is linked to three other carbon atoms. There are many forms of fullerenes which are prepared and contain around 30–3000 carbon atoms. The various kinds of fullerenes are briefly explained and shown below. Different fullerenes are made up of isomers with homologues which range from the lower homologue $C_{60}$ and $C_{70}$ up to higher homologues like $C_{240}$, $C_{540}$, and $C_{720}$. Several fullerenes containing $C_{70}$, $C_{76}$, $C_{80}$, $C_{84}$, etc. also exist (Jarrold, 2000). Among all these, the most stable fullerene is $C_{60}$. $C_{60}$ fullerene molecule consists of 12 pentagonal and 20 hexagonal rings. Every carbon atom is $sp^2$ hybridized with a truncated icosahedral structure (Terrones et al., 1995). It is known as Buckminsterfullerene and has been named after the architect, Buckminster Fuller who has fabricated cage-like geodesic Domes in 1960s (Yadav, 2018). $C_{60}$, which is a spherical fullerene, is similar to that as an electron deficient entity (an electrophile). So, it is readily attacked by a nucleophile which is an electron rich entity. In order to make $C_{60}$ molecule completely soluble, various molecules or polar functional groups are attached on the core of the fullerene. In the meanwhile, the unique inherent properties of fullerenes are also retained as it is and reasonably the biological availability can be achieved (Yadav and Kumar, 2008). $C_{60}$ fullerene and its derivatives which have a unique electronic π-system enable them to absorb ultraviolet (UV) or visible light making them a very potential photosensitizer.

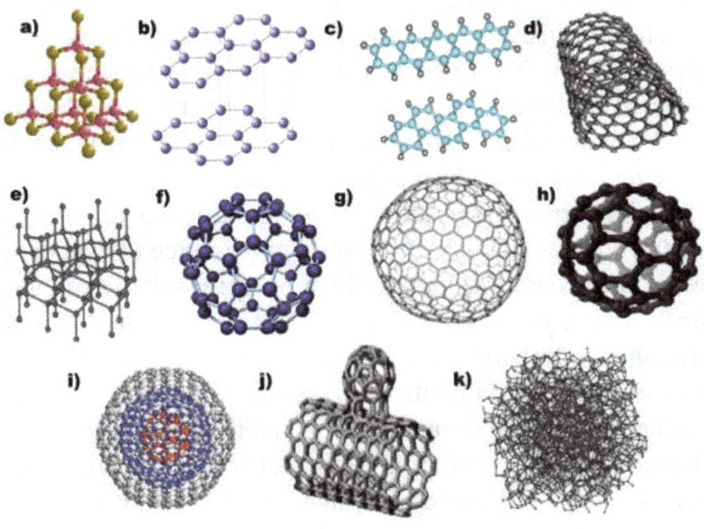


**FIGURE 4.1** a) Diamond, b) graphite, c) graphene, d) carbon nanotube, e) lonsdaleite, f) Buckminster fullerene ($C_{60}$), g) fullerite ($C_{540}$), h) $C_{70}$, i) nano-onions, j) nanobuds, k) and amorphous carbon.

## 4.2 HISTORY OF FULLERENES GLOOMY

Far back in 1966, when David Jones suggested the possibility of graphite balloons, this third allotrope of carbon had a very gloomy existence (Kroto and Walton, 1993; Jones, 1993). In 1970, some progress was made by Eiji Osawa who explored the vision of producing $C_{60}$ molecule (made up of 60 carbon atoms) in the structure of condensed icosahedron (Gund et al., 1970), followed by the enumeration by Eiji Osawa and Zensho Yoshida in 1971 about the possibilities of aromatic properties of the C60 molecule in a book titled "Aromatic Molecules" (Sheka et al., 2011). Subsequently, in order to determine its electronic structure, Huckel calculation of $C_{60}$ was carried out by Bochvar and Galpern with Stonkevich and coworkers in 1973 (Bochvar and Galpern, 1973; Stankevich et al., 1984). Later in 1980, Davidson used the general group theoretical techniques and calculated the distance between the highest occupied molecular orbital (HOMO) and the lowest unoccupied molecular orbital (LUMO) and suggested that, the closed hollow cell of fullerene $C_{60}$ molecule could be characterized by knowing the distance between HOMO and LUMO, indicating the chemical stability of the species. Thus, advancement in experimental technique of clustered molecules and planetary processes associated with inter/circum-stellar molecules led to the discovery of $C_{60}$ Buckminsterfullerene in 1985. This flagged the way for its extraction and characterization in 1990 (Kroto, 1988). During the investigations on cyanoacetylenes, which were assumed to exist in the molecular clouds surrounded by stars with carbon-rich atmospheres, Harry Kroto combined molecular spectroscopy with measurements made by radio astronomers on the same molecules in the material surrounding carbon stars, in particular the object known as IRC+10216 (Kroto, 1988). However, the existence of cyanoacetylenes in large quantity contradicted with the reported results of the interstellar chemistry existing model (Heath et al., 2016). Later in 1985, Kroto in collaboration with Robert Curl and Rick Smalley of Rice University, USA performed experiments to confirm the formation of long cyanopolyenes by the interaction of laser beams on graphite target, which evidenced the remarkably stable nature of $C_{60}$ molecule (Kroto et al., 1985; Kroto, 1997). Combination of mass spectral data with the circumstantial evidence provided the experimental support for the truncated Icosahedron structure of $C_{60}$. Initially, the proposed structure of cage of interconnecting pentagons and hexagons was received with understandable uncertainty (Kroto et al., 1991; Kroto, 1987). But later on, the uncertainty was cleared by W. Kratshchemer and D. R. Hutfirm in 1990 who made a breakthrough with the production of Isolable quantities of $C_{60}$ (Krätschmer et al., 1990; Kroto, 1992; Ajie et al., 1990; Howard et al., 1991). Further, using chromatographic technique the isolated material was separated into two components as $C_{60}$ and $C_{70}$. On analyzing these two individual components by 13C NMR, it was found that, $C_{60}$ component possesses single line resonance indicating the fact that the equivalent nature of all the carbon atoms is present. However, five lines were observed from $C_{70}$ indicating the existence of homologues of fullerene family (Anderson et al., 1997). Later, a technical model was adopted to describe the structure of $C_{60}$ by Smalley which had been previously used by 'Buckminster Fuller'. This led to the name Buckminsterfullerene which is similar to a soccer ball with a spherical shape having 12 pentagons and 20 hexagons like $C_{60}$ (Kroto et al., 1985). The expedition to study the physical and chemical properties of $C_{60}$ and its higher homologues along with the new substances produced from these compounds opened up a new outlook in the expansion of chemistry into diverse areas such as material chemistry/physics, superconductivity, and astrochemistry.

## 4.3 TYPES OF FULLERENE

Sir Harold W. Croto discovered the first fullerene in 1985 (Kroto et al., 1985). Subsequently, many different categories of fullerenes were explored based on their structural disparities. Some of them are listed below:

### *4.3.1 BUCKYBALL CLUSTERS*

The notably discovered nanoparticles were Buckyballs. In these buckyballs, each carbon atom is linked through covalent bonds to other three carbon atoms. Just like a soccer ball, the carbon atoms are attached in a specific pattern forming pentagons and hexagons. This type of outline gives a spherical structure to the Buckyball (Sinha and Sinha, 2001). Buckyball containing 60 carbon atoms is the most common one and is usually called as $C_{60}$. Depending upon the number of carbon atoms, there are also different sizes of buckyballs ranging from 20 to more than 100 carbon atoms. The strength of the buckyballs is due to the covalent bonds which is present between them (Hsu et al., 2013). In order to reinforce the material, these are often used in composites.

### 4.3.2 NANOTUBES

In 1991, Nanotubes were discovered by Iijima. The dimensions of nanotubes are very small. The diameter of nanotubes ranges from fraction to tens of nanometer whereas the length is up to some micrometers. The shape of nanotube looks like a graphene sheet which is rolled in the shape of a cylinder and is capped with fullerene like structures. Just like a tube, these nanotubes are hollow from inside. These may be single-walled or multiwalled which are formed by graphene lamella having single and multiple layers, respectively (Ganesh, 2013). The surface area of these nanotubes is found in the range of 150–1500 $m^2/g$. On this basis only, it acts as a good sorbent. Figure 4.1d shows structure of nanotube (Jarrold, 2000).

### 4.3.3 MEGATUBES

The word mega means large. So, these are one type of fullerenes whose diameter is larger than those of nanotubes. The walls of the megatubes are prepared by changing their thickness. The main use of megatubes is to transport various molecules having different sizes (Wang et al., 2007).

### 4.3.4 POLYMERS

An extensive variety of fullerenes containing polymers have been discovered with a number of technological applications in day-to-day life. These are formed by applying high temperature and under high pressure. The polymers which are synthesized in the presence of fullerenes have anomalous electrochemical, photophysical, and structural properties. The synthesized polymers were having different shapes like 1-D, 2-D, and 3-D chains (Giacalone and Martin, 2006). Various well-established materials of these classes are cross-linked, end-capped, star-shaped, double cable, $C_{60}$ dendrimer, etc. (Scarel and Alonso, 2013). Figure 4.1f shows the shape of polymers.

### 4.3.5 NANO-ONIONS

Carbon nano-onions are multilayered fullerenes which are spherical in shape. These consist of closed spherical carbon shells, polyhedral and multiple layered quasi-spherical shaped shells, whose structure resembles with that of an onion. These are simply known as carbon onions. These carbon nano-onions have a number of applications. These are used as potential drug transporters and also as delivery vehicles. They are used as good lubricants. But, they are poorly soluble in organic as well as aqueous solvents (Plonska et al., 2019; Xu, 2008). Figure 4.1i shows the shape of carbon nano-onions.

### 4.3.6 BUCKMINSTERFULLERENE ($C_{60}$)

Buckminsterfullerene is the naturally occurring most common fullerene. These are present in less quantity in soot. It is represented by a formula $C_{60}$ having 60 carbon atoms. These carbon atoms are arranged in a closed cage-like fused ring structure which looks like a soccer ball or simply football. The carbon atoms form polygons having 20 hexagonal and 12 pentagonal rings with 60 vertices and 32 faces. In $C_{60}$, each carbon atom is covalently bonded to three other carbon atoms having one double bond and two single bonds. The structure of Buckminsterfullerene is obtained by putting carbon atoms at all the 60 vertices of the rings. On comparing $C_{60}$ molecule with that of the size of a soccer ball, then the actual size of the soccer ball can be compared with the size of the earth. It is the most symmetric molecule. It is very much stable which can withstand high temperature and high pressure (Zheng et al., 2005; Shibuya et al., 1999). It can be used for photovoltaic applications. The structure of Buckminster-fullerene is presented in Figure 4.1f.

## 4.4 $C_{60}$ FULLERENE

$C_{60}$ form of fullerene was named after the American architect, who was well known for designing a large geodesic dome, called Buckminster Fuller. That's why it is called Buckminsterfullerene. The geodesic dome resembles the molecular structure of $C_{60}$. It is a compound which is composed of even number of carbon atoms which forms a three-dimensional cage-like fused ring polycyclic system. The 60 carbon atoms in $C_{60}$ fullerene are equivalent in nature. Hence, a single line in 13C NMR spectrum is observed for $C_{60}$ at $\delta = 143$ ppm (Diederich and Robert, 1992). It is a three-dimensional polyolefin which is electron deficient. Single bonds are present in the pentagonal structures whereas double bonds are present between pentagonal and hexagonal rings in the form of bridging bonds. Among all fullerenes, the stable fullerenes are those which obey isolated pentagon rule (Chiang et al., 1992). Fullerene $C_{60}$ is almost insoluble in water and other solvents which are polar, but sparingly soluble in organic solvents like toluene and benzene. However, it is found to be soluble in orthodichlorobenzene, dimethylnaphthalene, and 1-chloronaphthalene like solvents (Ruoff et al., 1993). The chemical properties and reactivity of fullerene $C_{60}$ are based on the fact that the bonding in $C_{60}$ has delocalized $\pi$ molecular orbitals which extends throughout the structure and contains both $sp^2$ and $sp^3$ hybridized C atoms. Fullerene $C_{60}$ is not "superaromatic" because it tends to evade double bonds in the pentagonal rings by which the electron delocalization becomes very poor. Therefore, $C_{60}$ acts as an electron deficient alkene and reacts rapidly with electron-rich species. $C_{60}$ shows many types of chemical reactions and some of them are nucleophilic addition reactions, electrophilic addition reactions, radical addition reactions, halogenations, pericyclic reactions, oxidation reactions, and the formation of endohedral complexes $M@C_{60}$, where M refers to a metal atom (Djordjevic et al., 2015). Figure 4.2 represents some of the important chemical reactions of fullerene $C_{60}$. The basic and common reactions are electrophilic addition reactions and are mostly found to be exothermic in nature (these reactions are accompanied by a change in hybridization of the C-atoms from $sp^2$ to $sp^3$, as a result of that the angular strain decreases in the cage). The number of addendums reduces the amount of evolved heat of the reaction. Therefore, the stability decreases with the increase in the number of adducts. Hence, one of the biggest problems during the synthesis of only one derivative of fullerene is the formation of a large number of its isomeric forms. In case of two addendums $C_{60}X_2$, total eight regioisomers along with around 23 stereoisomers are formed. Different chemical properties of $C_{60}$ fullerene (nucleophilic and electrophilic additions, radical additions, and pericyclic reactions) make the covalent bonding of a number of organic compounds and functional groups on its cage. Fullerene-based derivatives which are water soluble are very important to study different biological applications (Mojica et al., 2013).

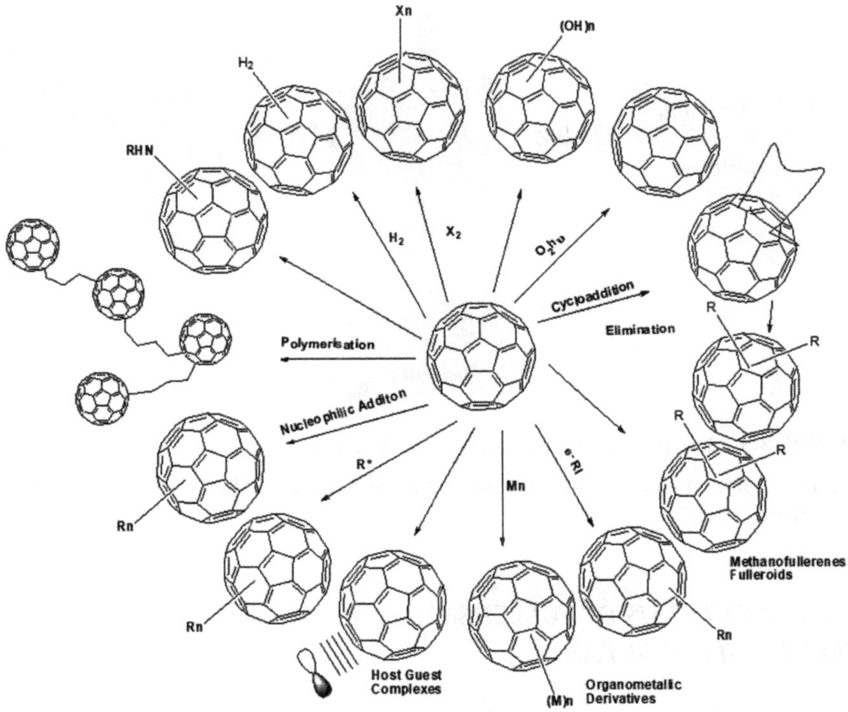


**FIGURE 4.2** Chemical reactions of fullerenes.

*Source*: Reprinted with permission from Djordjevic et al., 2015. Copyright © 2015 Aleksandar Djordjevic et al. https://creativecommons.org/licenses/by/3.0/

## 4.5 SYNTHESIS OF FULLERENES

Initially, synthesis of fullerenes was done by using laser vaporization of carbon in a non-reactive atmosphere. But by going through this method, that is, laser irradiation of carbon, very less quantity of fullerenes was produced (Nimibofa et al., 2018). However, the amount of $C_{60}$ fullerene can be increased by synthesizing through electrical arc heating of graphite and by going through laser irradiation method of Poly Aromatic Hydrocarbons (PAHs) (Fig. 4.3).

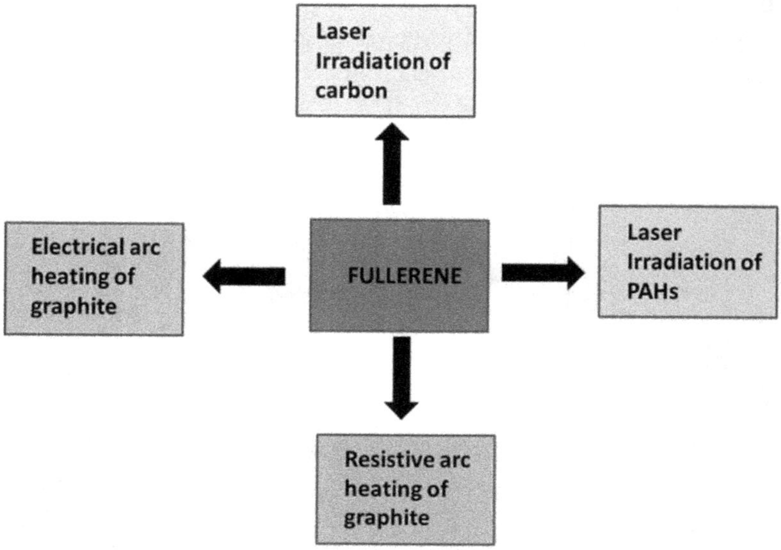


**FIGURE 4.3** Important routes for synthesis of fullerenes.

*Source*: Reprinted with permission from Nimibofa et al., 2018, https://creativecommons.org/licenses/by/4.0/

### *4.5.1 SYNTHESIS OF FULLERENES BY USING LASER VAPORIZATION OF CARBON*

This method of synthesis of fullerenes need an inert atmosphere and for this purpose helium gas has been used. Fullerenes are formed through a supersonic expansion nozzle by using a pulsed laser (Fig. 4.4). This pulsed laser is focused on a target of graphite that is kept in an inert atmosphere. From a solid disc of graphite, carbon is vaporized into a high-density helium flow which is done by using a focused pulsed laser (Geckeler and Samal, 1999). This method of synthesis is limited by the low yield of fullerene (< 1%). This method also requires very harsh conditions of temperature around 1300 °C and a pressure of around 1000 bar. This method produces $C_{60}$ and $C_{70}$, but it gives very low quantities of larger fullerenes such as $C_{84}$, and requires great effort for purification (more than 20 high-performance liquid chromatography cycles for $C_{84}$) (Whetten et al., 1990). Because of the uncontrollable character of the reaction, it is not possible to obtain a single fullerene or a specific isomer.

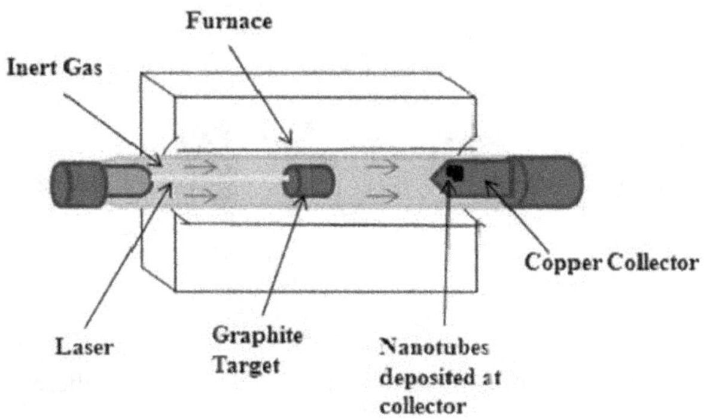


**FIGURE 4.4** Syntheses by laser vaporization of carbon.

*Source*: Reprinted with permission from Choudhary et al., 2013. Copyright © 2013 The Author(s). Licensee IntechOpen. https://creativecommons.org/licenses/by/3.0/

### 4.5.2 SYNTHESIS OF FULLERENES BY USING ELECTRIC ARC HEATING OF GRAPHITE

Microscopic quantity of fullerene was conceptualized and synthesized by Kratchmer and Huttmabn in the year of 1990. First time they had introduced the arc discharge method for the synthesis of fullerene. This process involves the formation of an electric arc in an inert atmosphere between the graphite rods (Fig. 4.5). When the inert atmosphere is created between the graphite rods it generates a fluffy condensate (soot) (Geckeler and Samal, 1999; Whetten et al., 1990). The soot contained the fullerenes which can be extracted by using small amount of toluene. After the extraction process the solvent was removed by using a high vacuumed rotary evaporator. The extracted samples were dried which contains mostly $C_{60}$. The high pure fullerene was obtained by the help of liquid chromatography process (Nimibofa et al., 2018; Chai et al., 1991; Diederich et al., 1991).

### 4.5.3 SYNTHESIS OF FULLERENES BY USING RESISTIVE ARC HEATING OF GRAPHITE

The synthesis of fullerene by using the resistive arc heating involves the vaporization of the carbon rods under partial helium atmosphere (Fig. 4.6). In this process, the heating of carbon rod generates soot-like substance which is faint grey white in color. The soot is the pure form of fullerene and collected around the carbon rods (Kyesmen et al., 2016; Parker et al., 1991).

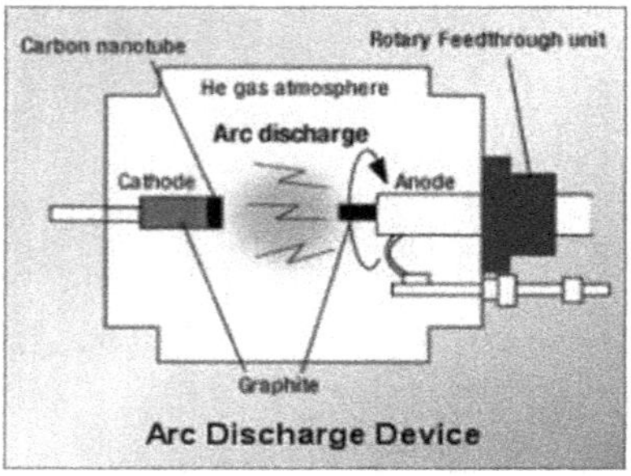


**FIGURE 4.5** Electric arc heating of graphite for the synthesis of fullerenes.

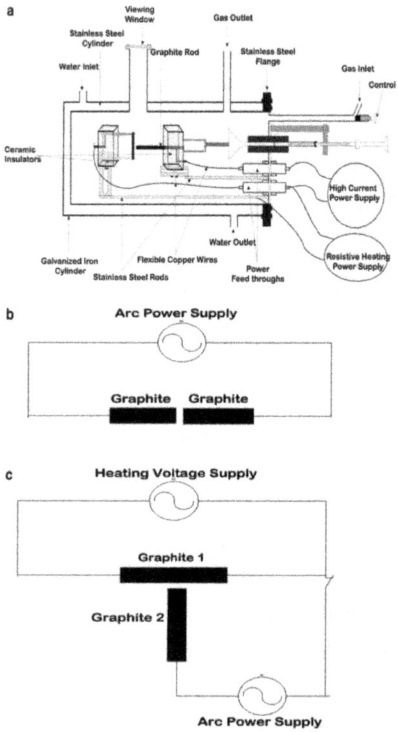


**FIGURE 4.6** (a) Arc discharge system for fullerenes synthesis, (b) basic circuit diagram for direct arc discharge, and (c) the basic circuit diagram that allows for resistive heating of one of the electrodes.

*Source*: Reprinted with permission from Kyesmen et al., 2016 https://creativecommons.org/licenses/by/4.0/

### 4.5.4 *SYNTHESIS OF FULLERENE WITH IRRADIATION OF POLYCYCLIC HYDROCARBONS (PAHS) BY LASER TREATMENT*

Fullerenes can be directly synthesized to obtain new homologues of fullerene family. These may not be obtained in better quantity by adopting a process of uncontrolled evaporation of graphite. The above approach to synthesize fullerene depends upon PAHs which require frameworks of carbon. The PAH molecules so obtained are "rolled up" under flash vacuum pyrolysis (FVP) conditions to form fullerenes (Kabdulov et al., 2010; Joshipura and Joshipura, 2000). It has been reported that under laser irradiation at 337 nm wavelength, a polycyclic aromatic hydrocarbon consisting 60 carbon atoms will form $C_{60}$ fullerene (Boorum et al., 2001). It is shown in Figure 4.7.

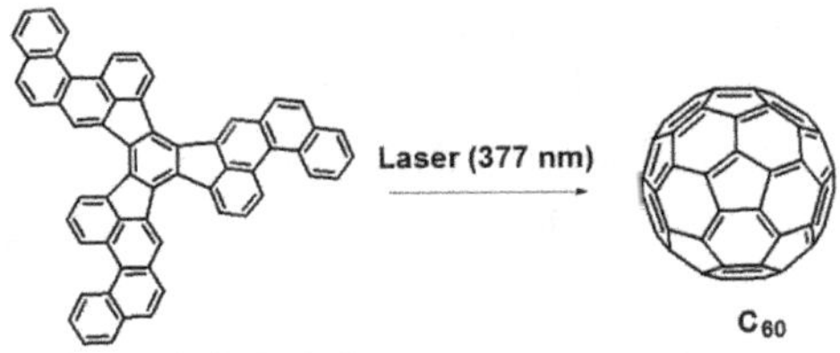


**FIGURE 4.7** Schematic for the laser irradiation of PAHs to synthesize fullerene.

*Source*: Reprinted with permission from Nimibofa et al., 2018, https://creativecommons.org/licenses/by/4.0/

## 4.6 REACTIVITY AND STRUCTURE OF FULLERENE

Because of its excellent electronic structure, the fullerene shows exceptional chemical properties (Mauter and Elimelech, 2008). One single carbon atom of a fullerene is connected to three other adjoining carbon atoms on the apex of a polyhedron, so that it forms one double bond and two single bonds. The basic structure of most abundant fullerene, Buckminsterfullerene, is as shown in Figure 4.1f.

### 4.6.1 *3D SHAPE OF FULLERENE*

In a $C_{60}$ fullerene, there are two types of bonds: one bond which behaves like a single bond is shared between a pentagon and a hexagon, and the other one which behaves like a double bond is shared between two hexagons. These single and double bonds in fullerenes cover their whole surface with a conjugated double bond system. Fullerenes having lesser number of hexagons show more $sp^3$ bonding features like they have more reactive carbon sites and have more strain. While the fullerenes having adjacent pentagons are usually less stable and are relatively more abundant than having isolated pentagons in which the $\pi$ (pie) electrons are delocalized throughout the structure (Campbell et al., 1996; Taylor and Walton, 1993). Fullerene has a curved surface.

So, it is quite natural to have pyramidal carbon atoms, and thus the atomic orbitals of carbon atoms have different hybridization such that the $\pi$ (pie) orbitals have different s and p characters providing the way for different chemical reactivity (Brettreich and Hirsch, 1998; Payne et al., 1992). Fullerenes do not contain replaceable hydrogen atoms as that found in classical aromatic compounds though they have conjugated $\pi$-system (Brettreich and Hirsch 1998). When an external chemical reagent and carbon atom interacts together, then either the hybridization of orbital changes from $sp^2$ to $sp^3$ which results in the breaking of $\pi$-bond or the free $\pi$-orbital can interact with the external chemical reagent (Park et al., 2001). Fullerenes undergo two main chemical transformations which are addition reactions and redox reactions. When fullerenes undergo addition reaction, they form exohedral products and during redox reactions, they form salts. In spite of having extreme conjugation in fullerenes, they act like electron-deficient alkene as they show addition reaction with nucleophiles, free radicals, and also homolytic reagents to form stable products (Tatiana et al., 1996). According to the reports, oxidation of $C_{60}$ and $C_{70}$ fullerenes was successfully done by electrochemical oxidation, photo-oxidation, ultrasound-induced oxidation, and dimethyl dioxirane addition. The unique nature of fullerene is that it contains carbon in soluble form because of which carbon films can be prepared in solutions. The solubility of $C_{60}$ as a function of solvent properties like refractive index, dielectric constant, hydrogen bonding strength, and size of the molecule is assessed which confirms the "like dissolving like" principle (Rai et al., 2007). Due to the nonpolar nature of fullerenes, they are hydrophobic and therefore are soluble mostly in aromatic hydrocarbons, carbon disulfide, etc. which are nonpolar solvents. On substituting fullerene with electron withdrawing group like nitro, aldehydic group, nitrile group, etc. the solubility decreases, while there is increase in solubility on substituting fullerene with electron releasing groups like methoxy group, methyl group, etc. (Bhoi et al., 2012).

## 4.7 APPLICATIONS

Fullerenes have unique chemical and physical properties which lead to the exploration of the different applications by many researchers. The functionalized derivatives of fullerene are broadly applied in areas like photovoltaics, medicines, gas storage/adsorption, pharmaceuticals, etc.

### 4.7.1 *MEDICAL APPLICATION*

The most important characteristics of fullerenes which make them effective in the field of medical chemistry are its hydrophobic nature, electronic configuration, size, and three-dimensional nature. Fullerenes act as potential therapeutic agents because of their exceptional cage-like structures and wide scope for functionalization (Bakry et al., 2007; Lin and Tan, 2012).

### 4.7.2 *ANTIOXIDANT/BIOPHARMACEUTICALS*

The reaction of fullerenes with free radicals is very rapid due to the presence of large number of conjugated double bonds and also due to low lying LUMO (Lowest Unoccupied Molecular Orbitals). According to the reports, fullerene is considered as the world's most efficient radical scavenger as a single $C_{60}$ molecule can add almost 34 methyl radicals (Gharbi et al., 2005; Cataldo and Tatiana, 2008).

### 4.7.3 *ANTIBACTERIAL/ANTIMICROBIAL ACTIVITY*

The antibacterial activities of fullerenes are shown by some of its hydrophilic derivatives like aminofullerene and fullerols, and have attracted attention in water treatment systems (Kang et al., 2009; Fortner et al., 2005). According to the reports, attaching multiple functional groups like hydroxyl, glycolic oxide, and carboxylic acid on $C_{60}$ prompts its photodynamic cytotoxic property against some pathogenic microorganisms. It includes cytotoxicity against some bacteria which are multi-antibiotic-resistant (Zhu et al., 2006; Wang et al., 2012). $C_{60}$ fullerene shows biocidal activity because of the ability to react with organic solvents generating reactive oxygen species like superoxide or singlet oxygen by photosensitization (Samanta and Das, 2017).

### 4.7.4 ANTIVIRAL ACTIVITY

Fullerene $C_{60}$ derivatives exhibit antiviral properties as they are linked to antioxidant property and their matchless molecular cage structure (Hong et al., 2008). Based on an investigation, it is found that the derivatives of fullerene could restrain and are able to form a complex with HIV protease (Kornev et al., 2011). Some reports also reflect that fullerenes having two amino groups called fulleropyrolidines were active against both the HIVs, that is, HIV-1 and HIV-2 (Kornev et al., 2011).

### 4.7.5 DIAGNOSTICS

Endofullerene is formed on inserting a metal ion into a fullerene cage. The endohedral metallofullerene written as EMFs cage so formed serves as an isolated chamber which can separate the reactive atoms from the biological environment (Shultz et al., 2010). One emerging application is the encapsulation of gadolinium in EMFs which has now been announced as one of the best candidate or agents for the next generation MRI (Toth, 2012). Some studies on biodistribution reveal the localization of EMFs to microphages. The above information concludes that their species target selectively to tissues which are rich in microphage and is considered as a valuable chemotherapeutic agent, which is used for treating leukemia and bone cancer (Goodarzi et al., 2017; Liang et al., 2008).

### 4.7.6 DRUG DELIVERY

As fullerenes have better biocompatibility, selective targeted delivery, and can release the carried drugs in a controlled manner, so they are considered as the potential drug carriers for the cellular delivery. Fullerenes show exceptional behavior in drug delivery and gene cellular delivery as they become soluble in water by attaching hydrophilic species (Isobe et al., 2006). As the functionalized fullerenes have the capability to bind with mitochondria by crossing the cell membranes, so they deliver drugs slowly to gain maximum therapeutic results. According to the investigations, the fullerenes which are DNA functionalized are very effective in comparison with lipid-based vectors which are commercially available (Isobe et al., 2001; Nakamura and Isobe, 2003).

### 4.7.7 DISINFECTANT

Fullerene can be used as a disinfectant, as fullerenes have antimicrobial properties and for the same the research is going on. Although these are known for having the strong antibacterial properties, some fullerenes did not endorse the highly effective antibacterial property. Li et al. reported that fullerene was unsuccessful to show its strong toxicity or antibacterial property on testing it on mammalian cells as disinfectant (2008). According to Lyon et al. (2006), the aggregate of $C_{60}$ molecules form stable fullerene water suspension (FWS) which is having different properties than $C_{60}$ bulk solid by four different methods. Antibacterial activities were studied on the FWS so produced. A strong antibacterial activity was demonstrated by all suspensions to *Bacillus subtilis* whose fractions with smaller aggregates revealed a higher antibacterial activity, in spite of the increase in toxicity excessively higher than the rise in presumed surface area. Also, different morphologies of the suspensions provide different antibacterial activity. The separated suspension of $C_{60}$ contains both amorphous and crystalline aggregates. The separated suspensions are centrifuged and after that it provides smaller size fractions having larger amorphous aggregates. These larger aggregates are crystalline in nature. On the other hand, the smaller fraction is more amorphous due to the absence of repeating structure. The lack of repeating structure defines the crystallinity of the larger fractions.

### 4.7.8 PHOTOVOLTAIC

The peculiar electronic transfer capacity and small reorganization energies explain the electron transfer process which makes possible for the multistep electron charge separation and artificial photosynthesis process (Guldi and Prato, 2000; Luis and Echegoyen, 1998; Troshina et al., 2007). The numerous covalent and noncovalent connected in fullerenes contain suitable electron donor–acceptor (D-A) hybrid dyads for the applications like photovoltaic devices, organic electronic, and photocatalysis. These various applications are due to the donor–acceptor (D-A) dyads which generate multistep charge separation states by the tuning of the electronic coupling between length and the nature of spacer (Lebedeva et al., 2015; Fukuzumi and Kojima, 2008). Also, some fullerene cages like cyanine, phthalocyanines, ruthenium bipyridine complex, metallocene, boron-dipyrrin, and tetrathiafulvalene are linked by some organic and transition metals which have the capability of donating the pair of electrons to the large molecule (Cnops et al., 2015; Martin et al., 2000; Herranz et al., 2000; Kanev et al., 2006). The allotropes of carbon have been applied as polymer transistors like Organic Field Effect Transistors (OFETS) and photodetectors. This is due to the presence of synergy between the organic field effect transistors (OFETS) and Organic Photovoltaics (OPVs). Similarly, fullerenes have also been applied as electrochemical material, doping material for the conducting polymer films, gas sensors, superconducting materials, and photoelectric devices (Boris, 1996; Nicolaidis et al., 2019; Ingo et al., 2004; Hayashi et al., 2004; Haddon, 1992; Gupta et al., 2019).

### *4.7.9 FULLERENE-BASED POLYMERIC MATERIALS*

The fullerene-based polymeric material has so many applications in many fields like catalysis, adsorption-separation, electroactive battery material, hydrogen storage, electronics, solar energy storage material, energy accumulation, etc. The various applications are due to the highly conductive nature and exceptional electronic properties of these materials (Avent et al., 1997). The synthesis of fullerene-based polymeric materials is carried out by different methods. Among various methods, the most followed one is regular attachment of fullerene cage to a polystyrene in a Friedel–Crafts type reaction and another is the indirect linking of the fullerene $C_{60}$ with an organic/inorganic space group such as $Pd(dba)_3$, $CHCl_3$, Pt(dba) or $Pt(cad)_2$ where 'cad' represents cycloocta-1,5-diene and 'dba' represents as benzylideneacetones. The reaction of fullerenes with spacer group makes them insoluble for the effective application in aqueous or organic medium. The reaction process also involves the replacement of organic spacers from the cage of fullerene to make them stable (Kiyokatsu et al., 1993; Milan et al., 1995; Yeoseon et al., 2014; Chae et al., 2009). The electrochemical formation of redox-active fullerene polymer is particularly attractive because these are formed on the cathodic surface of the electrode directly. The formation of the film of these materials is uniform and also smooth. The smooth film of polymers is formed by the reduction of $C_{60}O$ (Krystyna, 2010).

### 4.7.10 WATER PURIFICATION/ENVIRONMENT

The unique physical, chemical, and photochemical properties of carbon-based material have great interest in the environmental applications. Because of its lower solubility and stability in water, it has been greatly employed for the purification of water. A research investigation revealed that different functionalized fullerenes with active hydrophilic groups are synthesized and used for photocatalytic water purification process. These materials are inactive pathogenic microorganisms in nature (Li et al., 2019; Lyon et al., 2005). Moreover, fullerenes are also used as adsorbent material for adsorption of organic compounds (Snow et al., 2014). The photocatalytic and adsorption process involve the reactive oxygen species (ROS) such as singlet oxygen ($^1O_2$) and superoxide ($O_2^-$) via type I and II reactive photosensitization pathway. This process involves the $C_{60}$ fullerene assisted oxidation for water purification and disinfection (Richardson et al., 2000).

### 4.7.11 HYDROGEN STORAGE

The unique properties and cage-like structure make the fullerene as suitable candidate for hydrogen storage. The hydrogenation of fullerene makes the medium for the sorption of hydrogen at different temperatures and pressures. It involves the transformation of C-C to C-H bonds which requires low bond energy and it returns to its original structure by heating at high temperature. These properties of fullerenes are favorable for the sorption of maximum of 6.1% of hydrogen due to the large cage-like structure (Pupysheva et al., 2007; Wang et al., 2019; Zhang and Cheng, 2018).

### 4.7.12 ENERGY STORAGE MATERIALS

Energy storage and management seem to be essential to today's technological needs. Among the different source of energy generation, there is a need to develop the storage technology to save electrical energy. The storage of electrochemical energy is a big challenge among the available energy storage technologies. Electrochemical energy storage has great demand due to its low carbon footprint, high shelf-life, high efficiency, flexibility in grid oration, and low cost. Super capacitors are usually made of conductive layer of carbon and their allotropes. The capacitance of these supercapacitors depends on pore size distribution, electrolyte, electrical conductivity, and surface area (Kumar et al., 2018). The less permeable capacity of the electrolyte decreases the performance of the supercapacitor. To overcome these problems, nanotechnology has given new opportunity via the use of wide number of carbon-based material and their allotropic forms to increase the conductivity and capacitance (Campbell et al., 2018; Gaddam et al., 2019).

#### 4.7.12.1 SUPER CAPACITORS

There has been tremendous number of carbon-based materials used for energy storage applications. There is less focus on the allotropes of carbon like fullerenes. Fullerenes are the best material for supercapacitor applications due to the higher electron affinity and conductivity. The recent research also focused on the use of this hybrid nanomaterial for the storage of electrochemical energy (Bairi et al., 2019). The hybrid material composed of fullerene and graphene molecules has been applied for the electrode material in supercapacitors. These materials show a specific capacitance of 135.36 F/g and 101.88 F/g, respectively. These carbon-based materials show a good retention time rate of 92.25% after the completion of 1000 charge and discharge cycles (Jia et al., 2015).

#### 4.7.12.2 HIGH-PERFORMANCE LITHIUM ION BATTERIES

The intercalation chemistry of carbon-based materials plays a key role in the Li-ion battery. Different carbon-based materials are used as anode material in the battery due to the low cost, longer life, and safety. Among the different carbon materials fullerenes are having the fcc packing of C60 peapods and a promising solid-state ion intercalation chemistry. Different research groups have given their investigation on the use of hydrogenated fullerenes as high-performance anode material for the rechargeable batteries. From the results it revealed that the use of fullerenes as an additive in the commonly used graphite show an unexpected reversible capacity which improve the capacity of the rechargeable Li-ion batteries (Bai et al., 2019; Partha et al., 2019; Li et al., 2013).

#### *4.7.12.3 MATERIALS AS SUPERCONDUCTORS*

A composite is formed when an alkali metal is introduced in the interstitial site of $C_{60}$ fullerene. The composite structure shows a superconductivity property. Researches have investigated that the $K_3C_{60}$ composites show a high superconductance at critical temperature. This material shows a perfect 3-dimensional superconductivity with high ductility, current density, and highly stable in nature (Hebard, 1992; Shrivastava, 2019; Cantaluppi et al., 2018).

#### *4.7.12.4 REINFORCED COMPOSITES*

An effective way to increase the mechanical properties of the materials is by the utilization of nanomaterials which are the different forms of carbon acting like filler to strengthening the parent materials (Hasobe et al., 2004). On reinforcing Al–Mg alloy with $C_{60}$ fullerene, the malleability, mechanical strength, and thermal stability of the aluminum-matrix composite have been increased (Evdokimov et al., 2018).

### ***4.7.13 TREATMENT OF WASTEWATER***

Water contains a lot of contaminants which should be removed before they are discharged into the water bodies. Fullerenes are a type of substance capable enough to treat the waste water. According to the study, the characterization of fullerene on removal of contaminants from waste water is already done. In this report, on the basis of hydrophobic membrane a new polymer membrane named poly (2, 6-dimethyl-1,4-phenylene oxide) (PPO) which is modified by $C_{60}$ was introduced. The above modified $C_{60}$ polymer was used to study the nature of adsorption of estrogenic compounds. These compounds act as contaminants or pollutants which are present in surface water (Jin et al., 2007). The structure of cross section of PPO membrane is like sponge, whereas it looks like finger when the membrane contains 2% and 10% $C_{60}$. This study concluded that 95% of estrone was removed 10% by weight of $C_{60}$-PPO membrane. After 12 hours from static adsorption test, it was seen that all membranes are capable to adsorb estrone having similar adsorption capacity. The adsorption rate was the fastest for the membrane containing 2% of $C_{60}$. This adsorption rate was followed by 10% of $C_{60}$-PPO membrane and PPO membrane only. There is difference in adsorption capacity due to the difference in the size of pores on the internal structure or surface.

The rate of adsorption was also hampered by diffusion of estrone in the membrane. In this study, as there is increase in the size of the pores, the diffusion of the contaminant, that is, estrone into the internal porous structure of membrane and the surface membrane also increases. Polyelectrolyte composite membranes were also prepared by the reaction of fullerene with sulfonated polystyrene. The result of this composite preparation shows that the fullerene which was present in polyelectrolyte membrane has enhanced the oxidation resistance and decreased the methanol crossover of the membrane. The mechanical strength of the PPO membrane has not improved on adding fullerene. However, on adding fullerene to the membrane made it brittle. On increasing the amount of fullerene in the membrane; the maximum stress, rupture stress, and the Young's modulus of the membrane decreases. The composite membrane containing fullerene indicates less permeability of methanol as 50% more than 2.8% Flu-Ps and 1.4% Flu-Ps membrane. Also, the permeability was decreased by 30% in case of 2.8% Flu-Ps and 1.4% Flu-Ps than Nafion 117. The reason for the above behavior may be agglomeration of microscale fullerene in the membrane. From the above studies, it was concluded that the methanol crossovers can be decreased by inorganic fillers like $SiO_2$. Also, the methanol permeability of PS membranes was more than Nafion 117 because of the presence of micropores in the prepared membranes (Saga et al., 2008; Thines et al., 2017; Bassyouni et al., 2019).

## 4.8 CONCLUSIONS

This chapter discussed about Fullerenes, starting from their discovery to its applications. Fullerenes due to their unique chemistry found applications in various advanced fields of science, such as in nanotechnology, material science, pharmaceutical science and electronics. A brief discussion on various approaches such as Laser irradiation of carbon, PAHs, electrical arc heating, and resistive arc heating of graphite for the synthesis of fullerenes are reported. The unique chemical and physical properties of fullerenes lead to the exploration of the different applications of this molecule by many researchers. The functionalized derivatives of fullerene have also various applications in fast growing areas like photovoltaics, medicines, gas storage/adsorption, pharmaceuticals, and water treatment. Fullerenes due to the highly conductive nature and exceptional electronic properties are used in adsorption–separation, electroactive battery material, hydrogen storage, electronics, solar energy storage material, and energy accumulation. Furthermore, a multidisciplinary approach is highly pivotal to explore the profound performance of this unique material, Fullerene, in order to discover some new dimensions in its application in the near future.

## KEYWORDS

- **fullerene**
- **carbon nanotube**
- **carbon**
- **arc discharge**

## REFERENCES

Ajie, H.; Alvarez, M. M.; Anz, S. J.; Beck, R. D.; Diederich, F.; Fostiropoulos, K.; Huffman, D. R.; Kraetschmer, W.; Rubin, Y. Characterization of the Soluble All-Carbon Molecules $C_{60}$ and $C_{70}$. *J. Phys. Chem.* **1990,** *94*, 8630–8633.

Anderson, J. L.; West, M. A.; Foote, C. S. Extraction, Isolation, and Characterization of Fullerene $C_{60}$: A Safe and Reliable Separation Experiment. *J. Chem. Educ.* **1997,** *74*, 311.

Avent, A. G.; Benito, A. M.; Birkett, P. R.; Darwish, A. D.; Hitchcock, P. B.; Kroto, H. W.; Locke, I. W.; Meidine, M. F.; Odonovan, B. F.; Prassides, K.; Taylor, K.; Walton, D. R. M.; Wijnkoop, M. The structure of fullerene compounds. *J. Mol. Struct.* **1997,** *436*, 1–9.

Azzam, T.; Abraham, J. D. Current Developments in Gene Transfection Agents. *Curr. Drug Deliv.* **2004,** *1*, 165–193.

Bai, H.; Gao, H.; Feng, W.; Zhao, Y.; Wu, Y. Interaction in Li@ Fullerenes and Li+@ Fullerenes: First Principle Insights to Li-Based Endohedral Fullerenes. *Nanomaterials* **2019,** *9*, 630.

Bairi, P.; Maji, S.; Hill, J. P.; Kim, J. H.; Ariga, K.; Shrestha, L. K. Mesoporous Carbon Cubes Derived from Fullerene Crystals as a High Rate Performance Electrode Material for Supercapacitors. *J. Mater. Chem. A* **2019,** *7*, 12654–12660.

Bakry, R.; Vallant, R. M.; Najam-ul-Haq, M.; Rainer, M.; Szabo, Z.; Huck, C.; Günther K. B. Medicinal Applications of Fullerenes. *Int. J. Nanomed.* **2007,** *2*, 639.

Bassyouni, M.; Abdel-Aziz, M. H.; Sh Zoromba, M.; Abdel-Hamid, S. M. S.; Drioli, E. A Review of Polymeric Nanocomposite Membranes for Water Purification. *J. Indust. Eng. Chem.* **2019,** *73*, 19–46.

Bhoi, V. I.; Kumar, S.; Murthy, C. N. The Self-Assembly and Aqueous Solubilization of [60] Fullerene with Disaccharides. *Carbohydrate Res.* **2012,** *359*, 120–127.

Bochvar, D. A.; Galpern, E. G. Hypothetical Systems-Carbododecahedron, S-Icosahedrone And Carbo-S-Icosahedron. *Doklady Akademii Nauk Sssr* **1973,** *209*, 610–612.

Boorum, M. M.; Vasil'ev, Y. V.; Drewello, T.; Scott, L. T. Groundwork for a Rational Synthesis of $C_{60}$: Cyclodehydrogenation of a $C_{60}H_{30}$ Polyarene. *Science* **2001,** *294*, 828–831.

Boris, P. Transport and Dielectric Properties of Thin Fullerene ($C_{60}$) films. PhD diss.; Massachusetts Institute of Technology, 1996.

Brettreich, M.; Hirsch, A. A Highly Water-Soluble Dendro [60] Fullerene. *Tetrahedron Lett.* **1998,** *39*, 2731–2734.

Campbell, E. E. B.; Fowler, P. W.; Mitchell, D.; Zerbetto, F. Increasing Cost of Pentagon Adjacency for Larger Fullerenes. *Chem. Phys. Lett.* **1996,** *250*, 544–548.

Campbell, P. G.; Baumann, T. F.; Biener, J.; Merrill, M.; Montalvo, E.; Worsley, M. A.; Biener, M. M.; Hernández, M. R. C. Graphene Macro-Assembly-Fullerene Composite for Electrical Energy Storage. U.S. Patent 9,870,871, Issued January 16, 2018.

Cantaluppi, A.; Buzzi, M.; Jotzu, G.; Nicoletti, D.; Mitrano, M.; Pontiroli, D.; Riccò, M.; Perucchi, A.; Pietro, P. D.; Cavalleri, A. Pressure Tuning of Light-Induced Superconductivity in $K_3C_{60}$. *Nat. Phys.* **2018,** *14*, 837.

Cataldo, F.; Tatiana Da Ros, Eds. *Medicinal Chemistry and Pharmacological Potential of Fullerenes and Carbon Nanotubes*, Vol. 1; Springer Science & Business Media, 2008.

Chae, S. R.; Hotze, E. M.; Wiesner, M. R. Evaluation of the Oxidation of Organic Compounds by Aqueous Suspensions of Photosensitized Hydroxylated-$C_{60}$ Fullerene Aggregates. *Environ. Sci. Technol.* **2009,** *43*, 6208–6213.

Chai, Y.; Guo, T.; Jin, C.; Haufler, R. E.; Chibante, L. P. F.; Fure, J.; Wang, L.; Alford, J. M.; Smalley, R. E. Fullerenes with Metals Inside. *J. Phys. Chem.* **1991,** *95*, 7564–7568.

Chiang, L. Y.; Upasani, R. B.; Swirczewski, J. Versatile Nitronium Chemistry for $C_{60}$ Fullerene Functionalization. *J. Am. Chem. Soc.* **1992,** *114*, 10154–10157.

Choudhary, V.; Singh, B. P.; Mathur, R. B. Carbon nanotubes and Their Composites. Suzuki, Satoru, Ed. Syntheses and Applications of Carbon Nanotubes and Their Composites. *BoD–Books Demand* **2013,** *9*, 193–222.

Cnops, K.; Zango, G.; Genoe, J.; Heremans, P.; Martinez-Diaz, M. V.; Torres, T.; Cheyns, D. Energy Level Tuning of Non-Fullerene Acceptors in Organic Solar Cells. *J. Am. Chem. Soc.* **2015,** *137*, 8991–8997.

Delgado, J. L.; Herranz, M. A.; Martin, N. The Nano-Forms of Carbon. *J. Mater. Chem.* **2008,** *18*, 1417–1426.

Diederich, F.; Robert L. W. Beyond $C_{60}$: The Higher Fullerenes. *Acc. Chem. Res.* **1992,** *25*, 119–126.

Diederich, F.; Roland, E.; Rubin, Y.; Whetten, R. L.; Beck, R.; Alvarez, M.; Samir Anz.; Sensharma, D.; Wudl, F.; Khemani, K. C.; Koch, A. The Higher Fullerenes: Isolation and Characterization of $C_{76}$, $C_{84}$, $C_{90}$, $C_{94}$, and $C_{70}O$, an Oxide of $D_{5h}$-$C_{70}$. *Science* **1991,** *252*, 548–551.

Djordjevic, A.; Srdjenovic, B.; Seke, M.; Petrovic, D.; Injac, R.; Mrdjanovic, J. Review of Synthesis and Antioxidant Potential of Fullerenol Nanoparticles. *J. Nanomater.* **2015,** *16*, 280.

Evdokimov, I. A.; Perfilov, S. A.; Pozdnyakov, A. A.; Blank, V. D.; Bagramov, R.; Perezhogin, I. A.; Kulnitsky, B. A.; Kirichenko, A. N.; Aksenenkov, V. V. Nanostructured Composite Materials Based on Al–Mg Alloy Modified with Fullerene $C_{60}$. *Inorg. Mater. Appl. Res.* **2018,** *9*, 472–477.

Fortner, J. D.; Lyon, D. Y.; Sayes, C. M.; Boyd, A. M.; Falkner, J. C.; Hotze, E. M.; Alemany L. B.; Tao, Y. J.; Guo, W.; Ausman, K. D.; Kolvin, V. L.; Hughes, J. B. $C_{60}$ in Water: Nanocrystal Formation and Microbial Response. *Environ. Sci.Technol.* **2005,** *39*, 4307–4316.

Fukuzumi, S.; Kojima, T. Photofunctional Nanomaterials Composed of Multiporphyrins and Carbon-Based π-Electron Acceptors. *J. Mater. Chem.* **2008,** *18*, 1427–1439.

Gaddam, R. R.; Kumar, N. A.; Narayan, R.; Raju, K. V. S. N.; Zhao, X. S. Advanced Carbon Materials for Electrochemical Energy Storage. In *Nanomaterials Synthesis*; Elsevier, 2019; pp 385–418.

Ganesh, E. N. Single Walled and Multi Walled Carbon Nanotube Structure, Synthesis and Applications. *Int. J. Innov. Technol. Explor. Eng.* **2013,** *2*, 311–320.

Geckeler, K. E.; Samal, S. Syntheses and Properties of Macromolecular Fullerenes, a Review. *Polym. Int.* **1999,** *48*, 743–757.

Gharbi, N.; Pressac, M.; Hadchouel, M.; Szwarc, H.; Wilson, S. R.; Moussa, F. [60] Fullerene Is a Powerful Antioxidant In Vivo with No Acute or Subacute Toxicity. *Nano Lett.* **2005,** *5*, 2578–2585.

Giacalone, F.; Martin, N. Fullerene Polymers: Synthesis and Properties. *Chem. Rev.* **2006,** *106*, 5136–5190.

Goodarzi, S.; Tatiana, D. R.; Conde, J.; Sefat, F.; Mozafari, M. Fullerene: Biomedical Engineers Get to Revisit an Old Friend. *Mater. Today* **2017,** *20*, 460–480.

Guldi, D. M.; Prato, M. Excited-State Properties of $C_{60}$ Fullerene Derivatives. *Acc. Chem. Res.* **2000,** *33*, 695–703.

Gund, T. M.; Osawa, E.; Schleyer, P. R. A convenient, High-Yield Preparation of Diamantane (Congressane). *Tetrahedron Lett.* **1970,** *11*, 3877–3880.

Gupta, N.; Gupta, S. M.; Sharma, S. K. Carbon Nanotubes: Synthesis, Properties and Engineering Applications. *Carbon Lett.* **2019,** 1–29.

Haddon, R. C. Electronic Structure, Conductivity and Superconductivity of Alkali Metal Doped ($C_{60}$). *Acc. Chem. Res.* **1992,** *25*, 127–133.

Hasobe, T.; Hattori, S.; Kotani, H.; Ohkubo, K.; Hosomizu, K.; Imahori, H.; Kamat, P. V.; Fukuzumi, S. Photoelectrochemical Properties of Supramolecular Composite of Fullerene Nanoclusters and 9-Mesityl-10-Carboxymethylacridinium Ion on $SnO_2$. *Org. Lett.* **2004,** *6*, 3103–3106.

Hayashi, A.; Yamamoto, S.; Suzuki, K.; Matsuoka, T. The First Application of Fullerene Polymer-Like Materials, $C_{60}Pd_n$, as Gas Adsorbents. *J. Mater. Chem.* **2004,** *14*, 2633–2637.

Heath, J. R.; Robert F. C.; Kroto, H. (1939–2016). **2016,** 470–470.

Hebard, A. F. Superconductivity in Doped Fullerenes. *Phys. Today* **1992,** *45*, 26–32.

Herranz, M. A.; Illescas, B.; Martin, N.; Luo, C.; Guldi, D. M. Donor/Acceptor Fulleropyrrolidine Triads. *J. Org. Chem.* **2000,** *65*, 5728–5738.

Hong, J.; Yang, Z.; Jiang, W.; Geng, C.; Gong, M.; Xiao, H.; Wang, Z.; Cheng, L. Antiviral Activity of Nano Carbon Fullerene Lipidosome Against Influenza Virus In Vitro. *J. Huazhong Univ. Sci. Technol. [Medical Sciences]* **2008,** *28*, 243–246.

Howard, J. B.; Thomas McKinnon, J.; Makarovsky, Y.; Lafleur, A. L.; Johnson, M. E. Fullerenes $C_{60}$ and $C_{70}$ in Flames. *Nature* **1991,** *352*, 139.

Hsu, C. P.; Jiang, D. H.; Lee, S. L.; Horng, J. L.; Ger, M. D.; Chang, J. K. Buckyball-, Carbon Nanotube-, Graphite-, and Graphene-Enhanced Dehydrogenation of Lithium Aluminum Hydride. *Chem. Commun.* **2013,** *49*, 8845–8847.

Iijima, S. Helical Microtubules of Graphitic Carbon. *Nature* **1991,** *354*, 56.

Ingo, R.; Parisi, J.; Dyakonov, V.; Lutsen, L.; Vanderzande, D.; Hummelen, J. C. Effect of Temperature and Illumination on the Electrical Characteristics of Polymer–Fullerene Bulk-Heterojunction Solar Cells. *Adv. Funct. Mater.* **2004,** *14*, 38–44.

Isobe, H.; Nakanishi, W.; Tomita, N.; Jinno, S.; Okayama, H.; Nakamura, E. Nonviral Gene Delivery by Tetraamino Fullerene. *Mol. Pharma.* **2006,** *3*, 124–134.

Isobe, H.; Tomita, N.; Jinno, S.; Okayama, H.; Nakamura, E. Synthesis and Transfection Capability of Multi-Functionalized Fullerene Polyamine. *Chem. Lett.* **2001,** *30*, 1214–1215.

Jarrold, M. F. Chemistry: The Smallest Fullerene. *Nature* **2000,** *407*, 26.

Jia, M.; Guo, Q.; Gao, H. L.; Qin, X. Synthesis of $C_{60}$/Graphene Composite as Electrode in Supercapacitors. *Fullerenes Nanotubes Carbon Nanostruct.* **2015,** *23*, 477–482.

Jin, X.; Hu, J. Y.; Tint, M. L.; Ong, S. L.; Biryulin, Y.; Polotskaya, G. Estrogenic Compounds Removal by Fullerene-Containing Membranes. *Desalination* **2007,** *214*, 83–90.

Jones, D. E. H. Dreams in a Charcoal Fire: Predictions About Giant Fullerenes and Graphite Nanotubes. *Phil. Trans. R. Soc. London: A* **1993,** *343*, 9–18.

Joshipura, K. N.; Joshipura, K. N. Exploring the Fullerenes-How Geometrical Ideas Explain Strange Fullerene Structures. *Resonance* **2000,** 5.

Kabdulov, M. A.; Amsharov, K. Y.; Jansen, M. A Step Toward Direct Fullerene Synthesis: $C_{60}$ Fullerene Precursors with Fluorine in á Key Positions. *Tetrahedron* **2010,** *66*, 8587–8593.

Kanev, St.; Nenova, Z.; Koprinarov, N.; Ivanova, K. Conductivity and Photoconductivity Peculiarities Observed in $C_{60}$ Layers. *Semiconduct. Phys. Quantum Electron. Optoelectron.* **2006**.

Kang, S.; Meagan S. M.; Elimelech, M. Microbial Cytotoxicity of Carbon-Based Nanomaterials: Implications for River Water and Wastewater Effluent. *Environ. Sci. Technol.* **2009,** *43*, 2648–2653.

Katsnelson, M. Graphene: Carbon in Two Dimensions. *Mater. Today* **2007,** *10*, 20–27.

Kiyokatsu, J.; Uemura, T.; Ohta, H.; Nagashima, H.; Itoh, K. Separation and Identification of Higher Molecular Weight Fullerenes by High-Performance Liquid Chromatography with Monomeric and Polymeric Octadecylsilica Bonded Phases. *Analyt. Chem.* **1993,** *65*, 2650–2654.

Kornev, A. B.; Peregudov, A. S.; Martynenko, V. M.; Balzarini, J.; Hoorelbeke, B.; Troshin, P. A. Synthesis and Antiviral Activity of Highly Water-Soluble Polycarboxylic Derivatives of [70] Fullerene. *Chem. Commun.* **2011,** *47*, 8298–8300.

Krätschmer, W.; Lamb, L. D.; Fostiropoulos, K.; Huffman, D. R. Solid $C_{60}$: A New Form of Carbon. *Nature* **1990,** *347*, 354.

Kroto, H. Space, Stars, $C_{60}$, and Soot. *Science* **1988,** *242*, 1139–1145.

Kroto, H. Symmetry, Space, Stars and $C_{60}$. *Rev. Modern Phys.* **1997,** *69*, 703.

Kroto, H. W. $C_{60}$: Buckminsterfullerene, the Celestial Sphere That Fell to Earth. *Angew. Chem. Int. Ed. English* **1992,** *31*, 111–129.

Kroto, H. W. The Stability of the Fullerenes Cn, with n= 24, *28*, 32, *36*, 50, 60 and 70. *Nature* **1987,** *329*, 529.

Kroto, H. W.; Allaf, A. W.; Balm, S. P. $C_{60}$: Buckminsterfullerene. *Chem. Rev.* **1991,** *91*, 1213–1235.

Kroto, H. W.; Heath, J. R.; Sean C. O'Brien, Curl, R. F.; Smalley, R. E. $C_{60}$: Buckminsterfullerene. *Nature* **1985,** *318*, 162–163.

Kroto, H. W.; Walton, R. M. D. Stable Derivatives of Small Fullerenes. *Chem. Phys. Lett.* **1993,** *214*, 353–356.

Krystyna, P. Carbon Nanostructures for Separation, Preconcentration and Speciation of Metal Ions. *TrAC Trends Analy. Chem.* **2010,** *29*, 718–727.

Kumar, P.; Senthil, K.; Pavithra, G.; Ramalingam, S. Carbon Based Nanomaterials for Energy Storage. *Carbonaceous Compos. Mater.* **2018,** *42*, 33–56.

Kyesmen, P. I.; Onoja, A.; Amah, A. N. Fullerenes Synthesis by Combined Resistive Heating and Arc Discharge Techniques. *Springer Plus* **2016,** *5*, 1323.

Lebedeva, M. A.; Chamberlain, T. .; Khlobystov, A. N. Harnessing the Synergistic and Complementary Properties of Fullerene and Transition-Metal Compounds for Nanomaterial Applications. *Chem. Rev.* **2015,** *115*, 11301–11351.

Li, Q.; Sun, X.; Yang, Z.; Wang, X.; Wang, Q.; He, D. Network Structures of Fullerene-Like Carbon Core/Nano-Crystalline Silicon Shell Nanofibers as Anode Material for Lithium-Ion Batteries. *Carbon* **2013,** *54*, 29–35.

Li, Qilin.; Mahendra, S.; Lyon, D. Y.; Brunet, L.; Liga, M. V.; Li, D.; Alvarez, P. J. J. Antimicrobial Nanomaterials for Water Disinfection and Microbial Control: Potential Applications and Implications. *Water Res.* **2008,** *42*, 4591–4602.

Li, Z.; Zhang, Y.; Zhu, R.; Wen, G.; Dong, C.; Li, H. Self-Assembled Palladium Nanoflowers Supported on Fullerene: Electrochemical Catalytic Performance for the Reduction of 4-Nitrophenol. *Electrochem Commun.* **2019,** 106484.

Liang, X.; Chen, C.; Zhao, Y.; Jia, L.; Wang, P. C. Biopharmaceutics and Therapeutic Potential of Engineered Nanomaterials. *Curr. Drug Metabol.* **2008,** *9*, 697–709.

Lin, C. M.; Tan, Y. L. $C_{60}$ Fullerene Derivatized Nanoparticles and Their Application to Therapeutics. *Recent Patents Nanotechnol.* **2012,** *6*, 105–113.

Lopez, A. M.; Aurelio, M. A.; Prato, M. Materials Chemistry of Fullerene $C_{60}$ Derivatives. *J. Mater. Chem.* **2011,** *21*, 1305–1318.

Luis, E.; Echegoyen, L. E. Electrochemistry of Fullerenes and Their Derivatives. *Acc. Chem. Res.* **1998,** *31*, 593–601.

Lyon, D. Y.; Adams, L. K.; Falkner, J. C.; Alvarez, P. J. J. Antibacterial Activity of Fullerene Water Suspensions: Effects of Preparation Method and Particle Size. *Environ. Sci. Technol.* **2006,** *40*, 4360–4366.

Lyon, D. Y.; Fortner, J. D.; Sayes, C. M.; Colvin, V. L.; Hughes, J. B. Bacterial Cell Association and Antimicrobial Activity of a $C_{60}$ Water Suspension. *Environ. Toxicol. Chem.* **2005,** *24*, 2757–2762.

Martin, N.; Sánchez, L.; Herranz, M. A.; Guldi, D. M. Evidence for Two Separate One-Electron Transfer Events in Excited Fulleropyrrolidine Dyads Containing Tetrathiafulvalene (TTF). *J. Phys. Chem. A* **2000,** *104*, 4648–4657.

Mauter, M. S.; Elimelech, M. Environmental Applications of Carbon-Based Nanomaterials. *Environ. Sci.Technol.* **2008,** *42*, 5843–5859.

Milan, F.; Costa, D. A.; Balch, A. L.; Fawcett, W. R. Electrochemical Synthesis of a Redox-Active Polymer Based on Buckminsterfullerene Epoxide. *Angew. Chem. Int. Ed. English* **1995,** *34*, 194–196.

Mojica, M.; Alonso, J. A.; Méndez, F. Synthesis of Fullerenes. *J. Phys. Org. Chem.* **2013,** *26*, 526–539.

Nakamura, E.; Isobe, H. Functionalized Fullerenes in Water. THE first 10 Years of Their Chemistry, Biology, and Nanoscience. *Acc. Chem. Res.* **2003,** *36*, 807–815.

Nicolaidis, N. C.; Al-Mudhaffer, M. F.; Holdsworth, J. L.; Zhou, X.; Belcher, W. J.; Dastoor, P.C. Contribution of Fullerene Photocurrent Generation to Organic Solar Cell Performance. *J. Phys. Chem. C* **2019,** *123*, 11950–11958.

Nimibofa, A.; Newton, E. A.; Cyprain, A. Y.; Donbebe, W. Fullerenes: Synthesis and Applications. *J. Mater. Sci. Res* **2018,** *7*, 22.

Park, S.; Srivastava, D.; Cho, K. Local Reactivity of Fullerenes and Nano-Device Applications. *Nanotechnology* **2001,** *12*, 245.

Parker, D. H.; Wurz, P.; Chatterjee, K.; Lykke, K. R.; Hunt, J. E.; Pellin, M. J.; Hemminger, J. C.; Gruen, D. M.; Stock, L. M. High-Yield Synthesis, Separation, and Mass-Spectrometric Characterization of Fullerenes $C_{60}$ to $C_{266}$. *J. Am. Chem. Soc.* **1991,** *113*, 7499–7503.

Partha, B.; Maji, S.; Hill, J. P.; Kim, J. H.; Ariga, K.; Shrestha, L. K. Mesoporous Carbon Cubes Derived from Fullerene Crystals as a High Rate Performance Electrode Material for Supercapacitors. *J. Mater. Chem. A* **2019,** *7*, 12654–12660.

Payne, M. C.; Michael, P. T.; Allan, D. C.; Arias, T. A.; Joannopoulos, J. D. Iterative Minimization Techniques for Ab Initio Total-Energy Calculations: Molecular Dynamics and Conjugate Gradients. *Rev. Modern Phys.* **1992,** *64*, 1045.

Peter, S.; Wirz, L/N.; Avery, J. The Topology of Fullerenes. *Wiley Interdiscip. Rev.* **2015,** *5*, 96–145.

Plonska-Brzezinska, Marta E. Carbon Nano-Onions: A Review of Recent Progress in Synthesis and Applications. *Chem Nano Mat* **2019,** *5*, 568–580.

Pupysheva, O. V.; Farajian, A. A.; Yakobson, B. I. Fullerene Nanocage Capacity for Hydrogen Storage. *Nano Lett.* **2007,** *8*, 767–774.

Rai, P. K.; Nicholas A.; Parra-Vasquez, G.; Peng, H.; Hauge, R. H.; Pasquali, M. Solubility and Size Separation of Large Fullerenes in Concentrated Sulfuric Acids. *J. Phys. Chem. C* **2007,** *111*, 17966–17969.

Richardson, C. F.; Schuster, D. I.; Wilson, S. R. Synthesis and Characterization of Water-Soluble Amino Fullerene Derivatives. *Org. Lett.* **2000,** *2*, 1011–1014.

Ruoff, R. S.; Doris S. T.; Malhotra, R.; Lorents, D. C. Solubility of Fullerene ($C_{60}$) in a Variety of Solvents. *J. Phys. Chem.* **1993,** *97*, 3379–3383.

Saga, S.; Matsumoto, H.; Saito, K.; Minagawa, M.; Tanioka, A. Polyelectrolyte Membranes Based on Hydrocarbon Polymer Containing Fullerene. *J. Power Sour.* **2008,** *176*, 16–22.

Samanta, P. N.; Das, K. K. Noncovalent Interaction Assisted Fullerene for the Transportation of Some Brain Anticancer Drugs: A Theoretical Study. *J. Mol. Graphics Modell.* **2017,** *72*, 187–200.

Scarel, F.; Alonso, A. M. Fullerene $C_{60}$ Architectures in Materials Science. *Carbon Nanomater.* **2013,** 47–88.

Sheka, E. F.; Razbirin, B. S.; Nelson, D. K. Continuous Symmetry of $C_{60}$ Fullerene and Its Derivatives. *J. Phys. Chem. A* **2011,** *115*, 3480–3490.

Shibuya, M.; Kato, M.; Ozawa, M.; Fang, P. H.; Ōsawa, E. Detection of Buckminsterfullerene in Usual Soots and Commercial Charcoals. *Fullerene Sci. Technol.* **1999,** *7*, 181–193.

Shrivastava, S. K. Superconductivity in Organic Materials (2019).

Shultz, M. D.; Duchamp, J. C.; Wilson, J. D.; Shu, C. Y.; Ge, J.; Zhang, J.; Gibson, H. W.; Fillmore, H. L.; Hirsch, J. I.; Dorn, H. C.; Fatouros, P. P. Encapsulation of a Radiolabeled Cluster Inside a Fullerene Cage, 177Lu x Lu (3− x) N@ $C_{80}$: An Interleukin-13-Conjugated Radiolabeled Metallofullerene Platform. *J. Am. Chem. Soc.* **2010,** *132*, 4980–4981.

Sinha, A.; Sinha, A. Buckyball C60–the Story So Far. *Resonance* **2001,** *6*, 56–65.

Snow, S. D.; Park, K. E.; Kim, J. H. Cationic Fullerene Aggregates with Unprecedented Virus Photoinactivation Efficiencies in Water. *Environ. Sci. Technol. Lett.* **2014,** *1*, 290–294.

Stankevich, Ivan V.; Nikerov, M. V.; Bochvar, D. A. The Structural Chemistry of Crystalline Carbon: Geometry, Stability, and Electronic Spectrum. *Russian Chem. Rev.* **1984,** *53*, 640.

Tatiana, D. R.; Prato, M.; Novello, F.; Maggini, M.; Banfi, E. Easy Access to Water-Soluble Fullerene Derivatives via 1, 3-Dipolar Cycloadditions of Azomethine Ylides to $C_{60}$. *J. Org. Chem.* **1996,** *61*, 9070–9072.

Taylor, R.; Walton, D. R. M. The Chemistry of Fullerenes. *Nature* **1993,** *363*, 685.

Terrones, H.; Terrones, M.; Hsu, W. K. Beyond C $_{60}$: Graphite Structures for the Future. *Chem. Soc. Rev.* **1995,** *24*, 341–350.

Thines, R. K.; Mubarak, N. M.; Nizamuddin, S.; Sahu, J. N.; Abdullah, E. C.; Ganesan, P. Application Potential of Carbon Nanomaterials in Water and Wastewater Treatment: A Review. *J. Taiwan Inst. Chem. Eng.* **2017,** *72*, 116–133.

Toth, K. (Endo) fullerene Functionalization: From Material Science to Biomedical Applications, PhD diss.; Strasbourg, 2012.

Troshina, O. A.; Troshin, P. A.; Peregudov, A. S.; Kozlovskiy, V. I.; Balzarini, J.; Lyubovskaya, R. N. Chlorofullerene $C_{60}Cl_6$: A Precursor for Straightforward Preparation of Highly Water-Soluble Polycarboxylic Fullerene Derivatives Active Against HIV. *Org. Biomol. Chem.* **2007,** *5*, 2783–2791.

Wang, C. Y.; Zhang, Y. Y.; Wang, C. M.; Tan, V. B. C. Buckling of Carbon Nanotubes: A Literature Survey. *J. Nanosci. Nanotechnol.* **2007,** *7*, 4221–4247.

Wang, M.; Huang, L.; Sharma, S. K.; Jeon, S.; Thota, S.; Sperandio, F. F.; Nayka, S.; Chang, J.; Hamblin, M. R.; Chiang, L. Y. Synthesis and Photodynamic Effect of New Highly Photostable Decacationically Armed [60]-and [70] Fullerene Decaiodide Monoadducts to Target Pathogenic Bacteria and Cancer Cells. *J. Med. Chem.* **2012,** *55*, 4274–4285.

Wang, Y.; Luo, K.; Ye, W.; Du, S.; Francisco, J. S.; Yin, J.; Gao, P. Mg@$C_{60}$ Nano-Lamellae and Its 12.50 wt% Hydrogen Storage Capacity. *Int. J. Hydrog. Energy* **2019,** *44*, 15239–15245.

Whetten, R. L.; Alvarez, M. M.; Anz, S. J.; Schriver, K. E.; Beck, R. D.; Diederich, F. N.; Rubin, Y. Spectroscopic and Photophysical Properties of the Soluble $C_n$ Molecules, n= 60, *70*, 76/78, 84. *MRS Online Proceedings Library Archive*, 1990; p 206.

Xu, B. Prospects and Research Progress in Nano Onion-Like Fullerenes. *New Carbon Mater.* **2008,** *23*, 289–301.

Yadav, B. C.; Kumar, R. Structure, Properties and Applications of Fullerenes. *Int. J. Nanotechnol. App.* **2008,** *2*, 15–24.

Yadav, J. Fullerene: Properties, Synthesis and Application. *Res. Rev.: J. Phys.* **2018,** *6*, 1–6.

Yeoseon, C.; Ye, Y.; Mackeyev, Y.; Cho, M.; Lee, S.; Wilson, L. J.; Lee, J.; Alvarez, P. J. J.; Choi, W.; Lee, J. $C_{60}$ Aminofullerene-Magnetite Nanocomposite Designed for Efficient Visible Light Photocatalysis and Magnetic Recovery. *Carbon* **2014,** *69*, 92–100.

Zhang, Y.; Cheng, X. Hydrogen Storage Property of Alkali and Alkaline-Earth Metal Atoms Decorated $C_{24}$ Fullerene: A DFT Study. *Chem. Phys.* **2018,** *505*, 26–33.

Zheng, G.; Irle, S.; Morokuma, K. Towards Formation of Buckminsterfullerene $C_{60}$ in Quantum Chemical Molecular Dynamics. *J. Chem. Phys.* **2005,** *122*, 014708.

Zhu, S.; Eva, O.; Mary, L. H. Toxicity of an Engineered Nanoparticle (Fullerene, $C_{60}$) in Two Aquatic Species, Daphnia and Fathead Minnow. *Marine Environ. Res.* **2006,** *62*, S5–S9.

CHAPTER 5

# Biochar: An Advanced Carbon Material for Mitigation of Environmental Pollution

SANGHAMITRA MOHAPATRA[1,2] and CHINMAYEE ACHARYA[1,3]

[1]*CSIR—Institute of Minerals and Materials Technology, Bhubaneswar, India*

[2]*AcSIR—Academy of Scientific and Industrial Research, New Delhi, India*

[3]*North Orissa University, Baripada, India*

## ABSTRACT

Environmental pollution, causing serious harmful concerns, is increasing day by day. Various chemical and biological remediation strategies are being adopted to mitigate pollution. Out of those remediation strategies, biochar is one of the most investigated topics since the last decade. Biochar is a carbon derivative obtained from the carbonization of various organic biomasses under limited supply of or without oxygen, which yields a porous, carbon rich material with low density. Biochars have several physicochemical properties of interest such as high specific surface area and cation exchange abilities that are strongly affected by the source materials and pyrolysis temperatures, which enhance the sorption of a variety of contaminants to their surfaces. The sorption of these contaminants occurs through various processes, namely, ion exchange, electrostatic interaction, complex formation, membrane filtration, physisorption, and precipitation.

New Forms of Carbon: Nanocarbons. Aneeya Kumar Samantara & Satyajit Ratha (Eds.)

To increase the pollutant removal capacity, further modification of the biochars may also be done by adding some metals like iron, calcium, magnesium, sulfur, silver, zinc, etc., through acid treatment, with the help of clay particles. Such modifications increase the surface area, affect the polarity, alter the pH, and form new functional groups in biochar which increases the adsorption capacity. In this chapter, we will be discussing about the production methods of biochars and various biochar composites, properties, and their utilization for remediation of environmental pollution.

## 5.1 INTRODUCTION

The presence of heavy metals, organic and inorganic pollutants in environmental systems such as soil, air, and water pose challenges in environmental remediation. Many workers reported the presence of the following pollutants in soil; polycyclic aromatic hydrocarbons (PAHs), pesticides, polychlorinated biphenyls (PCB), explosives, metals, metalloids, and radionuclides (Zhu and Shaw, 2000; Kumar et al., 2012; Testiati et al., 2013; Vane et al., 2014). Once these pollutants enter the soil system, they contaminate the underground as well as surface water by leaching and runoff, respectively. Mobility of these heavy metals aided by several atmospheric events, for example, runoff water and wind, drastically increase their content in the topsoil, resulting in both air and water contamination/pollution, which could lead to several chronic health-related disorders in the inhabited living bodies. Presence of these contaminants in soil and water resources is undesirable due to their toxicity and anomalies they create. Increase in the concentration of these pollutants beyond their permissible limits has primarily resulted from rapid industrialization, urbanization, and agricultural activities (using pesticides and artificial fertilizers). Unlike most of the organic pollutants (which gradually degrade when disposed), inorganic pollutants, specifically the heavy metals, are nonbiodegradable and may be passed along the food chain through bioaccumulation. Therefore, removal of these toxic substances from the soil and water is one of the focus areas in the field of environmental sciences.

In this context, remediation of contaminated areas for pollution reduction and minimum downstream damage is essential (Powlson et al., 2011). Several strategies and methods have been employed to combat soil contamination, for example, soil washing, soil vapor extraction, farming, soil flushing, and ion exchanges (Zhou and Song, 2004; Prasad and Nakbanpote, 2015). However, these traditional methods are usually expensive and might create further glitches, such as loss of soil fertility and soil erosion (Khan et al., 2004; Kumpiene et al., 2008). Various removal techniques for wastewater treatment (e.g., ion exchange, precipitation, electrocoagulation, activated carbon adsorption, and packed bed filtration) have been employed by many researchers. For both soil and water treatments, most of these techniques are expensive and time consuming. Therefore, a more cost-effective alternative is required in this context, which would make use of low-cost adsorbents for soil and water treatment (Kailash et al., 2010). Such amendments include red mud (Ma et al., 2007), nanoparticles (Araújo et al., 2015; Mitra et al., 2017), carbon-based nanoparticles such as graphene, CNTs, activated carbons (Kemp et al., 2013; Ahmad et al., 2018; Karnib et al., 2014; Anastopoulos et al., 2017), biomass, and biochar (Tan et al., 2015). The word, "biochar," is a combination of "bio" as in "biomass" and "char" as in "charcoal". Biochar is a black carbon derived by pyrolysis of carbon-rich biomass in an oxygen-limited environment (Lehmann and Joseph, 2009). The International Biochar Initiative (IBI) (http://www.biochar-international.org/biochar) defines biochar as a solid material obtained from the carbonization of biomass. Various degrees of carbonization produce infinite varieties of biochars for use as fuels and adsorbents. The primary constituents of biochar include carbon (C), hydrogen (H), oxygen (O), nitrogen (N), sulfur (S), and ash in different proportions. These biochars possess both high specific surface area, and large number of pores, abundant surface functional groups and mineral components, which makes them quite useful as adsorbent to remove pollutants from aqueous solutions (Zhang et al., 2018; Zhu et al., 2018a).

Biochars have drawn significant research interest recently, primarily due to their multifunctionality such as carbon sequestration and soil fertility enhancement (Santín et al., 2017), bioenergy production (Zhu et al., 2018), and environmental remediation (Mohan et al., 2014; Inyang et al., 2016). Several recent publications reported biochar's excellent ability to immobilize organic (Dai et al., 2019) and inorganic pollutants (Yuan et al., 2018; Mohan et al., 2014) in both soil and water systems. Many researchers have reported that biochars are potential low-cost adsorbents, as they can immobilize or bind chemical compounds including some of the most common pollutants, for example, heavy metals, pesticides or herbicides (Cao and Harris, 2010; Chen and Yuan, 2011; Jiang et al.,

2012). Therefore, biochar could serve as an in situ alternative material for amending soils contaminated with heavy metals (Zhao et al., 2016).

## 5.2 BIOCHAR PRODUCTION

Biochar production is mostly done by utilizing the waste biomass. Utilization of waste biomass for biochar production is economical and beneficial as it provides clean alternatives of fossil fuels, and also the biochar products can be used to mitigate pollution (Barrow, 2012).Waste biomass from various sources has been extensively used to produce biochar (Cantrell et al., 2012), such as crop residues, forestry waste, animal manure, animal bones, food processing waste, paper mill waste, municipal solid waste, and sewage sludge (Li et al., 2013; Ahmad et al., 2014). Various "biomass to biochar" conversion methods have been developed to produce biochar (Czernik and Bridgwater, 2004; Mohan et al., 2006). Both biological (anaerobic digestion, hydrolysis, and fermentation) and thermal (combustion, pyrolysis, liquefaction, torrefaction, and gasification) conversion methods are used to transform biomass into fuel and byproducts. The most popular method for biochar production is pyrolysis. Pyrolysis is the process of charring in the absence or limited presence of oxygen. Pyrolysis technique involves various thermochemical conversion processes with limited/no supply of oxygen to yield different products, for example, bio-oil, biochar, and pyrogas. During pyrolysis, thermal decomposition of cellulose, hemicellulose, and lignin occurs. These components undergo their own reaction pathways at their own reaction temperatures which include crosslinking, depolymerization, and fragmentation. The end products are generally solid, liquid, and gaseous products. The solid and liquid products generated from the pyrolysis process are biochar and bio-oil, respectively, while the gaseous mixture containing CO, $CO_2$, $H_2$, and $C_1$–$C_2$ hydrocarbons, are called pyrogas/syngas.

Pyrolysis can be categorized as slow pyrolysis and fast pyrolysis depending on the heating rate and residence time. Conventional slow pyrolysis method has been mainly utilized for charcoal production. It is a thermodynamically controlled processing of biomass within a temperature range of 500–800°C, with very slow heating rate that varies within 5–10°C, for long residence time (may be for hours or days). The main product of slow pyrolysis is biochar (50–60%) along with bio-oil and gases. The yield of these products depends upon both biomass feedstock and temperature. Biochar produced by slow pyrolysis, generally has high surface area and found to be advantageous. Conventional carbonization of biomass is done by slow pyrolysis. This method has been used to generate charcoal for centuries. Slow pyrolysis favors high yield of biochar. The major limitation of slow pyrolysis is long production time and low potential of value-added utilization of coproducts, which may require expensive and complex purification and fractionation step (Kambo and Dutta, 2015). Fast pyrolysis involves heating of biomass for very short residence times, at very high heating rates of about ~1000°C $s^{-1}$ and fast quenching of vapors to maximize production of bio-oil. Mostly, the residence time in fast pyrolysis is shorter (less than 10 s). The fast pyrolysis process is based on kinetically controlled reactions, where vapors are separated rapidly to prevent its condensation and carbonization as char and noncondensable gases. This is achieved through restricting thermodynamic equilibrium by using very high heating rate that causes both drying and volatilization steps, spontaneously. Depending on the biomass properties, fast pyrolysis of biomass generally produces 60–75% of bio-oil, 15–25% biochar, and 10–20% of noncondensable gases. The nature of the end product, from pyrolysis process, depends on the operating parameters involved and the type of biomass (source material). The effect of operating parameters on the product yield is an essential feature of the pyrolysis technique, which needs to be explored, in order to get insights about the underlying optimization steps. The factors affecting the pyrolysis process and sorption properties of biochar such as pyrolysis temperature, feedstock, etc. are discussed below.

### *5.2.1 TEMPERATURE*

The pyrolysis temperature plays an important role for biochar quality. The temperature range varies for the pyrolysis of major biocomponents of biomass, that is, hemicellulose (150–350°C), cellulose (275–350°C), and lignin (250–500°C), which determines the end products of pyrolysis. Productions of condensable volatiles are relatively dominant during pyrolysis of cellulose than hemicellulose, whereas pyrolysis of hemicellulose produces relatively more noncondensable gases and less tarry compounds than cellulose. During pyrolysis, thermal decomposition of lignin occurs to form aromatic compounds and the char yield is more than cellulose. In addition, lignin pyrolysis also contributes to produce bio-oil (approximately 35%, mostly contains phenol) and approx. 10% of gases. The pyrolysis reactions for hemicellulose and lignin are of exothermic nature that leads to charring process. Whereas, pyrolysis reactions for cellulose are endothermic in nature which results in fast devolatilization reactions and contributes less char formation. Overall, the char yield mainly depends upon the lignin content of the biomass, which contributes more during charring process than cellulose and hemicellulose. The temperature range varies for the pyrolysis of major biocomponents of biomass, that is, hemicellulose (150–350°C), cellulose (275–350°C), and lignin (250–500°C), which determines the end products of pyrolysis. Production of condensable volatiles is relatively dominant during pyrolysis of cellulose than hemicellulose, whereas pyrolysis of hemicellulose produces relatively more noncondensable gases and less tarry compounds than cellulose. During pyrolysis, thermal decomposition of lignin, to form aromatic compounds and char, is more than cellulose. In addition, lignin pyrolysis also contributes to produce bio-oil (approx. 35%, mostly contains phenol) and approximately 10% of gases.

A number of reactions (primary and secondary) and devolatilization of biomass take place during the pyrolysis. At low temperature, the primary reactions predominate, and with the increase in the reaction temperature, vapor formation increases. However, with an increase in the temperature, the incidence of secondary reactions also increases. With increasing temperature, more volatiles are formed resulting residual biomass, that is, decreased biochar yield. The yield of biochar always decreases as temperature and heating rate increases, which is due to the significant loss of volatile matter or secondary decomposition of char at a higher temperature (Chutia et al., 2014). Secondary decomposition of the char, at a higher temperature, produces noncondensable gases, which contribute to the increase in gaseous product yield. The increased pyrolysis temperature results in low biochar yield, and the mineral ash content becomes more concentrated in the biochar. Consequently, biochars produced at higher pyrolysis temperatures have more total, soluble and exchangeable base cations and carbonates, less oxygenated functional groups, and higher pH (Fletcher et al., 2014; Qian et al., 2013; Yuan et al., 2011). Thus, more alkaline biochars are produced at higher pyrolysis temperature, which can lead to greater precipitation of metal cations (Kim et al., 2013).

Many literatures reported decreased VM content in biochar when pyrolyzed at high temperature. This is also associated with rapid loss of VM (Keiluweit et al., 2010; Spokas, 2010); as volatile organic compounds undergo the formation of organic, cyclic, and aromatic molecules with increased pyrolysis temperature. Sun et al. (2014) reported that, increasing pyrolysis temperature causes lower O, H, and N contents, whereas increases C, P, K, Ca, and Mg contents. Moreover, the increase in the pyrolysis temperature and residence time lead to dehydration, deoxidation, and decomposition of volatile materials in biomass. The increase in the C content and decrease in the O/H content in biochar lead to a low O/C and H/C ratio (Brewer et al., 2014; Cantrell et al., 2012). Thus, increasing pyrolysis temperature increases the aromaticity and decreases the polarity of the biochar, making the biochar surface less hydrophilic. Lower O and H contents have been reported when increasing the pyrolysis temperature, due to loss of surface functional groups. The decrease in the oxygen-containing functional groups can affect the metal sorption capacity of biochars, as reported by Ding et al. (2014), who found high Pb adsorption capacity of biochar when produced at low temperature. Formation of aromatic functional groups makes the biochar more stable; however, these groups are less important in adsorption of pollutants. Carboxyl and hydroxyl functional groups are important for adsorption and are abundantly available in low temperature biochars. Temperature also affects the surface area of biochar and higher temperature always results in larger surface area with micropores. Large surface area helps in better adsorption of pollutants.

### 5.2.2 FEEDSTOCK

The common feedstocks for biochar preparation are residues/wastes generated from forestry and agriculture. These are heterogeneous in nature, that is, they show great variations in physical and chemical structures, even within a single species. The general composition of every biomass is cellulose, hemicellulose, and lignin. Due to the differences in crystallinity and polymerization, lignin is the most stable component in an unaltered biomass, followed by cellulose and hemicellulose (Li et al., 2017a, b). Higher lignin content in feedstock commonly leads to a higher content of aromatic C in biochar and a slower rate of biochar mineralization (Ameloot et al., 2013; Windeatt et al., 2014). Biochars from wood are usually more stable than those from grasses, sludge, and husks (Cely et al., 2014; Hilscher et al., 2009; Zimmerman, 2010). The influence of feedstock type on biochar labile C content ($H_2O_2$ oxidizable fraction) decreased with increasing pyrolysis temperature (from 350°C to 650°C), and no significant effect was found in the case of biochar obtained at 650°C due to the dominant impact of pyrolysis temperature (Crombie and Mašek, 2015). Biochar approaches increasingly similar structures (i.e., mostly polyaromatic) with increasing temperature. And at higher temperatures, the formation of these structures becomes independent of feedstock type (McBeath et al., 2011). However, the structure changing rate in response to the increasing pyrolysis temperature varied for different biomass (Morales et al., 2015).

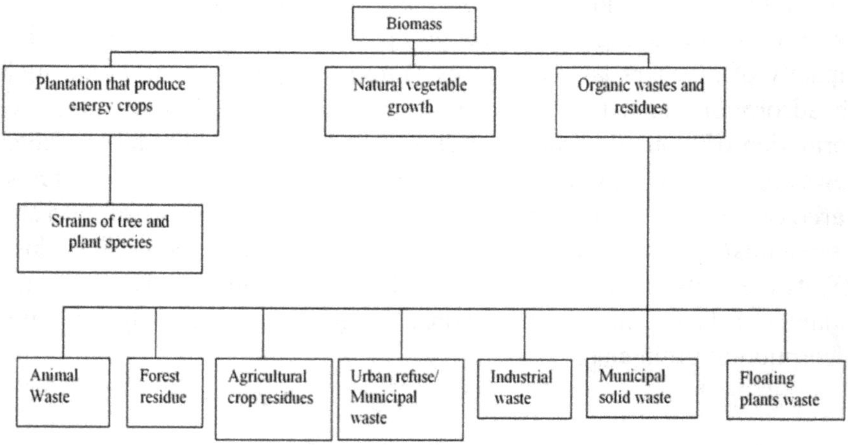


International Biochar Initiative has classified biomass into two types, that is, processed and unprocessed feedstocks. Unprocessed feedstocks generally correspond to the type of biomasses which are not contaminated and/or treated with any sources. Unprocessed feedstocks for biochar production comprise unutilized biomass residues or wastes of plant kingdom, fungi, and algae, cultivated in an uncontaminated environment. The processed feedstocks comprise biomass that were chemically and/or biologically processed such as wastewater sludge, paper pulp sludge, sewage sludge, animal/human manures, and compost. IBI emphasized that any combination of feedstocks can be used for biochar production, but the presence of contaminants should be limited to 2% on dry weight basis. Processed feedstocks also include the biomass cultivated in contaminated lands such as mining sites, wastewater irrigated sites, and plants used for phytoremediation process.

### 5.2.3 REACTION TIME

Reaction time is generally referred to the residence time of the process. Sometimes it is of few minutes and sometimes it varies up to hours. Biochar quality depends on reaction time, as large reaction time causes improved removal of volatile matters and magnification of carbon content. In the context of how the product composition, in pyrolysis, is affected by the residence time, Zhang et al. (2015) reported that the biochar yield decreased with increased residence time, at the same pyrolysis temperature. Longer reaction time not only increases the carbon content but also enhances several additional biochar characteristics such as porosity, surface area, and aromaticity. According to Lu et al. (1995), the specific surface area and pore area increased with increasing residence time up to 2 h at 500~900 °C, but they decreased when the residence time exceeded 2 h. Specifically, the specific surface area and pore area decreased rapidly when the residence time exceeded a time duration of 2 h, at high temperatures. Amarasinghe et al. (2016) investigated the effects of different residence time (15, 30, 45, and 60 min) on the preparation of biochar from refuse tea, including chemical, physical, and morphological properties of the products. As presented, temperatures in the range of 450–500°C, and residence time between 45 and 60 min showed the highest biochar mass recovery.

### 5.2.4 OTHER FACTORS

The pH of biochar increases with increase in the pyrolysis temperature due to the enrichment of ash content (Brewer et al., 2012; Gul et al., 2015; Windeatt et al., 2014). Several key features, such as hydrophobicity, aromaticity, and specific surface area have been found to increase, when the pyrolysis temperature is increased (>500°C) (Keiluweit et al., 2010). These properties impart a high degree of effectiveness on to the biochar towards the removal of organic pollutants. Biochar is often zwitterionic or dipolar and thus comprises both positively- and negatively-charged surfaces (Sohi et al., 2009; Tan et al., 2017). The negatively-charged functional groups are likely to attract the cations and contribute to increasing cation exchange capacity (CEC) of soils; moreover, anion exchange capacity (AEC) is also exhibited by O-containing functional groups (oxonium heterocycles) of biochar, dependent on the pH value (Brewer et al., 2012; Lawrinenko, 2014). Low pyrolysis temperature facilitates partial carbonization, which facilitates biochar yield with small pore size, low surface area, and high O-containing functional groups (Cox et al., 2012; Keiluweit et al., 2010; Lu et al., 2014), which increases the efficacy of biochar against potentially dangerous inorganic pollutants, due to increased ionic interactions with O-containing functional groups.

## 5.3 MODIFICATION OF BIOCHAR

Though biochar has the ability to remove pollutants from different mediums, its capacity is usually lower compared to other common biosorbents such as activated carbon. In order to enhance the adsorption capacity of biochar, various modifications have been used, for example, altering the specific surface area, pore structures, and content of functional groups. Attempts have been made to improve the surface area, porosity, pHPZC, and/or functional groups of biochar. Approaches to modify biochars include loading them with additives, such as minerals, reductants, organic functional groups, and nanoparticles, along with their activation with alkali solution. The commonly used modification methods can be generally divided into four types such as impregnation with minerals, nanoscale-metals assistance, surface oxidation, and surface reduction.

### *5.3.1 IMPREGNATION WITH MINERALS*

Minerals have their unique cation exchange capacity, surface charge, and mineral structure and are being widely used as low-cost adsorbents for the removal of various pollutants (Yao et al., 2014). Therefore, mineral impregnation can be used to enhance the functionality of modified biochar. Modification of biochar includes loading it with different minerals such as hematite (g-$Fe_2O_3$), magnetite, zero valent Fe, hydrous Mn oxide, calcium oxide, and birnessite (Han et al., 2016; Samsuri et al., 2013; Agrafioti et al., 2014; Van Vinh et al., 2015; Bakshi et al., 2018). The loading can be achieved before, during, or after pyrolysis of the feedstock.

Samsuri et al. (2013) studied the mechanisms of As(V) and As(III) sorption by Fe-coated biochars, obtained from empty fruit bunch and rice husk. The Fe coating significantly improved the As(III) sorption capacity of the biochar from 19 to 31 mg $g^{-1}$, while As(V) sorption increased from 5.5–7.1 to 15–16 mg $g^{-1}$. The possible sorption mechanism was through As complexation with Fe 3p on the biochar. However, another method has been reported for Cr(VI) and As(V) sorption by biochar modified with reductants like zero valent Fe or $Na_2SO_3$/$FeSO_4$, which can enhance metal reduction and surface complexation with the functional groups (Zhou et al., 2014; Pan et al., 2014). Bakshi et al. (2018) used low-cost hematite to impregnate low-ash biomass. They synthesized zero valent iron (ZVI)-biochar complexes by copyrolysis in a simple and effective method as compared to the traditional methods. Arsenic was effectively removed from drinking water by reduction of $As^{5+}$ to $As^{3+}$ and oxidation of $Fe^0$ to $Fe^{3+}$, where $As^{3+}$ and $Fe^{3+}$ were found to coexist on the surface of biochar in the forms of Fe(As)OOH and FeOOH. The γ-$Fe_2O_3$ particles present on the surface of biochar play a vital role in the activity of sorption sites, through electrostatic interaction (Wang et al., 2019). Agrafioti et al. (2014) investigated the production of calcium reagent ($Ca^{2+}$) and ferric reagent ($Fe^0$ and $Fe^{3+}$) modified biochar made from rice husk and municipal solid wastes. All modified biochar materials exhibited improved $As^{5+}$ and $Cr^{6+}$ removal capacities than their nonimpregnated counterparts. Furthermore, modified biochar with $Ca^{2+}$ and $Fe^{3+}$ was shown to remove more than 95% of $As^{5+}$ contamination, with the removal of $As^{5+}$ and $Cr^{6+}$. The sorption is mostly occurring via metal precipitation and electrostatic interactions between the modified adsorbents and pollutants. Coating biochar with $Fe^{3+}$ caused a change in the O/C ratios, (O + N)/C ratios, and polarity indices [(O + N)/C], which determine the capacity of biochar for $As^{5+}$ adsorption (Samsuri et al., 2013; Son et al., 2018a). Son et al. (2018b) synthesized engineered magnetic biochar by pyrolyzing waste marine macroalgae doped with iron oxide particles (e.g., magnetite, maghemite), which exhibited a greater capacity for the removal of $Cd^{2+}$, $Cu^{2+}$, and $Zn^{2+}$. Bakshi et al. (2018) investigated the main mechanisms for removal of arsenic from drinking water by magnetite-modified biochar, which included coprecipitation, surface adsorption, and intraparticle diffusion. Furthermore, Baig et al. (2014) reported that $Fe^{3+}$/$Fe^{2+}$ modified biochar could possess an improved affinity for $As^{3+}$ and $As^{5+}$, while $Al^{3+}$-modified biochar showed enhanced sorption of $As^{5+}$ (Qian et al., 2013). Mineral components such as montmorillonite, hematite ($\gamma$-$Fe_2O_3$), magnetite, calcium oxide, hydrous manganese oxide, and birnessite have been widely used to impregnate raw materials because of their magnetism, high number of adsorption sites, and widespread distribution. Biochar mineral composites are commonly synthesized by one-step oxygen-limited pyrolysis methods, in tubular or muffle furnaces, or they are directly impregnated with mineral components and dried under suitable conditions. However, the cost of synthesis of biochar–mineral complexes remain a major limitation and optimal impregnation ratios deserve further attention.

### *5.3.2 NANOSCALE-METALS ASSISTANCE*

Nanometal deposition on biochar improves the efficiency of pollutant adsorption by increasing the number of adsorption sites, the specific surface area, improving the reaction mechanism, thermal stability of biochar, and increasing its resistance to oxidation (Fang et al., 2011). There are different methods of preparing nanoscale-metal assisted biochars such as reduction/co-precipitation, carbothermal reduction and self-assembly, hydrothermal carbonization, and electro-derived/microwave-assisted pyrolysis (Fig. 5.1).

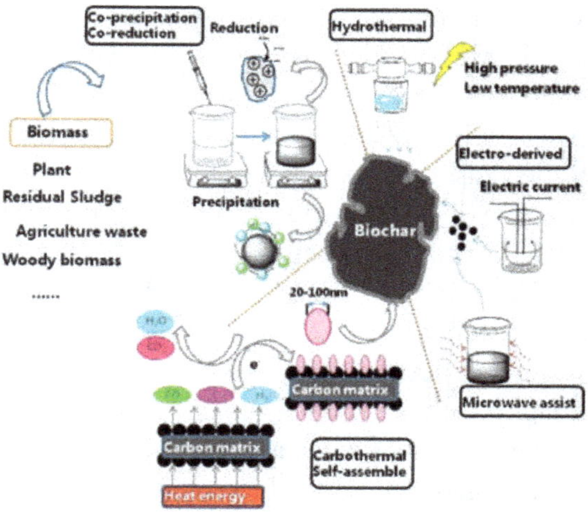


**FIGURE 5.1** Preparation of nanoscale-metal assisted biochar.

*Source*: Reprinted with permission from Ho et al., 2017.

One of the widely used nanoscale assisted biochar is nanoscale zero-valent iron (nZVI) modified biochar. It has gained significant attention, due to its strong adsorption capacity, affinity to pollutants, and reducibility. According to the recent reports, biochar can efficiently prevent the aggregation of nZVI with larger surface area and adsorption sites. nZVI-modified biochar has been applied in the removal of heavy metals from wastewater (Chen et al., 2012). Although the surface micropore structure and porosity of nZVI-modified biochar may be destroyed due to the generation or formation of NPs, more functional groups can be generated from the catalysis of iron ions. Dong et al. (2017) modified biochar with nZVI and treated it with HCl, which increased the contact area between nZVI and $Cr^{6+}$ and the sites for deposition of precipitates as well as reduced the agglomeration of nZVI particles. XRD and SEM identified the final products formed as FeOOH, $Cr_2O_3$, and $Cr_2FeO_4$, during the process of heavy metals removal by the reduction of $Cr^{6+}$ to $Cr^{3+}$. These findings coincided with the results of Zhuang et al. (2014), who synthesized nZVI-modified biochar for the removal of $Cr^{6+}$, and reported a similar Cr removal mechanism. Liu et al. (2015) reported that several heavy metals can be reduced directly by nZVI-modified biochar, such as $Pb^2$ to $Pb^0$, $Cu^{2+}$ to $Cu^0$, and $As^{5+}$ to $As^{3+}$. Lyu et al. (2017) synthesized FeS nanoparticle-assisted biochar by adding $Na_2S$ solution into biochar suspension loaded with $Fe^{2+}$-CMC complexation. FeS and $Fe_3O_4$ NPs crystals were present on the surface, while the granular-like particles (~256 nm) were observed to bond with biochar by C=C, C=O, and Si-O functional groups. Kim and Kan (2016) combined $TiO_2$ NPs with biochar for improved electron utilization efficiency. The $TiO_2$ granules dispersed well on the surface of the biochar in the anatase form. Recent studies have also demonstrated that sulfide-modified nanoscale zero-valent iron (S-nZVI) can increase electron conductivity, reducibility, and broaden the reactive lifetime of nZVI, as S-nZVI exhibits high electronegativity (5.02 V), roughly 1 V higher than that of nZVI (Kim et al., 2011; Levard et al., 2013).

### *5.3.3 SURFACE OXIDATION*

Yang and Jiang (2014) reported that at certain proportions, mixtures of concentrated sulfuric acid and nitric acid solutions produce nitronium ions ($NO^{2+}$), which undergoes electrophilic substitution to the aromatic ring of biochar, modifying nitro groups. Sodium dithionite ($N_2S_2O_4$) was used as a reducing agent for reduction of nitro groups to amino groups on the aromatic rings of biochar. The addition of $H_3PO_4$ to biochar at a suitable concentration impacts the physicochemical properties such as the degree of aromatization, functional groups and pH as well as zeta potential, exhibiting optimal adsorption properties due to the large concentration of acidic functional groups (Zhou et al., 2017a, 2017b). Song et al. (2014) produced $KMnO_4$-modified biochar with much higher adsorption capacity for $Cu^{2+}$, $Cb^{2+}$, and $Cd^{2+}$. After modification of the biomass with $KMnO_4$, ultrafine $MnO_x$ particles were produced on the biochar surface, enhancing the interaction of heavy metals with surface oxygen-containing functional groups, cation-π, and mineral components. Boudrahem et al. (2011) modified coffee residue biochar using chemical activation with phosphoric acid and zinc chloride. The resulted biochar was porous with high surface area and enhanced adsorption capacity for the removal of $Pb^{2+}$ and $Cd^{2+}$ from aqueous solution. Similarly, Zou et al. (2016) have modified biochar using hydrothermally carbonized *Cymbopogon schoenanthus* (L.) Spreng (HLG) material pretreated with different concentrations of $H_2O_2$. Biochar treated with 20% $H_2O_2$ resulted in the maximum amount of $Cu^{2+}$ uptake. Fan et al. (2018) reported the modification of aged wheat straw biochar by chemically treating it with $HNO_3$–$H_2SO_4$ and NaOH–$H_2O_2$ systems. The modification increased surface oxygen-containing functional groups in the resultant biochars, especially for carboxyl. A large number of well-developed mesopores were found in the aged biochar, and the specific surface area was found to increase by ~126% for biochar treated with NaOH–$H_2O_2$. The biochar, with much improved surface modifications, showed enhanced cadmium sorption capacity.

Gas activation methods are also available having the advantage of being environmentally friendly, without generating any secondary pollutants. However, the activation temperature and energy requirements are relatively high, with lower carbon yield than other methods, limiting the widespread application of this method. Except for hydrothermal treatment and conventional pyrolysis, the most commonly used biochar modification method is the microwave-assisted pyrolysis technique. Microwave heating makes only indirect contact with the heated materials, resulting in a rapid heating rate and reduced energy consumption, making this method more environmentally friendly than conventional heating methods (Mohamed et al., 2016).

### 5.3.4 SURFACE REDUCTION MODIFICATIONS

Surface reduction improves the content of basic functional groups (oxygen naphthalene and pyrone) on the surface of biochar, especially nitrogenous functional groups (i.e., amine, quaternary ammonium, and imidazole). The commonly used reducing agents are $NH_3 \cdot H_2O$, $Na_2SO_3$, $FeSO_4$, aniline, and $H_2$. Wu et al. (2017) produced biochar from coconut and activated it by ammonia treatment. The biochar specific surface area (BET) and basic functional groups improved after pyrolysis at 300°C. Pan et al. (2014) reported peanut straw biochar modification using $Na_2SO_3/FeSO_4$, which resulted with more oxygen containing functional groups on the biochar surface, after treatment.

## 5.4 APPLICATION OF BIOCHAR

### 5.4.1 APPLICATION OF BIOCHAR FOR ORGANIC POLLUTANT REMOVAL

In recent years, organic pollution has become a serious environmental issue for both water and soil. For the removal of organic pollutants, a number of treatment techniques have been developed, for example, flocculation (Guan and Tian, 2017), sedimentation (Wen et al., 2017), light treatment (Deng et al., 2017), membrane processes (Sarasidis et al., 2017), chemical oxidation (Gayathri et al., 2017), biological oxidation (Chen et al., 2017), photocatalytic oxidation/degradation (Janssens et al., 2017), adsorption (Huang et al., 2017), and combined methods (Sandoval et al., 2017; Zhou et al., 2015). Among these, adsorption is considered the most, because of its high removal rate and simple operation. Palma et al. (2016) reported carbonaceous material production from avocado skin by assessing the carbonization temperature and time effect, and adsorption of acid, alkali, and reactive dye contrast, to explore the feasibility of carbonized materials. Biochar-based nanocomposites also showed high affinity for organic pollutants. The concerned organic contaminants include crystal violet (Sun et al., 2015), methylene blue (Zhang and Gao, 2013; Inyang et al., 2014), phenanthrene (Tang et al., 2015), phenol (Karakoyun et al., 2011; Kong et al., 2014), sulfapyridine (Inyang et al., 2015), naphthalene (NAPH), and p-nitrotoluene (p-NT) (Chen et al., 2011a). The adsorption ability of these organic contaminants varied from 2.4 to 278.55 mg $g^{-1}$, which was found to be dependent on the type of nanomaterials, biochar substrates, and the target organic contaminants. Maneerung et al. (2016) activated the wood gasification biochar via steam, which showed a high adsorption capability (189.83 mg g-1) towards rhodamine B. Cattle-manure-derived low temperature biochar (CMB) has been investigated for its efficacy in removing methylene blue (MB) from aqueous solution by Zhu et al. (2018b) (Fig. 5.2). The zero-valent iron and graphitic $C_3N_4$ biochar composites were employed for the removal of various organic contaminants from aqueous solutions, including acid orange 7 (Quan et al., 2014), methylene blue (Pi et al., 2015), and methyl orange (Han et al., 2015).

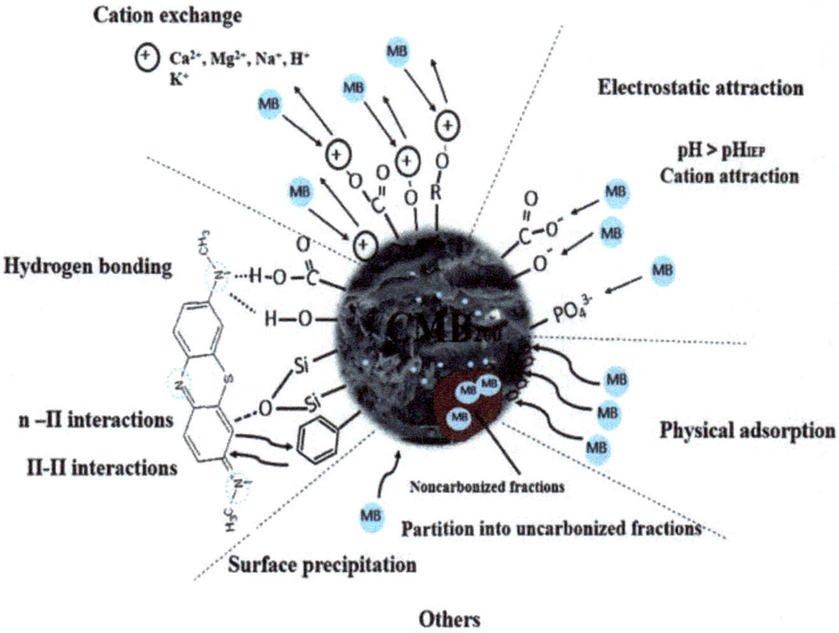


**FIGURE 5.2** Various mechanisms of methylene blue adsorption by biochar.

*Source*: Reprinted with permission from Zhu et al., 2018b, Copyright 2018. Royal Society of Chemistry

Biochar has been extensively used towards removing PAHs from aqueous solutions as it has high adsorptive capacity. Oleszczuk et al. (2012) reported about 57% removal of dissolved PAHs in sewage sludge.

Similarly, Beesley et al. (2010) used biochar in PAHs contaminated soil to remove >50% and > 40% of the HMW and LMW PAHs, respectively. They have also reported a reduction in the toxicity effect towards micro-organisms and plants due to immobilization of soil PAHs. Furthermore, Wang et al. (2006) demonstrated that using wood-based biochar, the PAHs (pyrene, phenanthrene, and benzo(a)anthracene) removal efficiency of ≥ 60% from aqueous solution could be achieved. Hale et al. (2011) demonstrated that particle size of biochar has little or no effect on sorption of PAHs. Contrasting results were reported by Zheng et al. (2010) in which they showed that smaller particle size provides shorter time duration for the sorption equilibrium. Sorption of organic contaminates by biochar is governed by the type of feedstock, production processes, sorbate concentration, and the age (Wang et al., 2006). Recent study revealed that wheat straw-based biochar produced at 400, 600, and 800°C, could achieve PAH removal efficiency of 71.8–98.6%, from soil washing effluent (Li et al., 2014). Chen et al. (2008) reported maximum adsorption of naphthalene (136.8 mg/g) by pine needle biochar produced at 700°C. It has been further demonstrated that the nonlinear relationship between production temperature and adsorption capacities of biochar could have resulted from the destruction of aliphatic alkyl and ester groups at the time of biochar production.

Biochars are considered as good adsorbents, for relatively mobile pesticides such as MCPA (2-methyl-4-chlorophenoxyacetic acid) and diuron. Adsorption improved when biochar was applied as a distinct layer in the adsorption column as opposed to when it is mixed with the substrate, as has been reported by Cederlund et al. (2017). Therefore, biochar is best used as a defined layer in a constructed wetland. Mandal et al. (2017) reported adsorption of atrazine and imidacloprid on different chars produced from eucalyptus bark, corn cob, bamboo chips, rice husk, and rice straw. Zhao et al. (2013) demonstrated the effect of $NH_4H_2PO_4$ treatment of corn stalk biochars for atrazine adsorption and reported that the treatment increased the adsorption capacity from 7.8 up to 53.9 mg $g^{-1}$ at 25°C. Xiao and Pignatello (2015) reported that the micro- and meso-porosity of biochars have strong impact on the adsorption of triazine herbicides. Cationic aromatic amines get adsorbed on the biochar surface by $\pi$–$\pi$ electron donor–acceptor interactions, which occur between the cation of the target molecule and the electron-rich polyaromatic surface of the biochar. There are different mechanisms by which biochar adsorb the organic pollutants, for example, electrostatic interaction, hydrophobic effect, hydrogen bonds, and pore-filling. The surface properties of biochar are important for the adsorption of organic pollutants. Surface of biochar is heterogeneous as the carbonized and non-carbonized fractions coexist. These carbonized and non-carbonized phases of biochar generally represent different adsorption mechanisms. The uptake of organic compounds is decided by partition into the non-carbonized organic matter and by adsorption onto the carbonized matter (Chen et al., 2008; Cao et al., 2009; Zheng et al., 2010).

### *5.4.2 APPLICATION OF BIOCHAR FOR INORGANIC POLLUTANT REMOVAL*

Biochars promise a low-cost alternative for the adsorption of heavy metal pollutants in aqueous media. Compared to activated carbon, biochar is a more efficient adsorbent with lower cost for metal removal. Metal sorption capacities of biochar are 2.4–147, 19.2–33.4, 0.3–39.1, and 3.0–123 mg $g^{-1}$ for Pb, Ni, Cd, and Cr, respectively (Inyang et al., 2016). Biochars are produced from various feedstocks (wood bark, dairy manure, pinewood, rice husk, etc.), under different pyrolysis conditions (governed by different parameters, such as temperature, rate of heat transfer, and residence time), to adsorb metals including arsenic (As), cadmium (Cd), chromium (Cr), mercury (Hg), and lead (Pb) from water (Qian et al., 2015; Xie et al., 2015; Inyang et al., 2016). Mohan et al. (2007) noted that removal efficiencies of lead (Pb) and cadmium (Cd) by biochar, obtained from the bark of an oak tree are comparable to that of Calgon F-400, a commercially available activated carbon. Chen et al. (2011b) reported that biochar produced from wood or corn straw can effectively adsorb copper (Cu) and zinc (Zn) in aqueous solutions. Similarly, Kong et al. (2011) reported 75–87% removal of mercury from aqueous solution by using soybean stalk biochar.

The implementation of wide range of techniques for the production and optimization of biochar, along with their surface modification, has yielded a large pool of highly efficient and low-cost engineered biochars, which show excellent adsorption capacities comparable to or even exceed some of the commercially available activated carbons. Previous studies reported the metal oxide NPs-modified biochars which can be used as efficient adsorbents for the removal of heavy metals from polluted environments. Such metal oxides are aluminum oxide, iron oxide, nickel oxide, manganese oxide, titanium oxide, cerium oxide, and magnesium oxide. Jung et al. (2015) synthesized nano-Mg-Al modified biochar by electromodification. Mg-modified biochars can also be used to adsorb various environmental pollutants such as heavy metals (Zhang et al., 2012; Usman et al., 2013), nitrate (Zhang et al., 2014), and phosphate (Dan et al., 2018). Gao et al. (2018) successfully synthesized sulfide-modified nZVI (S-nZVI/BC) biochar and utilized it for $Cr^{6+}$ removal. The maximum $Cr^{6+}$ removal capacity was observed at a pH value of 2.5. Sulfide-modified nZVI improved the electron transfer ability from the $Fe^0$ core to the $Cr^{6+}$ solution, due to the formation of FeS on the surface of nZVI. Alqadami et al. (2018) reported excellent adsorption capacities for $Pb^{2+}$, $Cd^{2+}$, and $Co^{2+}$ by magnetic nanocomposite biochar. Li et al. (2018) modified rice hulls with $La(NO_3)_3$ and $Ce(NO_3)_3$ to prepare biochar via one-step pyrolysis, with oxygen-rich functional groups and lanthanum/cerium oxide providing the main active sites for $As^{5+}$ removal. About five mechanisms have been proposed to govern metal sorption by biochar from aqueous solutions, that is, complexation, cation exchange, precipitation, electrostatic interactions, and chemical reduction. However, the role of each of these mechanisms played for each metal varies considerably depending on target metals.

## 5.5 CONCLUSIONS

This chapter summarizes the potential application of biochar for remediation of various environmental pollutants. Biochar has received significant attention during the past decade. However, studies are mostly at a lab scale, focusing on sorption of single pollutant from synthetic solutions. In natural waters, different pollutants coexist with other pollutants; thereby triggering a competition for sorption sites on biochar surface between the ions or organic pollutants. Further studies are required for the application of biochar in contaminated sites to measure its efficiency. All researches about the application of biochar pointed to the same conclusion that biochar can be a novel and feasible adsorbent. This is not only because of the biochars' excellent adsorption ability, but also the associated environmental and economic benefits. Despite the use of biochar as adsorbents is increasing, a number of research gaps and uncertainties still exist as identified in this review. To close these knowledge gaps, more relevant investigations are needed in further research.

## KEYWORDS


- **adsorption**
- **biochar**
- **composites**
- **pollution**
- **remediation**


## REFERENCES


Agrafioti, E.; Kalderis, D.; Diamadopoulos, E. Ca and Fe Modified Biochars as Adsorbents of Arsenic and Chromium in Aqueous Solutions. *J. Environ. Manag.* **2014,** *146*, 444–450.

Ahmad, A.; Lokhat, D.; Setapar, S. H. M.; Khatoon, A.; Shahadat, M.; Rafatullah, M. *Use of Carbon Nanotubes as Sorbents for Heavy Metal Remediation from Wastewater*; Nanocomposites for Pollution Control: Pan Stanford, 2018; pp 331–357.

Ahmad, M.; Rajapaksha, A. U.; Lim, J. E.; Zhang, M.; Bolan, N.; Mohan, D.; Vithanage, M.; Lee, S. S; Ok, Y. S. Biochar as a Sorbent for Contaminant Management in Soil and water: A Review. *Chemosphere* **2014,** *99*, 19–33.

Alqadami, A.; Khan, M.; Otero, M.; Siddiqui, M.; Jeon, B.; Batoo, K. A Magnetic Nanocomposite Produced from Camel Bones for an Efficient Adsorption of Toxic Metals from Water. *J. Clean. Prod.* **2018,** *178*, 293–304.

Amarasinghe, H. A. H. I.; Gunathilake, S. K.; Karunarathna, A. K. Ascertaining of Optimum Pyrolysis Conditions in Producing Refuse Tea Biochar as a Soil Amendment. *Procedia Food Sci.* **2016,** *6*, 97–102.

Ameloot, N.; Graber, E. R.; Verheijen, F. G. A.; De Neve, S. Interactions Between Biochar Stability and Soil Organisms: Review and Research Needs. *Eur. J. Soil Sci.* **2013,** *64*, 379–390.

Anastopoulos, I.; Anagnostopoulos, V. A.; Bhatnagar, A.; Mitropoulos, A. C.; Kyzas, G. Z. A Review for Chromium Removal by Carbon Nanotubes. *Chem. Ecol.* **2017,** *33*, 572–588.

Araújo, R.; Castro, A. C. M.; Fiúza, A. The Use of Nanoparticles in Soil and Water Remediation Processes. *Mater. Today: Proceedings*, **2015,** *2*, 315–320.

Bakshi, S.; Banik, C.; Rathke, S.; Laird, D. Arsenic Sorption on Zero-Valent Ironbiochar Complexes. *Water Res.* **2018,** *137*, 153.

Barrow, C. J. Biochar: Potential for Countering Land Degradation and for Improving Agriculture. *Appl. Geogr.* **2012,** *34*, 21–28.

Beesley, L.; Moreno-Jimenez, E.; Gomez-Eyles, J. L. Effects of Biochar and Green Waste Compost Amendments on Mobility, Bioavailability and Toxicity of Inorganic and Organic Contaminants in a Multi-Element Polluted Soil. *Environ. Pollut.* **2010,** *158*, 2282–2287.

Boudrahem, F.; Soualah, A.; Aissanibenissad, F. Pb(II) and Cd(II) Removal from Aqueous Solutions Using Activated Carbon Developed from Coffee Residue Activated with Phosphoric Acid and Zinc Chloride. *J. Chem. Eng. Data* **2011,** 56.

Cantrell, K. B.; Hunt, P. G.; Uchimiya, M.; Novak, J. M.; Ro, K. S. Impact of Pyrolysis Temperature and Manure Source on Physicochemical Characteristics of Biochar. *Bioresour. Technol.*, **2012,** *107*, 419–428.

Cao, X.; Harris, W. Properties of Dairy-Manure-Derived Biochar Pertinent to Its Potential Use in Remediation. *Bioresour. Technol.* **2010,** *101*, 5222.

Cao, X.; Ma, L.; Gao, B.; Harris, W. Dairy-Manure Derived Biochar Effectively Sorbs Lead and Atrazine. *Environ. Sci. Technol.* **2009,** *43*, 3285–3291.

Cederlund, H.; Börjesson, E.; Stenström, J. Effects of a Woodbased Biochar on the Leaching of Pesticides Chlorpyrifos, Diuron, Glyphosate and MCPA. *J. Environ. Manage.* **2017,** *191*, 28–34.

Cely, P.; Tarquis, A. M.; Paz-Ferreiro, J.; Méndez, A.; Gascó, G. Factors Driving the Carbon Mineralization Priming Effect in a Sandy Loam Soil Amended with Different Types of Biochar. *Solid Earth* **2014,** *5*, 585–594.

Chen, B. L.; Chen, Z. M. Sorption of Naphthalene and 1-Naphthol by Biochars of Orange Peels with Different Pyrolytic Temperatures. *Chemosphere* **2009,** *76*, 127–133.

Chen, B.; Chen, Z.; Lv, S. A Novel Magnetic Biochar Efficiently Sorbs Organic Pollutants and Phosphate. *Bioresour. Technol.* **2011a,** *102*, 716–723.

Chen, X.; Chen, G.; Chen, L.; Chen, Y.; Lehmann, J.; McBride, M. B.; Hay, A. G. Adsorption of Copper and Zinc by Biochars Produced from Pyrolysis of Hardwood and Corn Straw in Aqueous Solution. *Bioresour. Technol.* **2011b,** *102*, 8877–8884.

Chen, B.; Yuan, M. Enhanced Sorption of Polycyclic Aromatic Hydrocarbons by Soil Amended with Biochar. *J. Soils Sediments* **2011,** *11*, 62–71.

Chen, B.; Zhou, D.; Zhu, L. Transitional Adsorption and Partition of Nonpolar and Polar Aromatic Contaminants by Biochars of Pine Needles with Different Pyrolytic Temperatures. *Environ. Sci. Technol.* **2008,** *42*, 5137–5143.

Chen, J.; Qiu, X.; Fang, Z.; Yang, M.; Pokeung, T. Removal Mechanism of Antibiotic Metronidazole from Aquatic Solutions by Using Nanoscale Zero-Valent Iron Particles. *Chem. Eng. J.* **2012,** *181–182*, 113–119.

Chen, Y.; Lin, J.; Chen, Z. Remediation of Water Contaminated with Diesel Oil Using a Coupled Process: Biological Degradation Followed by Heterogeneous Fenton-Like Oxidation. *Chemosphere* **2017,** *183*, 286–293.

Cox, J.; Downie, A.; Hickey, M.; Jenkins, A.; Lines-Kelly, R.; McClintock, A.; Van Zwieten, L. Biochar in Horticulture: Prospects for the Use of Biochar in Australian Horticulture. *HAL Report*, **2012.** ISBN 978-1-74256-349-7.

Crombie, K.; Mašek, O. Pyrolysis Biochar Systems, Balance Between Bioenergy and Carbon Sequestration. *GCB Bioenergy* **2015,** *7*, 349–361.

Dai, Y.; Zhang, N.; Xing, C.; Cui, Q.; Sun, Q. The Adsorption, Regeneration and Engineering Applications of Biochar for Removal Organic Pollutants: A Review. *Chemosphere* **2019**.

Dan, J.; Bei, C.; Yoshimasa, A.; Motoi, M. Removal and Recovery of Phosphate from Water by mg-Laden Biochar: Batch and Column Studies. *Colloids Surf. A Physicochem. Eng. Asp.* **2018,** *558*, 429–437

Deng, F.; Zhong, F.; Lin, D. C.; Zhao, L. N.; Liu, Y. J., Huang, J. H.; Luo, X. B.; Luo, S. L.; Dionysiou, D. D. One-Step Hydrothermal Fabrication of Visible-Light-Responsive

AgInS2/SnIn4S8 Heterojunction for Highly-Efficient Photocatalytic Treatment of Organic Pollutants and Real Pharmaceutical Industry Wastewater. *Appl. Catal. B.* **2017,** *219*, 163–172.

Ding, W.; Dong, X.; Ime, I. M.; Gao, B.; Ma, L. Q. Pyrolytic Temperatures Impact 645 Lead Sorption Mechanisms by Bagasse Biochars. *Chemosphere* **2014,** *105*, 68–74.

Dong, H.; Deng, J.; Xie, Y.; Cong, Z.; Zhao, J.; Cheng, Y.; Zeng, G. Stabilization of Nanoscale Zero-Valent Iron (nZVI) with Modified Biochar for Cr(VI) Removal from Aqueous Solution. *J. Hazard. Mater.* **2017,** *332*, 79–86.

Fan, Q.; Sun, J.; Chu, L.; Cui, L.; Quan, G.; Yan, J.; Hussain, Q.; Iqbal, M. Effects of Chemical Oxidation on Surface Oxygen-Containing Functional Groups and Adsorption Behavior of Biochar. *Chemosphere* **2018,** *207*, 33–40.

Fang, Z.; Qiu, X.; Huang, R.; Qiu, X.; Li, M. Removal of Chromium in Electroplating Wastewater by Nanoscale Zero-Valent Metal with Synergistic Effect of Reduction and Immobilization. *Desalination* **2011,** *280*, 224–231.

Gao, J.; Yang, Li.; Liu, Y.; Shao, F.; Liao, Q.; Shang, J. Scavenging of Cr(VI) from Aqueous Solutions by Sulfide-Modified Nanoscale Zero-Valent Iron Supported by Biochar. *J. Taiwan Inst. Chem. Eng.* **2018,** *91*, 449–456.

Gayathri, P. V.; Yesodharan, S.; Yesodharan, E. P. Purification of Water Contaminated with Traces of Rhodamine B Dye by Microwave-Assisted, Oxidant-Induced and Zinc Oxide Catalyzed Advanced Oxidation Process. *Desalin. Water Treat.* **2017,** *85*, 161–174.

Guan, W.; Tian, S. C. The Modified Chitosan for Dyeing Wastewater Treatment via Adsorption and Flocculation. *Sci. Adv. Mater.* **2017,** *9*, 1603–1609.

Hale, S. E.; Hanley, K.; Lehmann, J.; Zimmerman, A.; Cornelissen, G. Effects of Chemical, Biological, and Physical Aging as Well as Soil Addition on the Sorption of Pyrene to Activated Carbon and Biochar. *Environ. Sci. Technol.* **2011,** *45*, 10445–10453.

Han, L.; Xue, S.; Zhao, S.; Yan, J.; Qian, L.; Chen, M. Biochar Supported Nanoscale Iron Particles for the Efficient Removal of Methyl Orange Dye in Aqueous Solutions. *PLoS One* **2015,** *10*, e0132067.

Han, Y. T.; Cao, X.; Ouyang, X.; Sohi, S. P.; Chen, J. W. Sorption Kinetics of Magnetic Biochar Derived from Peanut Hull on Removal of Cr (VI) from Aqueous Solution: Effects of Production Conditions and Particle Size. *Chemosphere* **2016,** *145*, 336–341.

Hilscher, A.; Heister, K.; Siewert, C.; Knicker, H. Mineralisation and Structural Changes During the Initial Phase of Microbial Degradation of Pyrogenic Plant Residues in Soil. *Org. Geochem.* **2009,** *40*, 332–342.

Ho, S. H.; Zhu, S.; Chang, J. S. Recent Advances in Nanoscale-Metal Assisted Biochar Derived from Waste Biomass Used for Heavy Metals Removal. *Bioresour. Technol.* **2017,** *246*, 123–134.

Huang, Z. H.; Li, Y. Z.; Chen, W. J.; Shi, J. H.; Zhang, N.; Wang, X. J.; Li, Z.; Gao, L. Z.; Zhang, Y. X. Modified Bentonite Adsorption of Organic Pollutants of Dye Wastewater. *Mater. Chem. Phys.* **2017,** 266–276.

Inyang, M. I.; Gao, B.; Yao, Y.; Xue, Y.; Zimmerman, A.; Mosa, A.; Pullammanappallil, P.; Ok, Y. S.; Cao, X. A Review of Biochar as a Low-Cost Adsorbent for Aqueous Heavy Metal Removal. *Crit. Rev. Environ. Sci. Technol.* **2016,** *46*, 406–433.

Inyang, M.; Gao, B.; Zimmerman, A.; Zhang, M.; Chen, H. Synthesis, Characterization, and Dye Sorption Ability of Carbon Nanotube–Biochar Nanocomposites. *Chem. Eng. J.* **2014,** *236*, 39–46.

Inyang, M.; Gao, B.; Zimmerman, A.; Zhou, Y.; Cao, X. Sorption and Cosorption of Lead and Sulfapyridine on Carbon Nanotube-Modified Biochars. *Environ. Sci. Pollut. Res.* **2015,** *22*, 1868–1876.

Janssens, R.; Mandal, M. K.; Dubey, K. K.; Luis, P. Slurry Photocatalytic Membrane Reactor Technology for Removal of Pharmaceutical Compounds from Wastewater: Towards Cytostatic Drug Elimination. *Sci. Total. Environ.* **2017,** *599–600*, 612–626.

Jiang, T. Y.; Jiang, J.; Xu, R. K.; Li, Z. Adsorption of Pb(II) on Variable Charge Soils Amended with Rice-Straw Derived Biochar. *Chemosphere* **2012,** *89*, 249–256.

Jung, K.; Hwang, M.; Jeong, T.; Ahn, K. A Novel Approach for Preparation of Modified Biochar Derived from Marine Macroalgae: Dual Purpose Electro-Modification for Improvement of Surface Area and Metal Impregnation. *Bioresour. Technol.* **2015,** *191*, 342–345.

Kailash, D.; Dharmendra, P.; Anil, V. Low Cost Adsorbents for Heavy Metal Removal from Wastewater: A Review. *Res. J. Chem. Environ.* **2010,** *14*, 100–103.

Karakoyun, N.; Kubilay, S.; Aktas, N.; Turhan, O.; Kasimoglu, M.; Yilmaz, S.; Sahiner, N. Hydrogel-Biochar Composites for Effective Organic Contaminant Removal from Aqueous Media. *Desalination* **2011,** *280*, 319–325.

Karnib, M.; Kabbani, A.; Holail, H.; Olama, Z. Heavy Metals Removal Using Activated Carbon, Silica and Silica Activated Carbon Composite. *Energy Procedia* **2014,** *50*, 113–120.

Kemp, K. C.; Seema, H.; Saleh, M.; Le, N. H.; Mahesh, K.; Chandra, V.; Kim, K. S. Environmental Applications Using Graphene Composites: Water Remediation and Gas Adsorption. *Nanoscale* **2013,** *5*, 3149–3171.

Khan, F. I.; Husain, T.; Hejazi, R. An Overview and Analysis of Site Remediation Technologies. *J. Environ. Manage*. **2004,** *71*, 95–122.

Kim, E.; Kim, J.; Azad, A.; Chang, Y. Facile Synthesis and Characterization of Fe/Fes Nanoparticles for Environmental Applications. *ACS Appl. Mater. Interfaces* **2011,** *3*, 1457–1462.

Kim, J. R.; Kan, E. Heterogeneous Photocatalytic Degradation of Sulfamethoxazole in Water Using a Biochar-Supported $TiO_2$ Photocatalyst. *J. Environ. Manage.* **2016,** *180*, 94–101.

Kim, W. K.; Shim, T.; Kim, Y. S.; Hyun, S.; Ryu, C.; Park, Y. K.; Jung, J. Characterization of Cadmium Removal from Aqueous Solution by Biochar Produced from a Giant Miscanthus at Different Pyrolytic Temperatures. *Bioresour. Technol.* **2013,** *138*, 266–270.

Kong, H.; He, J.; Gao, Y.; Wu, H.; Zhu, X. Cosorption of Phenanthrene and Mercury(II) from Aqueous Solution by Soybean Stalk-Based Biochar. *J. Agric. Food Chem.* **2011,** *59*, 12116–12123.

Kong, L.; Xiong, Y.; Sun, L.; Tian, S.; Xu, X.; Zhao, C.; Luo, R.; Yang, X.; Shih, K.; Liu, H. Sorption Performance and Mechanism of a Sludge-Derived Char As Porous Carbon-Based Hybrid Adsorbent for Benzene Derivatives in Aqueous Solution. *J. Hazard. Mater.* **2014,** *274*, 205–211

Kumar, B.; Gaur, R.; Goel, G.; Mishra, M.; Singh, S. K.; Prakash, D.; Sharma, C. S. Residues of Pesticides and Herbicides in Soils from Agriculture Areas of Delhi Region, India. *Electron. J. Environm. Agric. Food Chem.* **2012,** *11*, 328–338.

Kumpiene, J.; Lagerkvist, A.; Maurice, C. Stabilization of As, Cr, Cu, Pb and Zn in Soil Using Amendments—A Review. *Waste Manage.* **2008,** *28*, 215–225.

Lehmann, J.; Joseph, S. *Biochar for Environmental Management: Science and Technology*; Earthscan Ltd.: London, 2009.

Levard, C.; Hotze, E.; Colman, B.; Dale, A.; Truong, L.; Yang, X.; Bone, A. J.; Brown, G. E.; Tanguay, R. L.; Di Giulio, R. T.; Bernhardt, E. S.; Meyer, J. N.; Wiesner, M. R.; Lowry, G. V. Sulfidation of Silver Nanoparticles: Natural Antidote to Their Toxicity. *Environ. Sci. Technol.* **2013,** *47*, 13440–13448

Li, H.; Dong, X.; da Silva, E. B.; de Oliveira, L. M.; Chen, Y.; Ma, L. Q. Mechanisms of Metal Sorption by Biochars: Biochar Characteristics and Modifications. *Chemosphere* **2017b,** *178*, 466–478.

Li, H.; Qu, R.; Li, C.; Guo, W.; Han, X.; He, F.; Ma, Y.; Xing, B. Selective Removal of Polycyclic Aromatic Hydrocarbons (PAHs) from Soil Washing Effluents Using Biochars Produced at Different Pyrolytic Temperatures. *Bioresour. Technol.* **2014,** *163*, 193–198.

Li, J.; Li, Y.; Wu, M.; Zhang, Z.; Lü, J. Effectiveness of Low-Temperature Biochar in Controlling the Release and Leaching of Herbicides in Soil. *Plant Soil*, **2013,** *370*, 333–344.

Li, J.; Yan-Qun, Z.; Gang, L.; Guo-Xin, S. Mechanism of as (V) Removal from Water by Lanthanum and Cerium Modified Biochars. *Huan Jing Ke Xue* **2018,** *39*, 2211.

Li, X.; Liu, L.; Wang, X.; Ok, Y. S.; Elliott, J. A.; Chang, S. X.; Chung, H. J. Flexible and Self-Healing Aqueous Supercapacitors for Low Temperature Applications: Polyampholyte Gel Electrolytes with Biochar Electrodes. *Sci. Rep.* **2017a,** 7.

Liu, M.; Wang, Y.; Chen, L.; Zhang, Y.; Lin, Z. $Mg(OH)_2$ Supported Nanoscale Zero Valent Iron Enhancing the Removal of Pb(II) from Aqueous Solution. *ACS Appl. Mater. Interfaces* **2015,** *7*, 7961–7969

Lu, G. Q.; Low, J. C. F.; Liu, C. Y.; Lua, A. C. Surface Area Development of Sewage Sludge During Pyrolysis. *Fuel* **1995,** *74*, 344–348.

Lu, K.; Yang, X.; Shen, J.; Robinson, B.; Huang, H.; Liu, D.; Wang, H. Effect of Bamboo and Rice Straw Biochars on the Bioavailability of Cd, Cu, Pb and Zn to Sedum plumbizincicola. *Agr. Ecosys. Environ.* **2014,** *191*, 124–132.

Lyu, H.; Tang, J.; Huang, Y.; Gai, L.; Zeng, E. Y.; Liber, K.; Gong, Y. Removal of Hexavalent Chromium from Aqueous Solutions by a Novel Biochar Supported Nanoscale Iron Sulfide Composite. *Chem. Eng. J.* **2017,** *322*, 516–524.

Ma, H. W.; Hung, M. L.; Chen, P. C. A Systemic Health Risk Assessment for the Chromium Cycle in Taiwan. *Environ. Int.*, **2007,** *33*, 206–218.

Mandal, A.; Singh, N.; Purakayastha, T. J. Characterization of Pesticide Sorption Behaviour of Slow Pyrolysis Biochars as Low Cost Adsorbent for Atrazine and Imidacloprid Removal. *Sci. Total Environ.* **2017,** *577*, 376–385.

Maneerung, T.; Liew, J.; Dai, Y.; Kawi, S.; Chong, C.; Wang, C. H. Activated Carbon Derived from Carbon Residue from Biomass Gasification and Its Application for Dye Adsorption: Kinetics, Isotherms and Thermodynamic Studies. *Bioresour. Technol.* **2016,** *200*, 350–359.

McBeath, A. V.; Smernik, R. J.; Schneider, M. P. W.; Schmidt, M. W. I.; Plant, E. L. Determination of the Aromaticity and the Degree of Aromatic Condensation of a Thermosequence of Wood Charcoal Using NMR. *Org. Geochem.* **2011,** *42*, 1194–1202.

Mitra, S.; Sarkar, A.; Sen, S. Removal of Chromium from Industrial Effluents Using Nanotechnology: A Review. *Nanotechnol. Environ. Eng.*, **2017,** *2*, 11.

Mohamed, B. A.; Kim, C. S.; Ellis, N.; Bi, X. Microwave-Assisted Catalytic Pyrolysis of Switchgrass for Improving Bio-Oil and Biochar Properties. *Bioresour. Technol.* **2016,** *201*, 121–132.

Mohan, D.; Pittman Jr, C. U.; Bricka, M.; Smith, F.; Yancey, B.; Mohammad, J.; Steele, P. H.; Alexandre-Franco, M. F.; Gómez-Serrano, V.; Gong, H. Sorption of Arsenic, Cadmium, and Lead by Chars Produced from Fast Pyrolysis of Wood and Bark During Bio-Oil Production. *J. Colloid Interface Sci.* **2007,** *310*, 57–73.

Mohan, D.; Sarswat, A.; Ok, Y. S.; Pittman Jr, C. U. Organic and Inorganic Contaminants Removal from Water With Biochar, a Renewable, Low Cost and Sustainable Adsorbent—A Critical Review. *Bioresour. Technol.* **2014,** *160*, 191–202.

Oleszczuk, P.; Hale, S. E.; Lehmann, J.; Cornelissen, G. Activated Carbon and Biochar Amendments Decrease Pore-Water Concentrations of Polycyclic Aromatic Hydrocarbons (PAHs) in Sewage Sludge. *Bioresour. Technol.* **2012,** *111*, 84–91.

Pan, J.; Jiang, J.; Xu, R. Removal of Cr(VI) from Aqueous Solutions by Na2SO3/FeSO4 Combined with Peanut Straw Biochar. *Chemosphere* **2014,** *101*, 71–76.

Panwar, N. L.; Pawar, A.; Salvi, B. L. Comprehensive Review on Production and Utilization of Biochar. *SN Appl. Sci.* **2019,** *1*, 168.

Pi, L.; Jiang, R.; Zhou, W.; Zhu, H.; Xiao, W.; Wang, D.; Mao, X. G-$C_3N_4$ Modified Biochar as an Adsorptive and Photocatalytic Material for Decontamination of Aqueous Organic Pollutants. *Appl. Surf. Sci.* **2015,** *358*, 231–239.

Prasad, M. N. V.; Nakbanpote, W. Integrated Management of Mine Waste Using Biogeotechnologies Focusing Thai Mines. In *Environmental Sustainability*; Springer: New Delhi, 2015; pp 229–249.

Qian, K. Z.; Kumar, A.; Zhang, H. L.; Bellmer, D.; Huhnke, R. Recent Advances in Utilization of Biochar. *Renew. Sustain. Energy. Rev.* **2015,** *42*, 1055–1064.

Qian, W.; Zhao, A. Z.; Xu, R. K. Sorption of As(V) by aluminum-Modified Crop Strawderived Biochars. *Water Air Soil Pollut.* **2013,** *224*, 1610.

Quan, G.; Sun, W.; Yan, J.; Lan, Y. Nanoscale Zero-Valent Iron Supported on Biochar: Characterization and Reactivity for Degradation of Acid Orange 7 from Aqueous Solution. *Water Air Soil Pollut.* **2014,** *225*, 1–10.

Samsuri, A. W.; Sadegh-Zadeh, F.; Seh-Bardan, B. J. Adsorption of As(III) and As(V) by Fe Coated Biochars and Biochars Produced from Empty Fruit Bunch And Rice Husk. *J. Environ. Chem. Eng.* **2013,** *1*, 981–988.

Sandoval, A.; Hernández-Ventura, C.; Klimova, T. E. Titanate Nanotubes for Removal of Methylene Blue Dye by Combined Adsorption and Photocatalysis. *Fuel* **2017,** *198*, 22–30.

Santín, C.; Doerr, S. H.; Merino, A.; Bucheli, T. D.; Bryant, R.; Ascough, P.; Gao, X.; Masiello, C. A. Carbon Sequestration Potential and Physicochemical Properties Differ Between Wildfire Charcoals and Slow-Pyrolysis Biochars. *Sci. Rep.* **2017,** *7*, 11233.

Sarasidis, V. C.; Plakas, K. V.; Karabelas, A. J. Novel Water-Purification Hybrid Processes Involving In Situ Regenerated Activated Carbon, Membrane Separation and Advanced Oxidation. *Chem. Eng. J.* **2017,** *328*, 1153–1163.

Son, E. B.; Poo, K. M.; Chang, J. S.; Chae, K. J. Heavy Metal Removal from Aqueous Solutions Using Engineered Magnetic Biochars Derived from Waste Marine Macro-Algal Biomass. *Sci. Total Environ.* 2018a, *615*, 161.

Son, E. B.; Poo, K. M.; Mohamed, H. O.; Choi, Y. J.; Cho, W. C.; Chae, K. J. A Novel Approach to Developing a Reusable Marine Macro-Algae Adsorbent with Chitosan and Ferric Oxide for Simultaneous Efficient Heavy Metal Removal and Easy Magnetic Separation. *Bioresour. Technol.* **2018b,** 259.

Song, Z.; Lian, F.; Yu, Z.; Zhu, L.; Xing, B.; Qiu, W. Synthesis and Characterization of a Novel MnOx -Loaded Biochar and Its Adsorption Properties for $Cu^{2+}$ in Aqueous Solution. *Chem. Eng. J.* **2014,** *242*, 36–42.

Sun, P.; Hui, C.; Azim Khan, R.; Du, J.; Zhang, Q.; Zhao, Y. H. Efficient Removal of Crystal Violet Using $Fe_3O_4$-Coated Biochar: The Role of the $Fe_3O_4$ Nanoparticles and Modeling Study Their Adsorption Behavior. *Sci. Rep.* **2015,** *5*, 12638.

Tan, X.; Liu, Y.; Zeng, G.; Wang, X.; Hu, X.; Gu, Y.; Yang, Z. Application of Biochar for the Removal of Pollutants from Aqueous Solutions. *Chemosphere* **2015,** *125*, 70–85.

Tang, J.; Lv, H.; Gong, Y.; Huang, Y. Preparation and Characterization of a Novel Graphene/Biochar Composite for Aqueous Phenanthrene and Mercury Removal. *Bioresour. Technol.* **2015,** *196*, 355–363.

Testiati, E.; Parinet, J.; Massiani, C.; Laffont-Schwob, I.; Rabier, J.; Pfeifer, H. R.; Prudent, P. Trace Metal and Metalloid Contamination Levels in Soils and in Two Native Plant Species of a Former Industrial Site: Evaluation of the Phytostabilization Potential. *J. Hazard. Mater.* **2013,** *248*, 131–141.

Usman, A. R. A.; Sallam, A. S.; Alomran, A.; Elnaggar, A. H.; Alenazi, K. K. H.; Nadeem, M.; Alwabel, M. I. Chemically Modified Biochar Produced from Conocarpus Wastes: An Efficient Sorbent for Fe(II) Removal from Acidic Aqueous Solutions. *Adsorpt. Sci. Technol.* **2013,** *31*, 625–640.

Van Vinh, N.; Zafar, M.; Behera, S. K.; Park, H. S. Arsenic (III) Removal from Aqueous Solution by Raw and Zinc Loaded Pine Cone Biochar: Equilibrium, Kinetics, and Thermodynamics Studies. *Int. J. Environ. Sci. Technol.* **2015,** *12*, 1283–1294.

Vane, C. H.; Kim, A. W.; Beriro, D. J.; Cave, M. R.; Knights, K.; Moss-Hayes, V.; Nathanail, P. C. Polycyclic Aromatic Hydrocarbons (PAH) and Polychlorinated Biphenyls (PCB) in Urban Soils of Greater London, UK. *Appl. Geochem.* **2014,** *51*, 303–314.

Wang, L.; Wang, Y.; Ma, F.; Tankpa, V.; Bai, S.; Guo, X.; Wang, X. Mechanisms and Reutilization of Modified Biochar Used for Removal of Heavy Metals from Wastewater: A Review. *Sci. Total Environ.* **2019,** *668*, 1298–1309.

Wang, X. L.; Sato, T.; Xing, B. S. Competitive Sorption of Pyrene on Wood Chars. *Environ. Sci. Technol.* **2006,** *40*, 3267–3272.

Wen, C.; Zheng, Z. Q.; Xu, X. Y. Process Optimization of Coagulation in Pretreatment of Seed Coating Wastewater. *J. Tianjin Polytech. Univ.* **2017,** *36*, 64–67.

Windeatt, J. H.; Ross, A. B.; Williams, P. T.; Forster, P. M.; Nahil, M. A.; Singh, S. Characteristics of Biochars from Crop Residues: Potential for Carbon Sequestration and Soil Amendment. *J. Environ. Manage.* **2014,** *146*, 189–197.

Wu, W.; Li, J.; Tian, L.; Müller, K.; Niazi, N. K.; Xin, C.; Li, J. Unraveling Sorption of Lead in Aqueous Solutions by Chemically Modified Biochar Derived from Coconut Fiber: A Microscopic and Spectroscopic Investigation. *Sci. Total Environ.* **2017,** *576*, 766–774.

Xiao, F.; Pignatello, J. J. $\pi$+-$\pi$ Interactions Between (Hetero)Aromatic Amine Cations and the Graphitic Surfaces of Pyrogenic Carbonaceous Materials. *Environ. Sci. Technol.* **2015,** *49*, 906–914.

Xie, T.; Reddy, K.; Wang, C. W.; Yargicoglu, E.; Spokas, K. Characteristics and Applications of Biochar for Environmental Remediation: A Review. *Crit. Rev. Environ. Sci. Technol.* **2015,** *45*, 939–969.

Yang, G. X.; Jiang, H. Amino modification of biochar for enhanced adsorption of copper ions from synthetic wastewater. *Water Res.* **2014,** 48, 396–405

Yao, Y.; Gao, B.; Fang, J.; Zhang, M.; Chen, H.; Zhou, Y.; Creamer, A.; Sun, Y.; Yang, L. Characterization and Environmental Applications of Clay–Biochar Composites. *Chem. Eng. J.* **2014,** *242*, 136–143.

Yuan, P.; Wang, J.; Pan, Y.; Shen, B.; Wu, C. Review of Biochar for the Management of Contaminated Soil: Preparation, Application and Prospect. *Sci. Total Environ.* **2018.**

Zeng, Z. W.; Tan, X. F.; Liu, Y. G.; Tian, S. R.; Zeng, G. M.; Jiang, L. H.; Liu, J.; Liu, N.; Yin, Z. H. Comprehensive Adsorption Studies of Doxycycline and Ciprofloxacin Antibiotics by Biochars Prepared at Different Temperatures. *Front. Chem.* **2018,** *6*, 00080.

Zhang, J.; Liu, J.; Liu, R. Effects of Pyrolysis Temperature and Heating Time on Biochar Obtained from the Pyrolysis of Straw and Lignosulfonate. *Bioresour. Technol.* **2015,** *176*, 288–291.

Zhang, M.; Gao, B. Removal of Arsenic, Methylene Blue, and Phosphate by Biochar/A1OOH Nanocomposite. *Chem. Eng. J.* **2013,** *226*, 286–292

Zhang, M.; Gao, B.; Yao, Y.; Xue, Y.; Inyang, M. Synthesis of Porous MgO-Biochar Nanocomposites for Removal of Phosphate and Nitrate from Aqueous Solutions. *Chem. Eng. J.* **2012,** 21026–21032.

Zhang, P.; Sun, H.; Min, L.; Ren, C. Biochars Change the Sorption and Degradation of Thiacloprid in Soil: Insights Into Chemical and Biological Mechanisms. *Environ. Pollut.* **2018,** *236*, 158–167.

Zhang, X.; Zhang, S.; Yang, H.; Feng, Y.; Chen, Y.; Wang, X.; Chen, H. Nitrogen Enriched Biochar Modified by High Temperature $CO_2$–Ammonia Treatment: Characterization and Adsorption of $CO_2$. *Chem. Eng. J.* **2014,** *257*, 20–27.

Zhao, B.; Xu, R.; Ma, F.; Li, Y.; Wang, L. Effects of Biochars Derived from Chicken Manure and Rape Straw on Speciation and Phytoavailability of Cd to Maize in Artificially Contaminated Loess Soil. *J. Environ. Manag.* **2016,** *184*, 569–574.

Zhao, X.; Ouyang, W.; Hao, F.; Lin, C.; Wang, F.; Han, S.; Geng, X. Properties Comparison of Biochars from Corn Straw with Different Pretreatment and Sorption Behaviour of Atrazine. *Bioresour. Technol.* **2013,** *147*, 338–344.

Zheng, W.; Guo, M.; Chow, T.; Bennett, D. N.; Rajagopalan, N. Sorption Properties of Greenwaste Biochar for Two Triazine Pesticides. *J. Hazard. Mater.* **2010,** *181*, 121–126.

Zhou, N.; Chen, H.; Feng, Q.; Yao, D.; Chen, H.; Wang, H.; Lu, X. Effect of Phosphoric Acid on the Surface Properties and Pb(II) Adsorption Mechanisms of Hydrochars Prepared from Fresh Banana Peels. *J. Clean. Prod.* **2017a,** 165.

Zhou, N.; Chen, H.; Xi, J.; Yao, D.; Zhou, Z.; Tian, Y.; Lu, X. Biochars with Excellent Pb(II) Adsorption Property Produced from Fresh and Dehydrated Banana Peels via Hydrothermal Carbonization. *Bioresour. Technol.* **2017b,** *232*, 204–210.

Zhou, Q. X.; Song, Y. F. Remediation of Contaminated Soils: Principles and Methods. *Science* **2004,** 1–489.

Zhou, Y.; Zhang, L.; Cheng, Z. J. Removal of Organic Pollutants from Aqueous Solution Using 1198 Agricultural Wastes: A Review. *J. Mol. Liq.* **2015,** *212*, 739–762.

Zhu, L, Lei H, Zhang Y, Zhang X, Bu Q, Wei Y, et al. (2018a). A Review of Biochar Derived from Pyrolysis and Its Application in Biofuel Production. *SF J. Mater. Chem Eng.* **2018a,** *1* (1), 1007.

Zhu, Y.; Yi, B.; Yuan, Q.; Wu, Y.; Wang, M.; Yan, S. Removal of Methylene Blue from Aqueous Solution by Cattle Manure-Derived Low Temperature Biochar. *RSC Adv.* **2018b,** *8*, 19917–19929.

Zhu, Y. G.; Shaw, G. Soil Contamination with Radionuclides and Potential Remediation. *Chemosphere* **2000,** *41*, 121–128.

Zhuang, L.; Li, Q.; Chen, J.; Ma, B.; Chen, S. Carbothermal Preparation of Porous Carbon-Encapsulated Iron Composite for the Removal of Trace Hexavalent Chromium. *Chem. Eng. J.* **2014,** *253*, 24–33.

Zou, Y.; Wang, X.; Khan, A.; Wang, P.; Liu, Y.; Alsaedi, A.; Wang, X. Environmental Remediation and Application of Nanoscale Zero-Valent Iron and Its Composites for the Removal of Heavy Metal Ions: A Review. *Environ. Sci. Technol.* **2016,** *50*, 7290–7304.

CHAPTER 6

# Preparation and Properties of Activated Carbon

MIHIR RANJAN SAHOO

*School of Basic Sciences, Indian Institute of Technology Bhubaneswar, India*

## ABSTRACT

Activated carbon or activated charcoal (AC) is a porous and nongraphitizable form of carbon and considered as an efficient adsorbent due to its high porosity and large surface area per unit volume. AC can be prepared from various carbon-rich natural and synthetic precursors, agricultural byproducts, food wastes, domestic wastes, petrochemical products, and fruit shells through a two-stage process, that is, (i) carbonization of carbonaceous precursors and (ii) activation of the carbonized precursors by physically or chemically. Depending upon the starting raw materials and product selectivity, the selective activation process is required. Based on the physical characteristics, shape, and size of particles, activated carbons are broadly classified into various categories which have specific applications. ACs can adsorb a wide range of chemicals, organic compounds, and heavy metal ions and can be operated at a wide range of temperatures and humidity conditions. Specific surface area, pore size distributions, hardness, and iodine index are the major physical properties which qualitatively measure the adsorption efficiency of AC. Activated carbons show a wide range of applications such as water or air purifications, metal extractions, catalysis, gas storage, trapping heavy metal ions from wastewater, odor removal, gas storage, and many other industrial and medical applications.

New Forms of Carbon: Nanocarbons. Aneeya Kumar Samantara & Satyajit Ratha (Eds.)

## 6.1 INTRODUCTION

Carbon and its various allotropes have significantly influenced human life in various perspectives since many centuries. After hydrogen, it has the most number of compounds than any other element. The four valence electrons of the carbon atom are responsible for large number of interactions of carbon with the external world. The 2s and 2p orbitals of carbon atom intermix with each other to form various new hybrid orbitals which have different shapes and energies than the individual orbitals. This makes carbon a suitable candidate for forming bonds with other atoms. There are three types of hybrid orbitals seen in carbon atoms such as sp, $sp^2$, and $sp^3$ depending upon the number of 2p orbitals participating, resulting in the formation of different allotropes under different physical conditions. The cause of different possible hybridizations may be attributed to some interesting atomic features of carbon. First, from atomic configuration of C, it is observed that the energy levels of s and p-orbitals are neither too far nor too close to each other, which makes carbon-based compounds as potential candidates for various applications. Second, there is a chance that each electron may occupy any one of the four s–p hybrid states, as the number of available orbitals is same as the number of electrons. Another important factor is that the absence of core p-electrons in carbon leads to a stable planar structure and short bond lengths, while making covalent bonding with other C atoms. Hence, the allotropes formed may resemble each other chemically but their physical properties are quite different. The well-known allotropes, graphite and diamond, are formed because of $sp^2$ and $sp^3$ hybridizations in which C atoms form bond with other two and three carbon atoms, respectively. Other allotropes of carbon formed by aggregating C atoms include graphene (Geim, 2009), carbon nanotube (Dresselhaus et al., 1995), and activated carbons (AC) that are well known for their unique characteristics, suitable for wide range of applications. In this text, we will focus primarily on activated carbons, their classification, process of activation, and their use as catalysts in various reactions.

Basically, a pore or a space with size compared to the dimensions of a molecule and surrounded by carbon atoms is termed as activated carbon (AC), or popularly known as activated charcoal. The porous solid structures are usually taken as good adsorbents because the pore walls act as active sites for the attachment of various atoms and molecules. Activated carbon (AC), zeolites, activated alumina, silica gel, and synthetic resins are few examples of such porous materials, which are widely used as adsorbents. Among these adsorbents, AC is regarded as the most efficient due to its versatile applications in many fields, especially in removing pollutants (Mohammad-Khah and Ansari, 2009). The presence of highly hydrophobic graphene layers and hydrophilic functional groups on the surface of the activated carbon makes it a more promising material for adsorption and catalytic applications (Jia and Demopoulos, 2003; Soleimani and Kaghazchi, 2008). The pollutant particles (both gaseous and liquid) from local environment can be easily captured by the AC due to its high surface area and can also be used for the prevention of air and water pollutions (Mohammad-Khah and Ansari, 2009). AC can be used as an effective purifying agent for many water supplies (drinkable water and industrial wastewater) containing various toxins, contaminants, organic and inorganic compounds, and microbial and bio-refractive materials like heavy metal ions, chlorinated hydrocarbons, phenols, insecticides, herbicides, and so on (Ansari and Sadegh, 2007; Attia et al., 2010; Boopathy et al., 2013; Eba et al., 2011; El-Shafey et al., 2002; Mohamed et al., 2011; Mohan et al., 2005; Onundi et al., 2010). Apart from purification of water and air (filters out contaminants/pollutants like volatile organic compounds, dust, smokes, etc. and maintains a pleasant odor), AC can also be considered as an excellent catalyst, which has high adsorption potential and plays a significant role in several industrial uses like power plant and landfill gas emission, recovery of precious metals like gold from ore in carbon pulp, simple decolorization, hydrometallurgy, treatment of various liquids like sugar solutions, vegetable oils, glucose, fats and glutamates, and several medical processes (used in gas masks against toxic gases) too. Furthermore, AC shows high resistance to both acidic and basic media. They can be modified to obtain different physical forms, that is, granulates, fibers, pellets, and powders, as per the requirement for specific applications (Gálvez et al., 2011; Reinoso, 1997). In comparison to other microporous structures like activated alumina, silica, and zeolite, ACs are very cost-effective and have low toxicity. They can be prepared from any natural or synthetic solid carbonaceous precursors like wood, coconut husk, bamboo, sawdust, coal, petroleum pitch, fruit stones, cherry stones, date stones, tamarind seeds, domestic and agricultural waste products, and synthetic polymers like urea-formaldehyde resins, phenolic resins, and phenol-formaldehyde resins (Ahmed and Theydan, 2012; Mopoung et al., 2015; Nowicki et al., 2015; Ozbay and Yargic, 2016). The natural precursors are cheap and abundant with high-carbon contents and produce low ash, whereas the synthetic precursors are expensive but efficient. Thus, there is a high demand among researchers to use AC in most of the applications due to its simple design and easy operating process, excellent effluent treating capacity, and high selectivity toward specific products of interest (Tadda et al., 2016).

Based on the structural and mechanism of carbonizations involved, the family of carbons can be broadly classified into two groups: graphitizable (anisotropic carbons) and nongraphitizable (isotropic carbons) (Franklin, 1951). The solid carbon structures formed through the process of liquid phase carbonization are very well organized, ordered, and show good planar alignment with long-range stacking of graphene layers and are termed as graphitizable carbons in which the X-ray diffraction lines (h k l) of graphite lattice can appear by annealing beyond 2000. On the other hand, no internal recrystallization occurs during solid-phase carbonization, resulting in quite disordered structures with randomly formed graphitic alignment and termed as nongraphitizable carbons. This structural disorder creates nanopores inside the materials which have a tendency to form $sp^3$ bonds which prevents graphitic stacking and controls the properties of the materials (Dumanlı and Windle, 2012; Ghazinejad et al., 2017). In this case, no X-ray diffraction lines (h k l) appeared on heating these carbons at 2000 and beyond. Activated carbon belongs to the latter category of carbon, that is, nongraphitizable carbon, prepared from materials not passing through liquid-phase carbonization.

## 6.2 PREPARATION AND ACTIVATION

The versatile applications of activated carbons in everyday life have stimulated researchers to produce AC through different methods other than the traditional approach. As per the discussion in the previous section, various natural and synthetic precursors, agricultural by-products, domestic and food wastes, petrochemical products, polymers, and fruit shells can be used as the raw materials for the preparation of ACs (Tongpoothorn et al., 2011). However, in a fast-growing populated world, large-scale production of AC through economic means is highly essential in order to fulfill the rapid surge in the demands at individual level, industries, and for health sectors as well. In most of the cases, activation agents are required to convert organic precursors into activated carbon, which may be expensive. To optimize the production cost as well as quality of AC, it is very important to choose a specific activation method for a specific starting material. The nature of activating agents, raw materials, and the conditions of activation processes control the properties of the finally obtained AC (Buczek, 2016).

Basically, the preparation of activated carbon follows two major steps: (i) carbonization of the carbonaceous precursor (raw material) is completely decomposed at a temperature below 900°C in an inert environment, and (ii) activation of carbonized precursor to develop pore structures though physical (at a temperature between 800 and 900°C with carbon dioxide or steam as broadly used activating agents) or chemical (at room temperature with chemical reagents like phosphoric acid, potassium hydroxide, and zinc chloride) modifications (Tadda et al., 2016). The choice of activation process depends upon the starting raw materials and the desired form of the products, that is, whether powdered or granular or pellet form of the AC is required (Mohammad-Khah and Ansari, 2009).

### 6.2.1 PHYSICAL ACTIVATION

Physical activation involves two steps, that is, carbonization and activation. The first step includes the modification of raw materials into carbon in the absence of air and in the second step, the resulting char (carbon) is activated into activated carbon at an elevated temperature in the presence of suitable oxidizing agents such as carbon dioxide, air, or water vapor (steam) or a mixture of these gases (Yang et al., 2010) and can be represented through the following reactions:

$$C + CO_2 \rightarrow 2CO \tag{6.1}$$

$$C + H_2O \rightarrow CO + H_2 \tag{6.2}$$

This entire process of activation within the gas phase is termed as physical or thermal activation. The temperature range for carbonization of organic materials lies between 673 and 1073 K, whereas the temperature required for the activation process is within the range of 873 and 1173 K. Steam activation has a major impact on the surface area, volume, and the size distributions of pores and also enhances the oxygen contents in the system (Wang et al., 2017). The porosity formed in the porous carbon differs according to the method adopted. Producing carbon monoxide from carbon by using carbon dioxide or steam will not provide identical results. In steam activation, the pore size increases due to the removal of carbon from the pore walls, resulting from the steam–carbon chemical interactions held at the internal surfaces of carbon. In this activation process, pore size can be easily altered and the AC can be made suitable for specific applications. For example, in the case of water purification, microporous structure is required to adsorb small molecules in the solution, whereas in sugar decolorization, macroporous AC is used to adsorb large color molecules. The size of AC produced in this process is in between 1 and 3 mm. In general, physical activation for AC is preferred when raw materials such as agricultural biomass like rice husk, almond shell, sun flower shell, rice hull, and peanut hulls are used.

Although in some cases, physical activation is preferred to produce activated carbon due to its simplicity and cost-effectiveness, but it gives very low yield which cannot be adopted in industry for mass production (Lillo-Ródenas et al., 2003; Yorgun et al., 2009). The parameters that completely determine/control the amount of production are activation time and temperature. At higher temperature, during the carbonization process, most of the volatile compounds get detached from the organic materials resulting in a low yield of AC. For longer activation period, the time of partial burning of carbon and other organic compounds, due to the oxygen present in the water vapor (steam), increases which reduces the amount of product obtained (Baçaoui et al., 2001; Bergna et al., 2018).

### *6.2.2 CHEMICAL ACTIVATION*

In the chemical activation process, the chemical agents used for the activation of carbonized precursors are $ZnCl_2$, KOH, $H_3PO_4$, NaOH, $MgCl_2$, $K_2CO_3$, and $Na_2CO_3$. The procedure for this type of activation is simple: chemical reagents are mixed with the carbonized precursors to form a paste or solution which is impregnated on the surfaces of organic biomass and then heated in a furnace at temperatures ranging within 673–773 K for 1–4 h in nitrogen environment. To remove the chemical reagent, the product is treated with diluted mineral acid followed by filtration and drying in oven to yield the activated carbon. Various agricultural wastes such as olive seed, almond shells, nut shells, corn cob, rice husks, and rice straw and other waste materials like sawdust, wood, and peat can be considered as raw materials for the chemical activation.

The chemical reactions that take place between the precursors and activation agents, during the carbonizations, are not known clearly. However, Lillo-Ródenas et al. (2003) described that the reaction between carbon and alkaline metal hydroxide produces the corresponding alkaline metal, hydrogen, and metal carbonates and can be represented through the following set of equations:

$$2C + 6KOH \rightarrow 2K + 3H_2 + 2K_2CO_3 \quad (6.3)$$

$$2C + 6NaOH \rightarrow 2Na + 3H_2 + 2Na_2CO_3 \quad (6.4)$$

The main advantage of chemical activation over physical activation is higher final AC yields. In physical activation, to get a well-developed AC structure, a larger amount of carbon mass has to be removed. On the other hand, in chemical activation process, dehydrating chemical agent can go through pyrolytic decompositions which prevents the formation of tar and results in a higher yield of AC. Second, the temperature needed to carry out chemical activation is lower which leads to the production of better porous structures (Ahmadpour and Do, 1996). However, the limitation of using chemical activation is that the chemical impurities coming from the reagents are not eco-friendly and may also affect the chemical properties of AC (Benaddi et al., 1998).

## 6.3 PHYSICAL PROPERTIES

### *6.3.1 SURFACE AREA*

High surface area is a characteristic property of activated carbon which comes up with well-developed pores. Higher the surface area of the adsorbents, better the adsorption power due to the availability of more active sites for the adsorbate molecules. The interior surface area of AC can be calculated by using BET (Brunauer–Emmett–Teller) method (Brunauer et al., 1938). The interior surface area of AC is quite attractive, that is,

500–1500 $m^2$ or even more per unit gram. Generally, this large surface area is created during the activation process. In steam activation, at 1000°C, the holes in the carbonized precursors are burnt by steam molecules resulting pores developed inside the carbonaceous system. The chemical reagents like phosphoric acid can create similar porous structures at low temperatures.

### *6.3.2 PORE STRUCTURES*

The pore structures of AC can be determined from the type of raw materials used and the method of activation. As per the IUPAC classification system, pores of activated carbon are categorized into the following groups, according to the volume (or radius) (Fig. 6.1).

| Pores | Radius (in nm) |
|---|---|
| Micropores | 0.2-1 |
| Mesopores | 1-25 |
| Macropores | More than 25 |

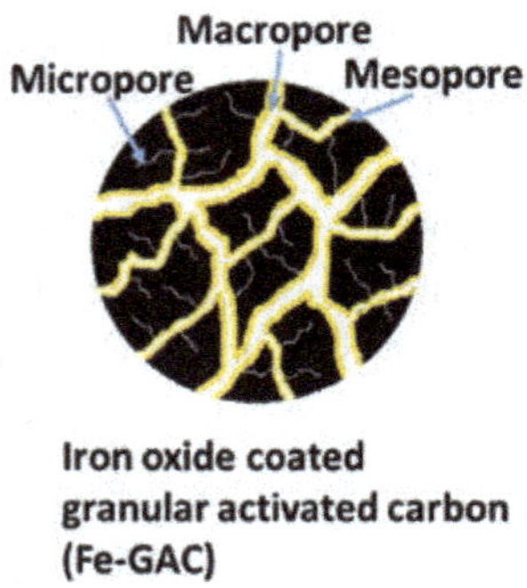


**FIGURE 6.1** Representation of different pore size in activated carbon.
*Source:* Reprinted with permission from Suresh Kumar et al. (2017). https://creativecommons.org/licenses/by/4.0/

Both geometry and volume of pores control the adsorption properties of AC. For example, large adsorbate molecules cannot be trapped by pores of very small size. Similarly, it is not possible for small molecules to be retained with macropores irrespective of their charges and polarities (Ahmedna et al., 2004). Macropores are usually developed in the activated carbon, prepared from the materials containing higher percentages of lignin such as cherry stones, grape seeds, and rice husk, while ACs prepared from raw materials containing higher amount of cellulose such as almond shells and coconut shells have microporous structures (Savova et al., 2001).

### 6.3.3 IODINE NUMBER

Iodine number is a key parameter to measure adsorption capacities of activated carbon and also regarded as a quality control parameter. It is generally used for the rough estimation of the available surface area and porosity of AC at room temperature. Iodine number is defined as the amount of iodine in milligram that can be adsorbed by 1 g of carbon when the residual filtrates have iodine concentration of 0.02 normal. In the adsorption mechanism, the surfaces of the adsorbents have various active sites where molecules of different size and functionality can be attached with different adsorption energies. To compare the adsorption capacities of pores of different size (micropores to macropores), chemists consider iodine atom as a good indicator due to its small size. Although the iodine number actually represents the carbon's capacity for iodine adsorption, increase in its value, however, indicates high porosity with good performance.

### 6.3.4 HARDNESS

Particle strength and resistance to attrition (breakdown of materials into smaller particles) can be measured through hardness/abrasion. Larger the hardness, higher the withstanding capacity against the frictional force imposed during backwashing. The activated carbon prepared from coconut shell shows the highest order of hardness.

### 6.3.5 APPARENT DENSITY

Apparent density is the weight of carbon per unit volume. In general, adsorption per unit volume will be affected by the apparent density. When density is higher, volume activity will be better- and high-quality AC can be obtained due to the availability of more volume.

### 6.3.6 ASH CONTENT

The amount of inorganic, unusable, and amorphous constituents (mineral oxides) on weight basis can be predicted from the ash content. The amount of ash content can be measured by converting the mineral to the corresponding oxides at temperature as high as 1173 K. The overall efficiency and the quality of activated carbon decrease, if the ash content increases. Hydrophilicity may increase due to presence of the ash content. In an ideal case, the ash content should be as low as possible to obtain desired AC. The typical values of inorganic constituents in coconut-based, wood-based, and coal-based ACs are 2–3% W/W, 5% W/W, and 8–15% W/W, respectively.

### 6.3.7 pH VALUE

The pH value has a strong impact on the adsorption process in the fluid phase as surface chemistry and surface charges are influenced by it. For example, when the pH value is low, the surface is positively charged. As a result, the efficiency of adsorbing cations will become less due to Coulombic repulsion. In some cases, if the pH value is less than 1 (strongly acidic), then the efficiency of adsorption will be maximum. But, such low pH value is not recommended because constituent materials suffer heavily due to corrosion. Thus, it is highly essential to neutralize the activated carbon before using it as an adsorbent. In highly acidic case, the activated carbon should be washed with deionized water until the pH value becomes zero. The pH value of AC depends on the chemical reagents used for the activation. The pH value of coconut shell-based AC lies in between 9 and 11.

## 6.4 ADSORPTION OF ACTIVATED CARBON

Adsorption is a pure surface phenomenon in which atomic and molecular species from a gas/vapor or solute dissolved in fluids are deposited on the surfaces of different species. The molecule can either enter into the solid (absorption) or attach to the surface (adsorption). When both phenomena occur simultaneously, it is called sorption. The species adsorbed and the surface on which adsorption takes place are termed as adsorbates and adsorbents, respectively. Examples of commonly used adsorbents are clay, silica, various synthetic resins, and activated carbons. In the case of AC, the specific surface area of adsorbent, size and structure of pores, and affinity and size of the adsorbate particle control the rate of adsorption. Depending upon the strength of adsorbate–adsorbent interactions, the process of adsorption can be categorized into three types: (i) electrical, (ii) chemical, and (iii) van der Waals or physical. When a permanent dipole moment is possessed by an adsorbate molecule (polar species), then the adsorption process is fully dominated by the electrical interaction due to the exchange of ions (Reucroft et al., 1971). The functional groups like carboxyl, ketones, phenols, and aldehydes control the acid–base character of AC. Adsorption of various adsorbate molecules on the surface of AC is affected by the change in the electrical properties of these functional groups. The adsorption power increases (decreases) if the charge of adsorbate species is opposite (same) to that of the surface of AC (Aygün et al., 2003). On the other hand, when the interaction between adsorbate and adsorbent is strong through the formation of chemical bonds, the process is termed as chemical adsorption. Organic molecules present in water bound to the surface of activated carbon at low temperature and this type of adsorption is called physical or van der Waals interaction. During physisorption, the chemical identity of the adsorbate species remains unchanged. It is a spontaneous process with $\Delta G < 0$, which implies an exothermic reaction with release of heat. The chemical reaction between reactive surface oxides of AC and the adsorbate species boosts this physical warming significantly. This adsorption heat gradually decreases as the accumulation increases. After some time, a balance between the amount of concentration of accumulated adsorbate species over the AC surfaces and the concentration of the same species in a solution (or local environment) will be achieved. This is known as adsorption equilibrium. The amount of accumulated adsorbate on the surface of adsorbent relative to the concentrations of the same in a local environment is termed as adsorption capacity. The adsorption power (capacity) of AC can be enhanced by increasing the surface area of the pores (>1000 $m^2/g$) (Tadda et al., 2016).

The complete adsorption process on activated carbon requires several steps: first, diffusion of adsorbate species (solute) will take place near the solid surface and then into the pore of particles, the pore walls and finally attached to the surface of pore walls. The contaminants dissolved in liquid or gaseous phase can be adsorbed by AC through various mechanisms such as Coulombic interaction, van der Waals interaction, hydrogen bonding, strong chemical bonds, dipole–dipole interaction, ligand exchange, and hydrophobic force (Mohammad-Khah and Ansari, 2009). Thus, the specific surface area, pore size and distribution, and polarity (hydrophobicity or hydrophilicity) of adsorbents, properties of adsorbate species, and physical and chemical conditions determine the adsorption capacities of AC. In general, the water molecule is more polar than the surfaces of carbonaceous matrix which leads to nonpolar contaminants like oil or hydrocarbons show a high tendency to be detached from the solution and can get adsorbed on the surface of AC. This phenomenon is termed as hydrophobic bonding which is very useful for removal of organic pollutants from the water.

The relationship between the concentrations of the chemicals (ions, molecules) adsorbed over the surface of AC and the concentration of same in the solution at adsorption equilibrium at constant temperature is known as the adsorption isotherm. Based on the sorption mechanism, there are three types of isotherms seen for AC: (i) linear, (ii) Freundlich, and (iii) Langmuir isotherm (Li et al., 2005; McKay et al., 1997). The linear isotherm is based on the fact that both the adsorbed and dissolved concentrations are directly proportional to each other. Freundlich adsorption isotherm was proposed by a German scientist, Freundlich, based on an empirical formula which relates the concentration of adsorbed gas on the surface of the solid adsorbent while keeping both the temperature and pressure at constant values. The relation can be expressed mathematically as

$$X / m = KC_e^{1/n} \left(n > 1\right) \tag{6.5}$$

where $X$ is the mass of adsorbed gas (in mg) and $m$ is the mass of the adsorbent (in g); $C_e$ represents the concentration at equilibrium (in g/L). $K$ and $n$ are the constants whose values depend on the nature of the adsorbed gas and adsorbent and can be determined from the experimental data. Taking the logarithm on both sides of the equation:

$$\log X / m = \log K + \frac{1}{n} \log C_e \tag{6.6}$$

A straight line will be obtained by plotting $X/m$ with $C_e$ in the log paper, and the slope, $1/n$, will define the adsorption intensity and $K$ is the adsorption capacity (L/mg). $1/n$ also indicates the adsorbate site's relative energy distribution and heterogeneity (Ayawei et al., 2017). The value of $1/n$ lies in between 2 and 10, for an effective AC. Also, this value is nearly equal to 1 for organic contaminants, whereas the value lies between 0.4 and 0.6 for heavy metal ions. If $n = 1$, the Freundlich isotherm will become a linear isotherm. The efficiency of activated carbon as an adsorbent can be determined from the isotherm. Steeper the isotherm, more efficient is the adsorbent. The adsorptive capacity of various AC in gas–solid phase adsorption can be described through Langmuir adsorption. Langmuir isotherm is based on the balanced condition between relative rates of adsorption and desorption. In this approach, adsorption and desorption are considered to be proportional to the fraction of bare and covered adsorbent surfaces, respectively (Günay et al., 2017). Langmuir adsorption method can be expressed mathematically through the following equation:

$$\frac{C_e}{q_e} = \frac{1}{K_l q_m} + \frac{C_e}{q_m} \tag{6.7}$$

where $C_e$ is the adsorbate concentration at equilibrium (mg/L), $q_e$ is the amount of the adsorbed species per unit mass of the adsorbent (mg/g), and $q_m$ is the maximum concentration of the adsorbate that can be attached to the adsorbate surface. $K_l$ defines the adsorption capacity of the activated carbon and is referred as the Langmuir constant. This constant can be correlated with the surface area and porosity variations in the activated carbon which implies that a higher adsorption capacity can be achieved by increasing both the surface area and pore volume. Langmuir isotherm can be expressed through an alternate equation as follows:

$$R_L = \frac{1}{1 + K_l C_0} \tag{6.8}$$

where $C_0$ is the initial adsorbate concentration (mg/L) and $R_L$ is a dimensionless constant termed as the separation factor and the physical significance of this factor is described below (Ayawei et al., 2017):

$R_L > 1$ implies the adsorption is unfavorable.
$R_L = 1$ implies the adsorption is linear.

$R_L = 0$ means the adsorption is irreversible.
And, $0 < R_L < 1$ implies, the adsorption is favorable.

## 6.5 CLASSIFICATIONS

Activated carbons are highly efficient and versatile adsorbents due to their large surface area and can be utilized in various fields. Based on the physical characteristics, shape, and size of particles, activated carbons are broadly classified into the following categories.

### 6.5.1 POWDERED ACTIVATED CARBON

If the constituent particles of activated carbon have powder or granular form with average diameter size laying in the range between 0.15 and 0.25 mm, this form is called the powdered activated carbon (PAC) resulting in high surface-to-volume ratio. Because of smaller particle size, adsorption power of PAC is very effective. These types of ACs can be made from materials like wood, ignite, and coal (Najm et al., 1991). Depending upon the preparation mechanism and the type of raw materials used, the densities may vary between 0.36 g/cm$^3$ and 0.74 g/cm$^3$. In general, 65–95% of the commercially available PAC can pass through a 325-mesh sieve. Depending on the concentration and the type of contactor, PAC can effectively remove various biocides. In addition to this, relatively low PAC (10–25 mg/L) can effectively remove compounds like 2,4-dichlorophenol, geosmin, and 2-methylisoborneol while high PAC with densities between 75 and 620 mg/L is used to remove *p*-nitrophenol and humic odor (Najm et al., 1991).

### 6.5.2 GRANULAR ACTIVATED CARBON

In the case of granular activated carbon, the particle size is large (average particle size ~1–5 mm) which leads to a smaller surface-to-volume ratio as compared to PAC. The particles of GAC can be retained through 50-mesh sieve. The apparent densities of GAC lie in the range of 400–500 kg/m$^3$ depending on the preparation process and raw materials used. GAC can be prepared by heating various organic materials such as wood, rice husk, sawdust, eucalyptus, and coconut shell without the presence of oxygen (carbonization) and then crushed and sieved into granules. The GAC can be divided into two subgroups: chopped carbon (no specific form) and formed carbon (cylindrical form). GAC can be used to remove contaminants from water and as column filter for gaseous effluents, effectively. Wood-based GACs can be extensively used as de-colorizing agents by removing color bodies like tannins.

### 6.5.3 EXTRUDED ACTIVATED CARBON/PELLETIZED ACTIVATED CARBON

Extruded activated carbons (EACs) can be prepared by fusing a mixture of PACs like pulverized anthracites or charcoals and a suitable binder, and then putting it under high pressure, these are extruded into cylindrical forms with average diameter size laying between 0.8 and 130 mm. To obtain pores with specific sizes, catalyst like potassium hydroxide (KOH) is mixed before extrusion. Due to their low-pressure drop, very less amount of dust contents, and high mechanical strength, EACs are suitable for gas phase applications.

### 6.5.4 IMPREGNATED ACTIVATED CARBON

These are the activated carbons in which internal surfaces are decorated with fine distribution of selected chemicals and metal particles. As a result, adsorptive power of AC is highly reinforced due to chemisorption between the chemicals and carbon (Henning and Schäfer, 1993). Examples of metallic cation impregnates attached on the adsorbents are Fe (Mondal et al., 2007; Shah et al., 2015), Mn (Liu et al., 2016; Zhang et al., 2008), Zn (Rezaee et al., 2008; Somy et al., 2009), Al (Ganiyu et al., 2016), and Cu (Rossin and Morrison, 1991). These types of ACs are generally used in gas purifications (treat flue gas in coal-fired generation plants) and protection from military gases. Various acid gases, mercury, ammonia, aldehydes, methylene blue, some radioactive methyl iodide, and inorganic gases such as arsine and phosphine can be removed effectively through specific impregnation (like metal oxides, triethylenediamine, etc.) on activated carbon.

### *6.5.5 POLYMER-COATED ACTIVATED CARBON*

If activated carbons are coated with biocompatible, smooth, and permeable polymers without blocking carbon pores, these are termed as polymer-coated activated carbons. These are extensively used for biomedical applications. Using bare AC in hemoperfusion leads to release carbon fines which can damage blood platelets. To overcome this problem, surfaces of charcoal granules are microcapsulated with the hexamethyledisiloxane polymer (Hasirci and Akovali, 1986). The AC, coated with polysulfide-rubber polymer, is also used for the removal of Hg ion from water (Kim et al., 2011).

### *6.5.6 ACTIVATED CARBON FIBER*

Activated carbons are also available in thin-fiber shapes and called activated carbon fibers (ACFs). In comparison to PAC and GAC, ACF shows fast intraparticle adsorption kinetics which enhances its suitability in the application of gas phase and aqueous phase adsorption (Suzuki, 1994). ACF can be applied in various fields such as removal of heavy metals, biomedical applications, catalytic applications, and natural gas storage (Lee et al., 2014).

## 6.6 CONCLUSION

Activated carbons are considered as effective adsorbents due to their large surface area and well-developed porous structures. Since agricultural and food wastes are considered as affluent sources for carbon, AC can be prepared from coconut shell, rice husk, fruit stones, almond shells, nutshells, cherry stones, and tamarind seeds. ACs can adsorb a wide range of chemicals simultaneously and can be operated at a wide range of temperatures and humidity conditions. In addition, they are inert and much safe to handle. We cannot underestimate the importance and requirements of ACs as they fulfill the demands in most of the sectors in removing the pollutants, toxic elements, and contaminants from liquid, gaseous, and solid matters.

## KEYWORDS


- **activated carbon**
- **gas storage**
- **energy**
- **classification**
- **activated carbon fiber**


## REFERENCES


Ahmadpour, A.; Do, D. D. The Preparation of Active Carbons from Coal by Chemical and Physical activation. *Carbon* **1996,** *34*, 471–479.

Ahmed, M. J.; Theydan, S. K. Physical and Chemical Characteristics of Activated Carbon Prepared by Pyrolysis of Chemically Treated Date Stones and Its Ability to Adsorb Organics. *Powder Technol.* **2012,** *229*, 237–245.

Ahmedna, M.; Marshall, W. E.; Husseiny, A. A.; Rao, R. M.; Goktepe, I. The Use of Nutshell Carbons in Drinking Water Filters for Removal of Trace Metals. *Water Res.* **2004,** *38*, 1062–1068.

Ansari, R.; Sadegh, M. Application of Activated Carbon for Removal of Arsenic Ions from Aqueous Solutions. *E-J. Chem.* **2007,** 4. https://doi.org/10.1155/2007/829187

Attia, A. A.; Khedr, S. A.; Elkholy, S. A. Adsorption of Chromium Ion (VI) by Acid Activated Carbon. *Braz. J. Chem. Eng*. **2010,** *27*, 183–193.

Ayawei, N.; Ebelegi, A. N.; Wankasi, D. Modelling and Interpretation of Adsorption Isotherms. *J. Chem*; Hindawi Limited **2017,** https://doi.org/10.1155/2017/3039817

Aygün, A.; Yenisoy-Karakaş, S.; Duman, I. Production of Granular Activated Carbon from Fruit Stones and Nutshells and Evaluation of Their Physical, Chemical and Adsorption Properties. *Micropor. Mesopor. Mater.* **2003,** *66*, 189–195.

Baçaoui, A.; Yaacoubi, A.; Dahbi, A.; Bennouna, C.; Phan Tan Luu, R.; Maldonado-Hodar, F. J.; Rivera-utrila, J.; Moreno Castilla, C. Moreno-Castilla, C. Optimization of Conditions for the Preparation of Activated Carbons from Olive-Waste Cakes. *Carbon* **2001,** *39*, 425–432.

Benaddi, H.; Legras, D.; Rouzaud, J. N.; Beguin, F. Influence of the Atmosphere in the Chemical Activation of Wood by Phosphoric Acid. *Carbon* **1998.**

Bergna, D.; Varila, T.; Romar, H.; Lassi, U. Comparison of the Properties of Activated Carbons Produced in One-Stage and Two-Stage Processes. *C* **2018,** *4*, 41.

Boopathy, R.; Karthikeyan, S.; Mandal, A. B.; Sekaran, G. Adsorption of Ammonium Ion by Coconut Shell-Activated Carbon from Aqueous Solution: Kinetic, Isotherm, and Thermodynamic Studies. *Environ. Sci. Pollut. Res*. **2013,** *20*, 533–542.

Brunauer, S.; Emmett, P. H.; Teller, E. Adsorption of Gases in Multimolecular Layers. *J. Am. Chem. Soc*. **1938,** *60*, 309–319.

Buczek, B. Preparation of Active Carbon by Additional Activation with Potassium Hydroxide and Characterization of Their Properties. *Adv. Mater. Sci. Eng.* **2016.**

Dresselhaus, M. S.; Dresselhaus, G.; Saito, R. Physics of Carbon Nanotubes. *Carbon* **1995,** *33*, 883–891.

Dumanlı, A. G.; Windle, A. H. Carbon Fibres from Cellulosic Precursors: A Review. *J. Mater. Sci.* **2012,** *47*, 4236–4250.

Eba, F.; Kouya Biboutou, R.; Nlo, J. N.; Bibalou, Y. G.; Oyo, M. Lead Removal in Aqueous Solution by Activated Carbons Prepared from Cola Edulis Shell (Alocacée), Pentaclethra Macrophylla Husk (Mimosaceae) and Aucoumea klaineana Sawdust (Burseraceae). *Afr. J. Environ. Sci. Technol.* **2011,** (Vol. 5). Retrieved from http://www.academicjournals.org/AJEST

El-Shafey, E. I.; Cox, M.; Pichugin, A. A.; Appleton, Q. Application of a Carbon Sorbent for the Removal of Cadmium and Other Heavy Metal Ions from Aqueous Solution. *J. Chem. Technol. Biotechnol.* **2002,** *77*, 429–436.

Franklin, R. E. Crystallite Growth in Graphitizing and Non-Graphitizing Carbons. *Proc. R. Soc. London A* 1951, *209* (1097), 196–218. https://doi.org/10.1098/rspa.1951.0197

Juntgen, H. Activated Carbons as Catalyst Supports. *Fuel* **2011,** *65*, 1436–1446.

Ganiyu, S. A.; Alhooshani, K.; Sulaiman, K. O.; Qamaruddin, M.; Bakare, I. A.; Tanimu, A.; Saleh, T. A. Influence of Aluminium Impregnation on Activated Carbon for Enhanced Desulfurization of DBT at Ambient Temperature: Role of Surface Acidity and Textural Properties. *Chem. Eng. J.* **2016,** *303*, 489–500.

Geim, A. K. Graphene: Status and Prospects. *Science (New York, N.Y.)* **2009,** *324*, 1530–1534.

Ghazinejad, M.; Holmberg, S.; Pilloni, O.; Oropeza-Ramos, L.; Madou, M. Graphitizing Non-Graphitizable Carbons by Stress-Induced Routes. *Sci. Rep.* **2017,** 7.

Günay, A.; Arslankaya, E.; Tosun, I. Lead Removal from Aqueous Solution by Natural and Pretreated Clinoptilolite: Adsorption Equilibrium and Kinetics. *J. Hazard. Mater.* **2017,** *146*, 362–371.

Hasirci, N.; Akovali, G. Polymer Coating for Hemoperfusion Over Activated Charcoal. *J. Biomed. Mater. Res.* **1986,** *20*, 963–970.

Henning, K. D.; Schäfer, S. Impregnated Activated Carbon for Environmental Protection. *Gas Sep. Purif.* **1993,** *7*, 235–240.

Jia, Y.; Demopoulos, G. P. Adsorption of Silver Onto Activated Carbon from Acidic Media: Nitrate and Sulfate Media. *Indust. Eng. Chem. Res.* **2003,** *42*, 72–79.

Kim, E. A.; Seyfferth, A. L.; Fendorf, S.; Luthy, R. G. Immobilization of Hg(II) in Water with Polysulfide-Rubber (PSR) Polymer-Coated Activated Carbon. *Water Res.* **2011,** *45*, 453–460.

Lee, T.; Ooi, C. H.; Othman, R.; Yeoh, F. Y. Activated Carbon Fiber—The Hybrid of Carbon Fiber and Activated Carbon. *Rev. Adv. Mater. Sci.* **2014,** *36*, 118–136.

Li, L.; Quinlivan, P. A.; Knappe, D. R. U. Predicting Adsorption Isotherms for Aqueous Organic Micropollutants from Activated Carbon and Pollutant Properties. *Environ. Sci. Technol.* **2005,** *39*, 3393–3400.

Lillo-Ródenas, M. A.; Cazorla-Amorós, D.; Linares-Solano, A. Understanding Chemical Reactions Between Carbons and NaOH and KOH: An Insight Into the Chemical Activation Mechanism. *Carbon* **2003,** *41*, 267–275.

Liu, H.; Hu, Z.; Liu, H.; Xie, H.; Lu, S.; Wang, Q.; Zhang, J. Adsorption of Amoxicillin by Mn-Impregnated Activated Carbons: Performance and Mechanisms. *RSC Adv.* **2016,** *6*, 11454–11460.

McKay, G.; El-Geundi, M.; Nassar, M. M. Adsorption Model for the Removal of Acid Dyes from Effluent by Bagasse Pith Using a Simplified Isotherm. *Adsorp. Sci. Technol.* **1997,** *15*, 737–752.

Mohamed, E. F.; Andriantsiferana, C.; Wilhelm, A. M.; Delmas, H. Competitive Adsorption of Phenolic Compounds from Aqueous Solution Using Sludge-Based Activated Carbon. *Environ. Technol.* **2011,** *32*, 1325–1336.

Mohammad-Khah, A.; Ansari, R. Activated Charcoal: Preparation, Characterization and Applications: A Review Article. *Int. J. ChemTech Res.* **2009,** *1*.

Mohan, D.; Singh, K. P.; Singh, V. K. Removal of Hexavalent Chromium from Aqueous Solution Using Low-Cost Activated Carbons Derived from Agricultural Waste Materials and Activated Carbon Fabric Cloth. *Indust. Eng. Chem. Res.* **2005,** *44*, 1027–1042.

Mondal, P.; Balomajumder, C.; Mohanty, B. A Laboratory Study for the Treatment of Arsenic, Iron, and Manganese Bearing Ground Water Using $Fe^{3+}$ Impregnated Activated Carbon: Effects of Shaking Time, pH and Temperature. *J. Hazard. Mater.* **2007,** *144*, 420–426.

Mopoung, S.; Moonsri, P.; Palas, W.; Khumpai, S. Characterization and Properties of Activated Carbon Prepared from Tamarind Seeds by KOH Activation for Fe(III) Adsorption from Aqueous Solution. *Sci. World J.* **2015**.

Najm, I. N.; Snoeyink, V. L.; Lykins, B. W.; Adams, J. Q. Using Powdered Activated Carbon: A Critical Review. *J—Am. Water Works Assoc.* **1991,** *83*, 65–76.

Nowicki, P.; Kazmierczak, J.; Pietrzak, R. Comparison of Physicochemical and Sorption Properties of Activated Carbons Prepared by Physical and Chemical Activation of Cherry Stones. *Powder Technol.* **2015,** *269*, 312–319.

Onundi, Y. B.; Mamun, A. A.; Al Khatib, M. F.; Ahmed, Y. M. Adsorption of Copper, Nickel and Lead Ions from Synthetic Semiconductor Industrial Wastewater by Palm Shell Activated Carbon. *Int. J. Environ. Sci. Technol.* **2010,** *7*, 751–758.

Ozbay, N.; Yargic, A. S. Comparison of Surface and Structural Properties of Carbonaceous Materials Prepared by Chemical Activation of Tomato Paste Waste: The Effects of Activator Type and Impregnation Ratio. *J. Appl. Chem.* **2016,** *2016*, 1–10.

Reinoso, F. R. Activated Carbon: Structure, Characterization, Preparation and Applications. In *Introduction to Carbon Technologies*, Reinoso, F. R., Heintz, E. A., Marsh, H., Eds.; University of Alicante, Secretariado de Publicaciones, **1997;** p 669.

Reucroft, P. J.; Simpson, W. H.; Jonas, L. A. Sorption properties of activated carbon. *Journal of Physical Chemistry* **1971,** *75*, 3526–3531.

Rezaee, A.; Godini, H.; Dehestani, S.; Khavanin, A. Application of impregnated almond shell activated carbon by zinc and zinc sulfate for nitrate removal from water. Retrieved December **2008,** *20*, 2019.

Rossin, J. A.; Morrison, R. W. Spectroscopic Analysis and Performance of an Experimental Copper/Zinc Impregnated, Activated Carbon. *Carbon* **1991,** *29*, 887–892.

Savova, D., Apak, E., Ekinci, E., Yardim, F., Petrov, N., Budinova, T., Razvigorova, M.; Minkova, V. Biomass Conversion to Carbon Adsorbents and Gas. *Biomass Bioenergy* **2001,** *21*, 133–142.

Shah, I.; Adnan, R.; Ngah, W. S. W.; Mohamed, N. Iron Impregnated Activated Carbon as an Efficient Adsorbent for the Removal of Methylene blue: Regeneration Kinetics Studies. *PLoS ONE* **2015,** *10*, e0122603.

Soleimani, M.; Kaghazchi, T. Adsorption of Gold Ions from Industrial Wastewater Using Activated Carbon Derived from Hard Shell of Apricot Stones—An Agricultural Waste. *Bioresour. Technol.* **2008,** *99*, 5374–5383.

Somy, A.; Mehrnia, M. R.; Amrei, H. D.; Ghanizadeh, A.; Safari, M. Adsorption of Carbon Dioxide Using Impregnated Activated Carbon Promoted by Zinc. *Int. J. Greenhouse Gas Control* **2009,** *3*, 249–254.

Stoeckli, H. F.; Rebstein, P.; Ballerini, L. On the Assessment of Microporosity in Active Carbons, a Comparison of Theoretical and Experimental Data. *Carbon* **1990,** *28*, 907–909.

Suresh Kumar, P.; Prot, T.; Korving, L.; Keesman, K. J.; Dugulan, I.; van Loosdrecht, M. C. M.; Witkamp, G. J. Effect of Pore Size Distribution on Iron Oxide Coated Granular Activated Carbons for Phosphate Adsorption—Importance of Mesopores. *Chem. Eng. J.* **2017,** *326*, 231–239.

Suzuki, M. Activated Carbon Fiber: Fundamentals and Applications. *Carbon* **1994,** *32*, 577–586.

Tadda, M. A.; Ahsan, A.; Shitu, A.; Elsergany, M.; Arunkumar, T.; Jose, B., Abdur Razzaque, M.; Nik, N. N. A review on activated carbon: process, application and prospects, *J. Adv. Civil Eng. Res.* **2016,** *2*, 7-13.

Tongpoothorn, W.; Sriuttha, M.; Homchan, P.; Chanthai, S.; Ruangviriyachai, C. Preparation of Activated Carbon Derived from *Jatropha curcas* Fruit Shell by Simple Thermo-Chemical Activation and Characterization of Their Physico-Chemical Properties. *Chem. Eng. Res. Design* **2011,** *89*, 335–340.

Wang, B.; Gao, B.; Fang, J. Recent Advances in Engineered Biochar Productions and Applications. *Crit. Rev. Environ. Sci. Technol.* **2017,** *47*, 2158–2207.

Yang, K.; Peng, J.; Srinivasakannan, C.; Zhang, L.; Xia, H.; Duan, X. Preparation of High Surface Area Activated Carbon from Coconut Shells Using Microwave Heating. *Bioresour. Technol.* **2010,** *101*, 6163–6169.

Yorgun, S.; Vural, N.; Demiral, H. Preparation of high-Surface Area Activated Carbons from Paulownia Wood by $ZnCl_2$ Activation *Micropor. Mesopor. Mate.* **2009,** *122*, 189–194.

Zhang, J.; Shi, Q.; Zhang, C.; Xu, J.; Zhai, B.; Zhang, B. Adsorption of Neutral Red onto Mn-Impregnated Activated Carbons Prepared from Typha Orientalis. *Bioresour. Technol.* **2008,** *99*, 8974–8980.

# CHAPTER 7

# Carbon Nanotubes: A New Dimension in Human Healthcare Applications

RASHMI REKHA SAMAL[1,2] and MADHABI MADHUSMITA BHANJADEO[3]

[1]*Academy of Scientific & Innovative research (AcSIR), New Delhi, India*

[2]*Environment & Sustainability Department, CSIR-Institute of Minerals and Materials Technology, Bhubaneswar, Odisha, India*

[3]*Department of Biochemistry, School of Life Sciences, Ravenshaw University, Cuttack, Odisha, India*

## ABSTRACT

Carbon nanotubes (CNTs) signify solitarily the most extensively experimented allotropes of carbon. Because of their unique physicochemical properties, CNTs are ideal candidates for a wide range of biomedical applications, including drug delivery, gene therapy, biosensors, and tissue engineering. This chapter summarizes the properties and functionalization of CNTs for application toward recent approaches in human health-care advancements against nondurable diseases. The use of various functionalized CNTs in chemotherapy for efficient drug loading, targeted drug delivery, and controlled release is one of the therapeutic applications. In the similar context, other therapeutic approaches like gene therapy, photodynamic, and photothermal therapy also evidence the usability of CNTs as the carrier of antisense oligo, small interference ribonucleic acids, and photosensitizers, respectively. Another dimension of therapeutic application includes the tissue engineering and wound-healing approaches.

New Forms of Carbon: Nanocarbons. Aneeya Kumar Samantara & Satyajit Ratha (Eds.)

Specifically for these applications, the great mechanical properties along with biocompatibility are exploited to design new scaffolds for the regeneration of neural and bone tissues. Moreover, the antibacterial effect of CNTs benefits its use in bandage materials for curing of long-term critical wounds. This will be an informative chapter for material science scholars as well as biomedical researchers.

## 7.1 INTRODUCTION

Persistence is to man's character what carbon is to steel. The above saying is widely attributed to Napoleon Hill in order to demonstrate that just as carbon is necessary for the existence of steel, so is persistence to a man's character. Carbon is one of the most abundant elements of living as well as nonliving spheres of the planet and is widely studied/explored for countless applications. And carbon is now not only limited for steel and other metallurgical production or thermal energy production but also for the regeneration of nervous tissue to visualize persistent tumors. Among others, the carbon nanotube (CNT) is the most extensively studied allotrope of carbon in recent advanced applications. CNTs are highly ordered, all-carbon hollow graphitic nanomaterials with a high aspect ratio, lengths ranging from hundreds to micrometers, and diameters ranging from less than 2 nm for single-walled carbon nanotubes (SWCNTs) to 2–100 nm for multi-walled carbon nanotubes (MWCNTs). SWCNTs and MWCNTs are conceptualized as rolled-up structures of single or multiple sheets of graphene, similar to Russian dolls. These one-dimensional carbon allotropes have a larger surface area, higher mechanical strength but a low weight, a wide range of electronic properties, and exceptional chemical and thermal stability (Tanaka et al., 2020).

CNTs have been experimented for their possible functions in nanocomposite structural materials, semiconductor devices, and sensors. The recently well-documented ability to chemically modify and functionalize nanofibrous carbon materials to improve their solubility and biocompatibility has resulted in the creation of an entirely new class of bioactive carbon nanostructures for biological applications. So far, there has been a growing interest among biomedical scientists in investigating all of the aforementioned characteristics and demonstrating the suitability of CNTs for health-care applications. CNTs, for example, are being researched as a suitable platform for cell growth in the field of tissue regeneration, as a nanovehicle to deliver a variety of diagnostic or therapeutic agents, vaccines, or for the purpose of gene transfection, it acted as vectors (El-Sayed and Kamel, 2020). The following sections concisely narrates some aspects of bio-functionalized CNTs for various therapeutic applications as well as their usability in in-vitro and in-vivo bio-imaging in both invasive and noninvasive approaches with some outstanding examples. These aspects are being explored by researches worldwide in an exponentially increasing pattern in the last three decades. A detailed discussion on this aspect as well as on the toxicological effects of the CNTs are been presented in this chapter.

## 7.2 STRUCTURAL AND FUNCTIONAL CHARACTERIZATION

### *7.2.1 MORPHOLOGY AND PROPERTIES*

Japanese physicist Sumio Iijima and his coworkers discovered the unique allotrope, CNTs in 1991. Later, many of the materials and biomedical researchers have explored its physicochemical properties and targeted transportation of biomolecules and drugs. Based upon their physical properties such as diameter, length, and structure, CNTs are categorized into two broad classes. One is SWCNTs, which consist of a single layer of cylinder graphene sheet capped at both ends in a hexagonal carbon network by maintaining the diameter range from 0.4 to 2 nm, and another one is MWCNTs which contain multiple layers of concentric graphene sheet layers with diameters in the range of 1–3 nm for the inner tubes and 2–100 nm for the outer tubes with approximately 0.34 nm of interlayer separation. They exhibit a well-ordered arrangement of hexagonal carbon atoms (C–C distance of ~1.4 Å) linked through $sp^2$ bonds, which makes them the stiffest and strongest filament to be known. The basic $sp^2$ hybridized carbon arrangement of SWCNTs is different from MWCNTs. The structural arrangement of SWCNTs is established in armchair, zigzag, chiral, or helical manner (Fig. 7.1) (Gao et al., 2012). On the other hand, based on arrangements of graphene sheets, MWCNTs can be categorized into two types: one is a "Russian-doll"-like structure where the graphite sheets are arranged in concentric layers and the other is a parchment-like model where the single sheet of graphite is rolled around itself (Dresselhaus,

2004). Particularly, the hollow cylindrical tube structure of CNTs is a key feature for drug delivery as the drugs can be encapsulated in their inner hollow core region and other molecules can be attached to the external surfaces which facilitate them to be biocompatible for targeting purposes. During the process of drug delivery, SWCNTs are known to be more efficient than MWCNTs due to its one-dimensional structure and ultrahigh surface area which provides efficient drug-loading capacity. Additionally, SWCNT–anticancer drug complex or functionalized SWCNT has longer blood circulation time than the anticancer drug on its own and releases the drug into a specific region of the target. This phenomenon results in more prolonged and sustained drug uptake by tumor cells, which is mediated by the increased permeability and retention effect. The above facts suggest that SWCNTs are suitable candidates for drug delivery and evolved as favorable and suitable nanoplatform for subsequent cancer therapeutics. In comparison with SWCNTs, MWCNTs are recognized as a more useful nanomaterial in the case of thermal treatment of cancer (Chen et al., 2017). The reason behind the fact that MWCNTs release substantial vibrational energy after exposure to near-infrared light as they have more freely available electrons and comprise more metallic tubes than SWCNTs and produce heat within a tissue which can be directed to destroy cancer cells. However, it has also been suggested that the functionalization of MWCNTs' outer shell depends upon the side wall of the tube. The advantages of CNTs over other nanomaterials rely on their unique combination of electromagnetic or electrochemical along with mechanical and optical properties. Including exceptionally high tensile strength and elastic modulus, rich surface chemical functionalities, and size stability at the Nanoscale label, CNTs are also excellent thermal and electrical conductors, with additional abilities to absorb optical intensity, photoluminescence, and generate strong Raman signals that enable their facile and nondestructive characterization.

### *7.2.2 SYNTHESIS*

Due to the growing demand of CNTs in biomedical applications, the synthesis and purification process emerging the attention of researchers to produce specific, effective, and viable nanotubes. The general growth mechanism of CNT is very simple. The process can be commonly

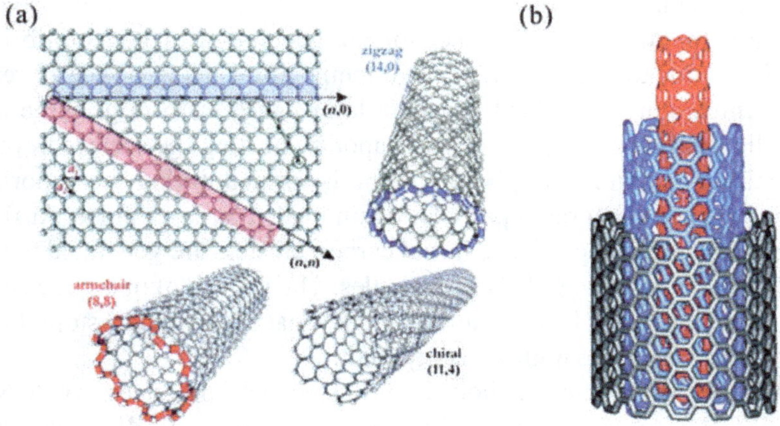


**FIGURE 7.1** Different structural forms of carbon nanotube: (a) different types of single-walled carbon nanotube forms formed by graphene sheet and (b) structure of a multiwalled carbon nanotube.

*Source*: Reproduced with permission from Gao, et al., 2012.

explained by the formation of a precursor on the surface of metal catalyst and rod-like structure of carbon formed rapidly followed by the slow graphitization of its wall (Anazawa et al., 2002). However, based on the particle size of the catalyst, single-walled nanotubes (SWNTs) or multiwalled nanotubes (MWNTs) are grown. CNTs are generally synthesized by three major techniques, namely, arc discharge or electric discharge using graphite cathodes, laser, or light ablation in addition to chemical vapor deposition (CVD) (Bianco et al., 2005; Dai, 2002). The arc discharge method is one of the oldest and common methods where vapor is created by an arc discharge between two carbon rods which act as electrodes and place in an enclosure filled with inert gases at low pressure (between 50 and 700 mbar) with or without catalyst. In the controlled atmosphere with high temperature and voltage (25–40 V), a constant gap is maintained for prolonged period between the carbon electrodes, corresponding to the position of anode after the plasma formation between carbon electrodes. The nanotubes along with the byproducts are formed inside the reaction chamber after all the cooling and depressurization process. In the laser ablation technique, the high-intensity light from a laser is used to vaporize the graphite sheet inside an oven at 1200°C. The laser source can be pulse or continuously depend upon the demand of higher light intensity. Moreover, favorable conditions are supplied by pulse laser for the generation of nonequilibrium conditions due to a short time scale of temperature changes. The extreme high temperatures provided by pulse laser can transform graphite and metallic catalyst into carbon–metal vapor by heating the surface beyond critical temperature. The plasma plume is formed during the vaporization where the evaporated particles again heated by laser pulse. At first, from that vaporized particles, initial carbon clusters are generated which then form the tubular SWCNT molecules. The growth of nanotube come to a termination by having much carbon coat layer which stops them from absorbing more high-intensity light.

Another important method is CVD, which produces MWCNTs or SWCNTs (with poor quality) based on thermal decomposition of a hydrocarbon vapor in the presence of a metal catalyst. The basic requirements of this method involve the carbon source which commonly includes methane ($CH_4$), carbon monoxide (CO), and acetylene ($C_2H_2$). Furthermore, plasma or heated coil treated as energy source to transfer energy to a gaseous carbon molecule. This synthesis process can be achieved by catalyst preparation followed by nanotube synthesis. After the deposition of transition metal on substrate, the thermal annealing process induces the catalyst particle nucleation which further leads to formation of cluster on the surface of substrate from which nanotube is generated. The entire process of nanotube synthesis through CVD is conducted by maintaining the temperature within 650–900°C. Different kinds of CVD are also used to synthesize CNTs such as plasma-enhanced CVD (PECVD), thermochemical-based CVD, catalytic-induced alcohol CVD, vapor phase growth, aerogel-based CVD, and laser-assisted CVD (Hu et al., 2009).

### *7.2.3 FUNCTIONALIZATION*

Despite many advantages of CNTs, there are certain limitations which restricted to their purification process during biomedical use. Meanwhile, the disagreement arises in solubility issues of CNTs present within aqueous media as they are easily contaminated with metal catalysts and amorphous carbons and are known to be generally insoluble and do not exhibit good biocompatibility. To resolve the nature of dispersion and solubilization, several methods have been explored by modifying the surface properties of CNTs. These methods can be simply described in terms of functionalization approach that is currently implemented to increase the bio-solubility of CNTs. This functionalization process depends upon the attachment of organic or inorganic moieties to their hexagonal tubular graphene sheet (Thakare et al., 2010). Moreover, the physicochemical properties can be modulated through the functionalization of CNTs. Presently, the two major functionalization techniques are available for the modification of CNTs named as noncovalent functionalization and covalent functionalization basing on the interaction of active material and CNTs. These methods have been extensively applied via using different chemical groups, whereas CNTs depend upon the reactivity of functional group with tips and side walls. It has been shown that CNT tips have a higher affinity for binding functional groups compared to that of the side walls (Pastorin et al, 2006).

#### *7.2.3.1 NONCOVALENT FUNCTIONALIZATION*

This functionalization process preserves the aromatic/electronic character of the surface carbon along with the structural properties of CNTs causing minimal structure damage. The noncovalent bonding is established between the hydrophobic chains of biocompatible functional groups and the hydrophobic surface of CNTs. The major driving forces like van der Waals interactions, $\pi$–$\pi$ interactions, and hydrophobic interaction act as key regulators in the stabilization of noncovalent functionalization. However, with the influence of chemical treatment, the charge on the surface of CNTs is able to adsorb the molecules through ionic interaction. Other biomolecules like nucleic acids, amphiphilic polymers, oligomers, surfactants, and peptide contributed to the structural backbone of functionalized biomolecules by establishing the $\pi$–$\pi$ interactions between aromatic or amino acid bases and surface of CNTs. It has been reported that polymers and biopolymers (nucleic acids and peptides) are very efficient in the dispersion process, whereas surfactants used widely due to their low cost and ease of availability (Lu and Chen, 2011). Concurrently, the weaker strength provided by the noncovalent bond in the noncovalent functionalization makes the CNTs inadequate nanocarrier for tumor target drug delivery.

#### *7.2.3.2 COVALENT FUNCTIONALIZATION*

The covalent functionalization or chemical functionalization approach can be achieved by modifying the CNT surface through different techniques. These techniques can be classified as surface oxidation and addition reaction to CNTs. In these techniques, the CNTs treated with strong acid solutions via surface oxidation process that generates cuts at the surface of CNTs and produce a number of carboxylic acid groups at the defect point, predominantly on the open ends. These incorporated carboxylate functionalities increase the hydrophilicity of the CNTs, thereby enhancing the solubility in polar solvents (water) and biocompatibility (Chen et al., 2017). The oxidized CNTs can be further coated with PEG (polyethylene glycol), a hydrophilic substance with the ability to make CNTs more bio-stable (Madani et al., 2011). Furthermore, the alternative side-wall functionalization process was accomplished by addition reactions where the activation of carboxylic group performed by treating with concentrated sulfuric and nitric acid and heated with ultrasonic wave. Thus, it purifies CNTs through the removal of metal catalyst and amorphous carbon (Saifuddin et al, 2013). This process allows for side-wall covalent functionalization, and carboxylic acid groups attached to the surface of CNTs to make it water soluble. However, the side wall of CNT got damaged and led to the alteration in structural properties. The covalent bonding of biocompatible functional groups of CNT in the surface region provides stability and makes CNT-suitable vehicle for drug delivery. The features of the carbon atoms in CNTs produce amazing properties that is suitable for a variety of applications in the electronics, photonics, renewable energy, drug delivery, and the biomedical sector.

#### *7.2.3.3 HYBRID FUNCTIONALIZATION*

Apart from carboxylated CNTs (MWNTs–COOH), covalently functionalized CNTs (MWNTs–$NH_2$), noncovalently functionalized CNTs (MWNTs–PPA), and a new strategy called hybrid functionalization have also been adopted to enhance the reproducibility and better understanding regarding the structure–property relationship of CNTs. Hybrid functionalized CNTs (MWNTs–COOH–PPA) are developed by modifying MWNTs to improve the dispersion of MWNTs within the matrix with controlled regulation of interfacial interactions (Li and Kim, 2011). A different kind of composites, polyurethane (PU), is prepared from these modifying MWNTs with improved thermal and mechanical properties which can be helpful in the prevention of the detachment of noncovalent wrapping. The hybrid functionalized CNTs (MWNTs–COOH–PPA/PU) have exceptional tensile strength. The increment rate in tensile strength is 104% higher than pure PU (Zhu and Wei, 2012). Other functional components like metal nanoparticles (NPs), quantum dots, inorganic oxides, and organic species can be incorporated to develop CNT-based hybrid nanomaterials with synergistic properties (Eder, 2010). Additionally, the efficient fabrication of CNT-based nanohybrids is the prime requirement for the activation of CNT graphitic surfaces. Earlier successful delivery of an anticancer drug, Doxorubicin, through nanohybrid carrier into tumor cells made CNT-based hybrid nanomaterials promising therapeutic approach (Jha et al., 2020). However, for biocompatibility and other safety and health issues, a detailed investigation is required to gain a clear toxicity profile of these hybrid materials.

## 7.3 APPLICATION IN HUMAN HEALTH CARE

### 7.3.1 *THERAPEUTIC APPLICATIONS*

Having the status of the most universally exploited nanomaterials, CNTs have drawn interest of numerous researchers as a desirable candidate material in the field of health care and therapeutics. CNTs are fabricated as hollow cylindrical tubes consisting of carbon with a high aspect ratio and stable hybridization, depending on the number of graphite layers, different classes of SWNTs, double-walled nanotubes, and MWNTs (Kumar et al., 2017). Owing to their excellent surface area, cell permeability, and biocompatibility, the CNTs are assumed to be as smart candidates for the targeted delivery of various drugs of interest. Basic applications include the use of functionalized CNTs as the nanocarrier for the delivery of therapeutic molecules. Another application explores the use of CNT to form scaffold in tissue engineering, wound healing, and regenerative medicines (Yang et al., 2007). The last three decades have principally focused on employing CNTs for drug delivery in cancer therapy as well as many other degenerative diseases. The CNT-based therapeutic applications in various aspects are schematically presented in Scheme 7.1.

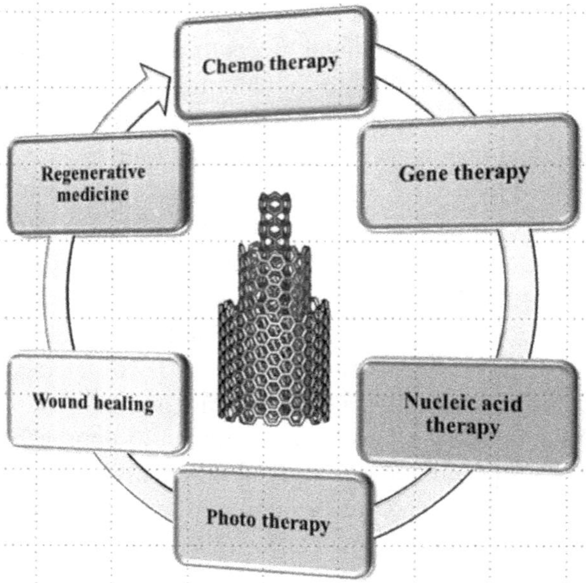


**SCHEME 7.1** Graphical representation of different types of CNT-based therapeutic applications.

### 7.3.1.1 CNTS IN CHEMOTHERAPEUTIC ADVANCES

In actuality, the archetypal attempt to treat such diseases is the administration of chemotherapeutic medicines that experience general toxicity, limited curative interface, drug resistance, and lesser penetrating ability (Kumar et al., 2017). In this perspective, CNTs present an advantageous opportunity as a smart vehicle for anticancer drugs by reducing their toxicity and enhancing local accumulation in the desired site. From the current status, several types of chemically active therapeutic compounds have been verified in relation to their interaction and functionality with the CNTs (Saliev, 2019). The utmost and vital efforts have been implemented toward the conjugation of standard chemotherapeutics with CNTs that have been extensively used for cancer treatment in the clinics as chemotherapy well-known drug which includes cisplatin, doxorubicin, paclitaxel, methotrexate, and flutamide. SWCNTs of 5.4–12.67 μg $ml^{-1}$ were used as nanovehicle for delivery of tamoxifen for human breast cancer cell line to show higher toxicity toward cancerous cells (Oskoueian et al., 2018). There are many literatures available where different anticancer chemotherapeutic agents have been delivered using CNT as the nanovehicle to treat bladder cancer, colon cancer, and other forms of cancer in both cell lines as well as animal models (Simon et al., 2019). Apart from cancer treatment, functionalized CNTs also find application in other diseases like metabolic syndrome, cardiac failure, and inflammation. Few examples include the study where SWCNT were used to administer prednisolone for collagen-induced arthritis (Nakamura et al., 2011). Some earlier reports regarding oral administration of SWCNT with nifedipine for hypertension in animal models suggest about lower cytotoxicity and effective drug delivery (Liu et al., 2009). Another study explored one example of MWCNT with clonidiene for topical application to treat tachycardia and hypertension (Strasinger et al., 2014). Transdermal delivery of MWCNT is also reported along with drug ditiazemhydrochloride for faster action to treat angaina pectoris and hypertension (Bhunia et al., 2013). These reports establish the exploration of CNT for effective drug delivery as well as shed light on the evaluation of toxicity effects on the cell. The cytotoxic effects may serve a primary challenge to translate the bench science to biomedical industry. Further research on development of nontoxic and bio-functionalized CNT will focus to reduce the cytotoxic effect as much as possible.

#### *7.3.1.2 CNTS IN GENE THERAPY AND NUCLEIC ACID THERAPEUTICS*

From several years, gene vectors are implemented for therapeutic applications but the real potential has yet to be realized. Researchers have developed their interest for inventing several innovative tools to deliver genes at the required specific site of action in mammalian cell lines with the advent of hollow cylindrical carbon nanomaterials such as CNTs. Research in the realm of nonviral gene therapy based on plasmid DNA and antisense oligonucleotides have advanced for decades and will resume toward improving target-specific delivery and transfection efficiencies to the levels required for in-vivo clinical trials. The diversified functionalization of CNTs has evolved as one of the most pliable nonviral vectors for gene therapy.

DNA and small interference ribonucleic acids (siRNA) are double-stranded nucleic acids and both contain anionic phosphodiester backbones with similar negative charge-to-nucleotide ratio and can interact electrostatically with cationic agents. Despite sharing similarities, it is to be considered that siRNA molecules possess distinct characteristics, and delivery technologies should be advanced to match the criteria specifically. There are a number of hindrances which need to be addressed in order to achieve the efficient delivery of siRNA molecules into the targeted cells. First is the molecules' tendency to degrade by serum and tissue nucleases. In contrast to DNA, ribose of RNA backbone where hydroxyl group situated at 2′ position of the pentose ring instead of hydrogen, making the RNA backbone more susceptible to hydrolysis by serum nucleases in the extracellular environment, which cleave along the phosphodiester backbone of nucleic acids. The remarkably small size of siRNA fragments (less than 21 base pairs) makes a less potent RNAi effect and causes rapid excretion via kidney even after stable transfection of siRNA and the inefficient endocytosis by targeted tumor cells and the inefficient release from endosomes. To address such challenges, viral capsid protein has been in use for transfection of siRNA. Viruses have evolved to efficiently overcome these blockages; however, the immunogenicity caused by the viral particles has limited the successful nucleic-acid-based therapy. In similar condition, CNT offers less immunogenic and efficient vector for delivery of siRNA inside living cells. Dai and coworkers demonstrated that SWCNTs are able to deliver siRNA into human T cells and primary cells by acting as nonviral molecular transporters (Liu et al., 2007). The role of CNTs as a novel gene-delivery vector system was first reported by Bianco et al. in 2005. They covalently functionalized CNTs using the Prato reaction, 1,3-dipolar cyclo-addition of azomethine to fabricate soluble material in aqueous condition. Both SWCNT and MWCNT were functionalized with a pyrrolidine ring bearing a free amine-terminal oligoethylene glycol moiety attached to the nitrogen atom which increases the solubility of CNTs in aqueous solutions. Small bundles of nanotubes were formed with a diameter of around 20 nm and length of around 200 nm. The delivery of plasmid DNA and the expression of β-galactosidase (marker gene) in CHO cells were studied. The expression was only 10 times higher than the naked pDNA alone, still much less effective than that of liposomes. However, they found that the DNA carbon nanotube (DNA-CNT) complexes were of effective transport vehicle without any toxic effect on the activated or nonactivated lymphocyte unlike dendrimers and liposomes (Cheung et al., 2010). These traditional nonviral genes that deliver vectors generally cause destabilization of the cell membrane and lead to pronounced cytotoxicity while achieving effective delivery of DNA. They attributed the lower cytotoxicity of the DNA–CNT complex to the capability of penetrating cell membrane. They studied the internalization mechanism of the amine functionalized CNTs and found that probably these CNTs entered the cells by a spontaneous mechanism in which they behaved like nanoneedles and passed through the cell membrane without causing cell death. A recent work involved the induction of suicide gene, iC9, to cause safe cell death in human breast cancer cell line using MWCNT (Mohseni-Dargah et al., 2019). Recently, many approaches involving CNTs were made extensively as a potential delivery agent of nucleic acids cell internalization. Various formulations of CNTs have been developed and proved to enhance the delivery and the uptake of different nucleic acid types such as plasmid DNA, siRNA, and miRNA. These specific results of nucleic acid delivery with CNTs represent a critical advance in the field of gene therapy with promising future formulations to be generated as a sanctioned medicine for systemic treatment of many diseases followed by required clinical studies.

#### *7.3.1.3 CNT-MEDIATED PTT AGAINST CANCER*

Over the past few decades, cancer has progressed at a phenomenal rate and provides a significant threat to human health. However, considerable efforts are being directed to selectively target the cancerous tissue with marginal impairment to normal tissue. In recent times, the arena of nanostructure studies has crossed roads with the foreground of cancer therapy and appeared as a promising therapeutic platform through NP-mediated therapy. This field provides a range of nanomaterial; particularly CNTs arise as a novel delivery system to improve the pharmacological performance of the delivered drug and mark a significant step in photothermal therapy (PTT) and radiofrequency-based thermal treatments which represent the potential form of noninvasive cancer therapy.

However, CNTs as one of the potential dynamic nanocarriers have received tremendous attention due to its versatile applications in biomedical fields, particularly in the areas of tissue engineering, biosensing, thermal ablation, and drug delivery (Simmons et al., 2009). Considering the fact that CNTs have the ability to enter cells independent of cell type and functional group through passive diffusion across lipid bilayer or endocytosis by attaching through surface of the cell and provide multiple sites for attachment of different molecules and experience the improved permeability along with retention validity, that is, they exhibit higher accumulation in tumor tissues (Fabbro et al., 2012). These features make the CNTs a competent tool for all kinds of diagnosis and therapeutics in drug delivery field. Hence, they have a predominant role as mediators and delivery vectors in PTT and photodynamic therapy which is recently considered as a new therapeutic in oncology to directly destroy cancer cells with the minimal toxicity to surrounding healthy tissues (Doughty et al., 2019). CNTs adopted extensively in PTT due to its impressive absorption ability under near-infrared (NIR) (700–1100 nm) and radio-frequency radiation where they undergo synchronized oscillations to convert NIR light that kills the targeted cancer cells. Furthermore, due to their remarkable physiochemical properties such as special electrical and thermal capabilities, these nanosized carbon materials are able to determine the capacity of heat transmission and cancer cell erosion. CNTs are able to target the malignant tumor tissues in a specific and selective manner where the cell death mechanism involves mostly three processes such as cell membrane destruction, irreversible protein, and DNA denaturation and angiogenesis blocking (Fig. 7.2) (Pinto and Pocard, 2018). NIR releases significant heat by using functionalized SWCNT/MWCNT where they are able to generate the heat through the excitation and relaxation of optical transitions by absorbing the incident photons as well as by transferring the heat to the surrounding matrix. The transmission of laser beam to the target tumor cell can be eventuated in nanosecond exposure for ablation of individual tumor cells with least damage to normal tissues and maximum temperature can be reached up to 300°C. On the other hand, primary malignant tumor cells can be treated by disabling the cell function with exposure of few minutes' laser beam where the maximum range is maintained within 85–90°C. The ablation temperature range by heating the nanotube is found to be 50–70°C through continuous laser irradiation at high-power density (3.5–35 W $cm^{-2}$) for a long time (3–4 min) inside the cell.

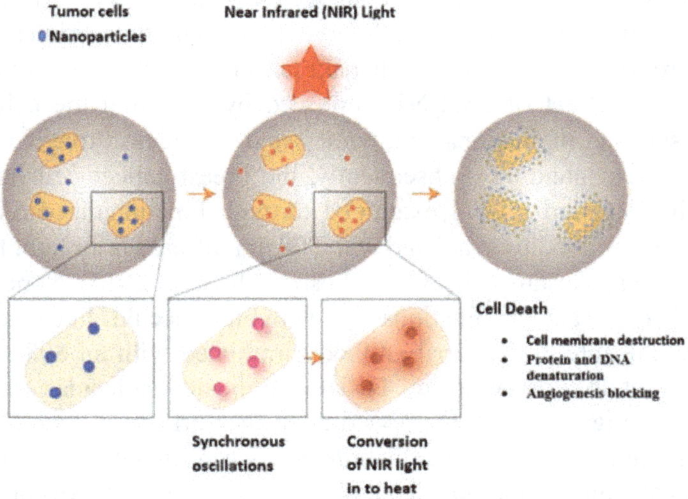


**FIGURE 7.2** Schematic presentation of photothermal therapy (PTT) using nanoparticles in tumor cells.

*Source*: Reproduced with permission from Pinto et al., 2018. Copyright 2018. Walter de Gruyter and Company

In agreement, the in-vivo as well as in-vitro experiments have been conducted by many researchers to explore the tumor destruction effect of SWCNTs and MWCNTs. A photo-ablative archetypal exploiting MWCNTs along with NIR spectral waves in order to annihilate kidney cancer cells were proposed by Torti et al. to augment the thermal destruction functioning of tumor cells by means of boron and nitrogen dopants (2007). The study confirmed the earlier proposed antenna theory to be truly effective with an efficient NIR coupling resulting at nanotube lengths. This was showed in exceeding half of the wavelength of the radiating NIR with increased temperature. The radiation resulted in elevated temperature of MWCNTs through induction processes; subsequently, the heat produced from MWCNTs was transmitted toward the neighboring area to carry on obliteration of the cancer cells at a relatively lower radiation doses. In their findings, they showed that lengths between 700 and 1100 nm were most desirable for killing the tumor cells (Levi-Polyachenko et al., 2009). In a similar way, a desirable uniform size (~0.81 nm) described for SWCNTs with a narrow absorption peak at 980 nm for selective PTT to kill cancer cells (Zhou et al., 2009).

Another captivating finding is uncovered by discovery of nanobombs in the field of CNT-based drug delivery which could be administrated for killing breast cancer cells (Panchapakesan et al., 2005). Moreover, the photothermal effect of SWCNT improved by attaching the cell-bound ligands such as monoclonal antibodies and peptides through covalent or noncovalent bonding. Subsequently, the breast cancer cell lines like HER2, IGF1R, and BT474 examined through functionalized SWCNTs with specific antibodies irradiated with 808 nm infrared laser light for only 3 min (Panchapakesan et al., 2005; Khosroshahi et al., 2015). Nonetheless, MWCNTs could also be functionalized with biomolecules like conjugated disialoganglioside monoclonal antibody irradiated with an 808-nm NIR laser to target disialoganglioside. Disialoganglioside molecules are tumor-associated antigens highly expressed upon the surface of neuroblastoma stNB-V1 tumor cell phenotypes. In order to trace the monoclonal antibodies or antidisialoganglioside-bound MWCNTs, Rhoda mine B was probed on carboxylated CNTs functionalized alongside anti-disialoganglioside as well as without them. MWCNTs bound with antibodies developed as a prospective coupling agent for the efficient PTT of neuroblastoma cells (Wang et al., 2009).

However, the release of drug can be achieved in a controlled and sustained manner with the use of induction heating of template-synthesized CNTs, because particularly, NIR absorption and energy transduction efficiency found to be high across a wide frequency range maintaining a flexibility in the choice of both material characteristics and excitation wavelengths (Yu et al., 2012). Recently, a fascinating development is found to be useful where an implantable bio-electronic device, a unique CNT-based photothermal–electrical converter, is able to convert the thermal energy generated by NIR light to electrical energy and the device can be operated from the outside of the body (Miyako et al., 2011). Furthermore, the advancement of CNTs in PTT is extended by developing an immuno-specific augmented nanotube system. This system was developed using an immune adjuvant, glaciated chitosan and SWCNTs where the stimulated thermal as well as immunogenic effects of tumor cells using GC–SWCNTs and NIR laser system were examined in animal cancer models in in-vitro and in-vivo environment conditions with minimal side effects (Zhou et al., 2012).

### 7.3.1.4 WOUND HEALING WITH CNTS

The advanced and biocompatible materials fabricated with CNTs may enable the improved recuperation of neuronal tissue damage as well as muscular tissue loss consequential from injury while precluding septicity. Reports have demonstrated that CNTs can be exploited effectually like platforms or scaffolds meant for the enhanced and efficient growth of various mammalian cells for instance nerve cells, otherwise difficult to grow pluripotent stem cells, cardiac cells including smooth muscles, and endothelial along with epithelial cells. These advances use CNTs for cell growth, with no apparent toxicity, and provide a platform for generation of wound-healing antiseptic nanofibers from CNTs. Simmons et al. in 2009 have created a novel nanocomposite material from the combination of SWCNTs with polyvinylpyrollidone in water. A novel nanocomposite scaffold based on polyvinyl alcohol (PVA) and MWCNT was synthesized and employed for antimicrobial bioactivity. The results have indicated that the nanocomposite membranes were obtained with thickness, mechanical properties, and swelling behavior comparable to those characteristics of skin tissues. Moreover, these designed hybrids have also presented preliminary biocompatibility based on the response of cell cultures using MTT assay (Santos et al., 2014). The filtration of that aqueous suspension synthesized a highly purified micropermeable layer which has the facility for antiseptic iodine on the surface of a complex of SWCNT wrapped in polymer. This material is intensely antiseptic; in addition, the control samples without iodine also had no evident antimicrobial activity. A novel three-phase disinfectant material ideal for wound dressing was reported from CNTs, silver nanoparticles (Ag NPs), and PVA nanofibers (Jatoi et al., 2019). Ag NPs were first assembled on CNT surfaces forming CNTs–Ag NP nanostructures which were combined to the PVA-based nanofibers in advance of electrospinning to synthesize PVA/carbon nanotubes–Ag NP nanocomposites. Every aspect of the physiochemical characterizations substantiated the tri-phasic structural architecture of nanofiber. Owing to the growth of Ag NPs on carbon nanotubes surfaces as well as entrenching into PVA-based nanofibers, the dressings could be employed aiming at safer and persistent wound-healing applications. The remarkable disinfection abilities along with sustained bacteriostatic features of the nanocomposite were confirmed with the antibacterial assays. Another application of CNTs was reported the slow release of isoniazid on tuberculosis wounds in a controlled manner. Transmission electron microscopy analysis revealed that the diameter of CNTs and NPs remained among 150 and 250 nm which noticeably extended the release time of isoniazid. In addition, the rate of release remained further continual in inhibiting the multiplication of remaining pathogenic *Mycobacterium tuberculosis* nearby surgical wound. In-vitro investigations exhibited that chitosan/CNT nanocomposites not only supplemented purpose of isoniazid but also were able to decrease the negative effects of cytotoxicity and inflammation. A further experiment on animal model of tuberculosis ulcer revealed that isoniazid/chitosan/CNT NPs stimulated the restoration of tuberculosis ulcer. Assessed with the isoniazid group and isoniazid/CNTs group, the surface area of wounds reduced by 94.6% and 89.8%, respectively (Chen et al., 2019a, b). Weakened wound healing is systemically concomitant with various health concerns, including diabetes, bedsores, and extensive burns. In such cases, healing often takes a longer time, subjecting patients to various complications. In this case, SWCNT or MWCNT complexes with chitosan hydrogel can serve as efficient material for enhancing the therapeutic molecule exposure and release into the targeted area.

#### *7.3.1.5 CNTs IN REGENERATIVE MEDICINES*

CNTs are fibrous yet robust nanostructures with a nanosized diameter and possess exceptional physical strength as well as chemical potency. These structures are used as fillers to improve the performance of materials in the industrial field. Since these groups of structures possess remarkable physicochemical properties, CNTs are exploited for their potential as biomaterials. For regenerative medicine, a supporting scaffold or matrix plays key role to maintain the continuous growth of new tissues. Scaffolds are prerequisite for tissue regeneration because of their capability to retain an unrelenting delivery of growth factors and also to offer a pulpit intended for cell proliferation to customize tissue organization. In recent years, diverse materials have been formulated and assessed as scaffold for regenerative medicine. Functionally specific scaffolds require excellent cell permeability, chemical properties, mechanical properties, and safety. Considering the excellent mechanical strength of CNTs, elaborate studies have been experimented aiming on exploration of these CNTs as reinforcing agents in composite biomaterials. Many reports regarding regeneration of bone tissues and nerve tissues are evident of CNTs as efficient scaffold material for regenerative medicine (Tanaka et al., 2020). While designing a bone scaffold, one of the primary criteria is the mechanical strength and CNTs stand out in comparison to other metallic- or ceramic-based bone scaffolds used in orthopedics (such as titanium, stainless steel, and alumina), SWCNTs being less dense and lighter scaffolds with very high tensile strength in the range of terapascals (Tran et al., 2009). Furthermore, it is also observed that the surface-free energies of material affect cell adhesion, hence take part in diverse forms of cells' steps toward greater tissue regeneration. Recent advancement in this field has reported the 3D printing of tough hydrogel-incorporated CNTs for bone regeneration (Cui et al., 2019). Another study of biodegradable PLGA–MWCNT used to regenerate bone tissues (Díaz et al., 2020). Apart from bone tissue regeneration, other emerging interdisciplinary field includes nerve tissue regeneration and stem-cell regeneration. Neural tissue engineering has triggered rising research interest to develop advanced biological scaffolds that rebuild, retain, or recuperate neural tissue functions. Since natural neural tissues have numerous nanostructured features like nanostructured extracellular matrices that neural cells interact with CNTs which also have such nanofeatures and exceptional electrical, mechanical and biocompatibility are ultimate candidates for neural tissue repair. Specifically, with the rapid progress of CNT production technologies, a variety of CNTs with nanometer to millimeter lengths and widths have been synthesized and widely investigated for various neural applications. An improved efficacy of peripheral nerve regeneration and neural regeneration was noticed by poly-L-lactic acid MWCNT conduit containing chitosan-encapsulated curcumin and electrospun hyaluronic acid–CNT nanofibers (Steel, 2018; Jahromi et al., 2020).

### 7.3.2 CNT-BASED BIO-IMAGING APPLICATIONS

The advanced and novel perspective of CNTs in the field of biomedical applications was explored through critical analysis. In recent years, CNTs have been widely studied as in-vitro and in-vivo imaging agents due to unique physical properties. CNT-based imaging probes widely investigated with efficient penetration in tissue layers along with precised spatial resolution that has shown a great promise toward biomedical application.

Different types of CNT-based bio-imaging techniques is featured in Scheme 7.2.

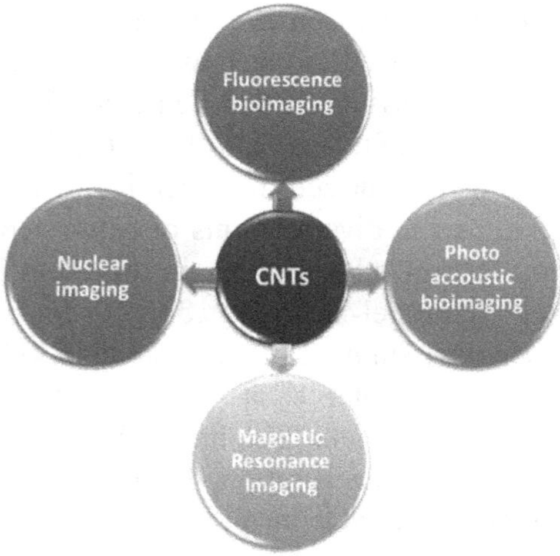


**SCHEME 7.2** Graphical representation of different types of CNT-based bio-imaging techniques.

### *7.3.2.1 FLUORESCENCE BIO-IMAGING*

Among imaging probes, MWCNTs evolved as the environment-friendly fluorescent probe which has a tremendous role in the scientific research and medical diagnosis. The exceptional potential of MWCNTs to replace conventional organic fluorescent dyes and metal-based quantum dots made them a convenient candidate for fluorescence bio-imaging. The potentiality of MWCNTs also involves the utilization of the broad absorption spectrum, low background, high signal-to-noise ratio, label-free detection, real-time monitoring, high sensitivity, their simple preparation, tuneability in emission, photochemical stability, high biocompatible, easily dispersibility, and intrinsic band-gap fluorescence near-infrared II (NIR-II) region (Gong et al., 2013). The modern emerging fluorescence imaging in NIR-II (1000–1700 nm) region empowers picturing of deep anatomical features with unprecedented spatial resolution. In comparison to the traditional NIR (NIR-I, 650–900 nm) fluorescence imaging, the minimized auto-fluorescence, optical scattering, and absorption of tissue with increased applicable power at longer wavelengths found to be more promising strategy. The development of activatable NIR-II fluorescence probes hold great promise for reducing the nonspecific activation and highly depends on high fluorescence quantum yield. This high fluorescence quantum yield can improve the resolution of in-vivo imaging by avoiding the loss of dynamic bio-information with nanoprobes dosage reduction (Tang et al., 2019). Mostly, the developed NIR-II fluorescence probes have a quantum yield below 5%. However, increasing the selectivity of probes for pathological parameters can be alternative way to develop precise activation and reduce false-positive signal. Recently, naphthalimide-derived carbon dots demonstrated its formidable ability to detect endogenous formaldehyde in lysosomes of HeLa cells via fluorescence imaging (Chen et al., 2019a, b). These CNT-based bio-imaging probes will further encourage to moving toward extensive investigation for improving the sensitivity and specificity toward target cells. Improvement in the real-time imaging of pathological parameters in living organisms is essential to understand the underlying pathological mechanism in order to optimize therapeutic interventions.

#### *7.3.2.2 PHOTOACOUSTIC IMAGING*

Photoacoustic (PA) imaging is biomedical imaging methodology to envision the object dependent on the capacity of the object to retain illuminated light. The features like high-contrast, high-spatial resolution with safety measurement made PA imaging technique one of the efficient bio-imaging modalities. PA imaging technique relied on principle where endogenous molecules or contrast agents absorbed the laser pulses in the biological sample and led to generation of heat followed by induction of transient thermoelastic expansion. The unique feature of PA imaging technique to acquire sound detection capacity shows great advantages over traditional optical imaging technique. The advantages proved over fluorescence imaging by avoiding absorbance and scattering of the emission light achieved advanced penetration properties in deep anatomical regions along with augmented three-dimensional resolutions in spatial arrangements. The role of both MWNTs and SWNTs was well demonstrated during cancer treatment by adopting PTT. The strong NIR absorbance builds CNTs as competent photothermal agent in PA imaging. The implementation of SWNTs with low (~7 pg) noise-equivalent detection sensitivity is able to allow for tissue analysis application. The effect of conjugated SWNTs is found to be more competent as PA contrast agent than plain SWNTs. This feature was evident in mice tumor cells where arginylglycylaspartic acid (RGD)–conjugated SWNTs provide strong PA signals than plain SWNT-injected group (Wu et al., 2013). The sensitivity of PA signal can further be upgraded by attaching gold layer in terms of golden nanotubes (GNTs) with SWNTs to enhance the intrinsic PA signals by elevating the intensity of absorbance and optical density near NIR region. Implementation of GNTs with antibody can be used for specific recognition and targeting of receptor with a better intensity of PA signal while extremely low laser fluency level at a few m J $cm^{-2}$ is obtained without antibody conjugation (Gong et al., 2013). The detection of circulating tumor cells is an extended application of GNTs under PA imaging (Galanzha et al., 2009). The intrinsic optical absorbance and coupling with other light-absorbing nanostructures or molecules with strong NIR absorbance made CNTs a promising contrast agent in PA imaging.

#### *7.3.2.3 MAGNETIC RESONANCE IMAGING*

In the current scenario, the techniques of magnetic resonance imaging (MRI) are considered as the mean for noninvasive modalities of imaging used universally in biomedical application. MRI emerged as frequently used technique due to its high spatial resolution and soft-tissue-imaging properties. Extensive investigations have been widely studied by exploring the application of CNTs for MRI. Considering that, MWCNTs are mostly used as excellent scaffold for contrast agents due to their magnetic and electronic properties with remarkable ability of cell membrane penetration. The strong confinement of rare earth metals like $Gd^{3+}$ ion clusters within SWNTs and amphiphilic gadolinium(III) chelate (GdL) conjugate with MWNTs through noncovalent strategy evolved as probable candidate for MRI (Richard et al., 2008). MWCNT hybrids and iron-rich MWCNTs with covalent functionalization propose a promising MRI candidate. Most importantly, the utilization of MWNTs in stem-cell labeling and tracking is one of the successful appliances which were detected through MWNT-labeled mesenchymal stem cells in animals during MRI (Vittorio et al., 2011). In MRI, the best results have shown through functionalization of CNTs with nonionic poloxamers (e.g., Pluronic®) or their "custom-made" counterparts, in terms of contrast effect and low cytotoxicity (Kuźnik and Tomczyk, 2016).

#### 7.3.2.4 *NUCLEAR IMAGING*

The implementation of CNTs in nuclear imaging has also gained attention in diagnosis and therapeutic aspects of medical engineering field. The intrinsic characteristics of SWNTs is enabled to expand the versatility as imaging probe by including external labels such as radio-isotopes. The labeled isotopes $^{125}I$ and $^{14}C$ are helpful in accessing the bio-distribution of SWNTs and MWNTs in animal (Wang et al., 2004; Deng et al., 2008). With advancement of having high sensitivity without tissue penetration depth limitation, PEGylated SWNT-coupled RGD peptide and CNTs–anti-CD20 antibody conjugates used in specific targeting of human tumor cells (McDevitt et al., 2007). Hence, in recent time, the development in CNT-based nuclear imaging has occupied the limited place compared to other advanced imaging techniques but by combing with other imaging techniques can build extraordinary interests in the improvement of CNT-based bio-imaging modality.

## 7.4 SUMMARY AND CONCLUSION

Besides, their application in the delivery of therapeutic agents, the bio-functionalized CNTs are currently being intensely explored due to their impressive ability as scaffolding materials, where they are able to provide the support for growth of bone cells, neurons, and cardiomyocytes and even direct or promote the differentiation of stem cells into specific lineages and hold a great promise for drug delivery in cancer therapy, as well as its diagnosis. These intrinsic properties open new avenues for the development of novel immune conjugates for cancer phototherapy with high performance and efficacy in the distinguishingly specific thermal ablation of cancer cells. In other respects, besides their role as carrier of chemotherapeutic drugs, antimicrobials, and anti-inflammatory agents, to more complex peptide-based vaccines, antibodies and siRNA have also been successfully delivered with CNTs using a multitude of strategies, admirable efficacy, and reduced toxicity (Varkouhi et al., 2011).

Apart from its role as a tremendous nanocarrier in drug delivery system, CNT has immense impact in the field of genetic engineering and tissue engineering by manipulating genomes and atoms in order to develop bio-imaging genomes (Harrison and Atala, 2007). The covalent bond between carbon atoms provides mechanical stability and thermal conductivity to the nanotubes which makes them potential transporter in diagnosis and therapeutics application. CNTs also identified as effective vector in gene therapy due to its tabular nature. Certainly, CNTs have a dominant role as catalyst due to the larger surface area where the amount of catalyst can be incorporated at molecular level; along with that, it also acts as biosensor for its fluorescence ability (Zhu et al., 2011). The advantages of CNTs have also been explored as artificial implants and preservative due to its high tensile strength and antioxidant properties. Nonetheless, recent studies explored the advantages and applications of CNTs in the field of therapeutics and diagnosis as a promising object which can be further implemented in the future of medicine. Many of the deadlocks have to be unlocked involving nanotubes in nanotherapeutics also where nanotube structures stand in connection with integration inside living systems or biological models. However, further concerns exist, for instance, what would be possible toxicological impacts of nanotubes on a natural or artificial living system are and how cells can be motivated to efficiently uptake nanotubes can create new horizons for human health-care systems with CNT in them.

## KEYWORDS

- **CNTs**
- **carbon nanotubes**
- **siRNA**
- **drug delivery**
- **therapeutics**

## REFERENCES

Anazawa, K.; Shimotani, K.; Manabe, C.; Watanabe, H.; Shimizu, M. High-Purity Carbon Nanotubes Synthesis Method by an Arc Discharging in Magnetic Field. *Appl. Phys. Lett.* **2002,** *81*, 739–741.

Bhunia, T.; Giri, A.; Nasim, T.; Chattopadhyay, D.; Bandyopadhyay, A. A Transdermal Diltiazem Hydrochloride Delivery Device Using Multi-Walled Carbon Nanotube/Poly (Vinyl Alcohol) Composites. *Carbon*, **2013,** *52*, 305–315.

Bianco, A.; Hoebeke, J.; Kostarelos, K.; Prato, M.; Partidos, C. D. Carbon Nanotubes: On the Road to Deliver. *Curr. Drug Delivery*, **2005,** *2*, 253–259.

Chen, G.; Wu, Y.; Yu, D.; Li, R.; Luo, W.; Ma, G.; Zhang, C. Isoniazid-Loaded Chitosan/Carbon Nanotubes Microspheres Promote Secondary Wound Healing of Bone Tuberculosis. *J. Biomater. Appl.* **2019a,** *33*, 989–996.

Chen, S.; Jia, Y.; Zou, G. Y.; Yu, Y. L.; Wang, J. H. A Ratiometric Fluorescent Nanoprobe Based on Naphthalimide Derivative-Functionalized Carbon Dots for Imaging Lysosomal Formaldehyde in HeLa Cells. *Nanoscale* **2019b,** *11*, 6377–6383.

Chen, Z.; Zhang, A.; Wang, X.; Zhu, J.; Fan, Y.; Yu, H.; Yang, Z. The Advances of Carbon Nanotubes in Cancer Diagnostics and Therapeutics. *J. Nanomater.* 2017.

Cheung, W.; Pontoriero, F.; Taratula, O.; Chen, A. M.; He, H. DNA and Carbon Nanotubes as Medicine. *Adv. Drug Deliv. Rev*. **2010,** *62*, 633–649.

Cui, H.; Yu, Y.; Li, X.; Sun, Z.; Ruan, J.; Wu, Z. Direct 3D Printing of a Tough Hydrogel Incorporated with Carbon Nanotubes for Bone Regeneration. *J. Mater. Chem. B* **2019,** *7*, 7207–7217.

Dai, H. Carbon Nanotubes: Synthesis, Integration, and Properties. *Acc. Chem. Res*., **2002,** *35*, 1035–1044.

Deng, X.; Yang, S.; Nie, H.; Wang, H.; Liu, Y. A Generally Adoptable Radiotracing Method for Tracking Carbon Nanotubes in Animals. *Nanotechnology* **2008,** *19*, 075101.

Díaz, E.; Puerto, I.; Sandonis, I.; Ribeiro, S.; Lanceros Mendez, S. Hydrolytic Degradation and Cytotoxicity of Poly (Lactic Coglycolic Acid)/Multiwalled Carbon Nanotubes for Bone Regeneration. *J. Appl. Polym. Sci*. **2020,** *137*, 48439.

Doughty, A. C.; Hoover, A. R.; Layton, E.; Murray, C. K.; Howard, E. W.; Chen, W. R. Nanomaterial Applications in Photothermal Therapy for Cancer. *Materials* **2019,** *12*, 779.

Dresselhaus, M. S. Nanotube Antennas. *Nature* **2004,** *432*, 959–960.

Eder, D. Carbon Nanotube—Inorganic Hybrids. *Chem. Rev*. **2010,** *110*, 1348–1385.

El-Sayed, A.; Kamel, M. Advances in Nanomedical Applications: Diagnostic, Therapeutic, Immunization, and Vaccine Production. *Environ. Sci. Pollut. Res*. **2020,** *27*, 19200–19213.

Fabbro, C.; Ali-Boucetta, H.; Da Ros, T.; Kostarelos, K.; Bianco, A.; Prato, M. Targeting Carbon Nanotubes Against Cancer. *Chem. Comm*. **2012,** *48*, 3911–3926.

Galanzha, E. I.; Shashkov, E. V.; Kelly, T.; Kim, J. W.; Yang, L.; Zharov, V. P. In Vivo Magnetic Enrichment and Multiplex Photoacoustic Detection of Circulating Tumour Cells. *Nat. Nanotechnol*. **2009,** *4*, 855–860.

Gao, C.; Guo, Z.; Liu, J. H.; Huang, X. J. The New Age of Carbon Nanotubes: An Updated Review of Functionalized Carbon Nanotubes in Electrochemical Sensors. *Nanoscale* **2012,** *4*, 1948–1963.

Gong, H.; Peng, R.; Liu, Z. Carbon Nanotubes for Biomedical Imaging: The Recent Advances. *Adv. Drug Deliv. Rev*. **2013,** *65*, 1951–1963.

Harrison, B. S.; Atala, A. Carbon Nanotube Applications for Tissue Engineering. *Biomaterials* **2007,** *28*, 344–353.

Hu, C. Y.; Xu, Y. J.; Duo, S. W.; Zhang, R. F.; Li, M. S. Non-Covalent Functionalization of Carbon Nanotubes with Surfactants and Polymers. *J. Chin. Chem. Soc.* **2009,** *56*, 234–239.

Jahromi, H. K.; Farzin, A.; Hasanzadeh, E.; Barough, S. E.; Mahmoodi, N.; Najafabadi, M. R. H. Enhanced sciatic nerve regeneration by poly-L-Lactic Acid/Multi-Wall Carbon Nanotube Neural Guidance Conduit Containing Schwann Cells and Curcumin Encapsulated Chitosan Nanoparticles in Rat. *Mater. Sci. Eng. C*, **2020,** *109*, 110564.

Jatoi, A. W.; Kim, I. S.; Ni, Q. Q. Cellulose Acetate Nanofibers Embedded with Ag NPs Anchored $TiO_2$ Nanoparticles for Long Term Excellent Antibacterial Applications. *Carbohydr. Polym.*, **2019,** *207*, 640–649.

Jha, R.; Singh, A.; Sharma, P. K.; Fuloria, N. K. Smart Carbon Nanotubes for Drug Delivery System: A Comprehensive Study. *J. Drug Deliv. Sci. Technol.*, **2020,** 101811.

Khosroshahi, M. E.; Hassannejad, Z.; Firouzi, M.; Arshi, A. R. Nanoshell-Mediated Targeted Photothermal Therapy of HER2 Human Breast Cancer Cells Using Pulsed and Continuous Wave Lasers: An In Vitro Study. *Lasers Med. Sci.*, **2015,** *30*, 1913–1922.

Kumar, S.; Rani, R.; Dilbaghi, N.; Tankeshwar, K. Kim, K. H. Carbon Nanotubes: A Novel Material for Multifaceted Applications in Human Healthcare. *Chem. Soc. Rev.*, **2017,** *46*, 158–196.

Kuźnik, N.; Tomczyk, M. M. Multiwalled Carbon Nanotube Hybrids as MRI Contrast Agents. *Beilstein J. Nanotechnol.* **2016,** *7*, 1086–1103.

Levi-Polyachenko, N. H.; Merkel, E. J.; Jones, B. T.; Carroll, D. L.; Stewart IV, J. H. Rapid Photothermal Intracellular Drug Delivery Using Multiwalled Carbon Nanotubes. *Mol. Pharma.* **2009,** *6*, 1092–1099.

Li, H.; Kim, I. Fabrication and Applications of Carbon Nanotube-Based Hybrid Nanomaterials by Means of Non-Covalently Functionalized Carbon Nanotubes. In *Carbon Nanotubes–From Research to Applications*; 2011; pp 211–228.

Liu, H.; Bu, Y.; Mi, Y.; Wang, Y. Interaction Site Preference Between Carbon Nanotube and Nifedipine: A Combined Density Functional Theory and Classical Molecular Dynamics Study. *J. Mol. Struct.*, **2009,** *901*, 163–168.

Liu, Z.; Winters, M.; Holodniy, M.; Dai, H. siRNA Delivery Into Human T Cells and Primary Cells with Carbon Nanotube Transporters. *Angew. Chem.* **2007,** *119*, 2069–2073.

Lu, L.; Chen, W. Supramolecular Self-Assembly of Biopolymers with Carbon Nanotubes for Biomimetic and Bio-Inspired Sensing and Actuation. *Nanoscale* **2011,** *3*, 2412–2420.

Madani, S. Y.; Naderi, N.; Dissanayake, O.; Tan, A.; Seifalian, A. M. A New Era of Cancer Treatment: Carbon Nanotubes as Drug Delivery Tools. *Int. J. Nanomed.*, **2011,** *6*, 2963.

Malekzad, H.; Zangabad, P. S.; Mirshekari, H.; Karimi, M.; Hamblin, M. R. Noble Metal Nanoparticles in Biosensors: Recent Studies and Applications. *Nanotechnol. Rev.*, **2017,** *6*, 301–329.

Martinelli, V.; Cellot, G.; Toma, F. M.; Long, C. S.; Caldwell, J. H.; Zentilin, L.; Mestroni, L. Carbon Nanotubes Promote Growth and Spontaneous Electrical Activity in Cultured Cardiac Myocytes. *Nano Lett.* **2012,** *12*, 1831–1838.

McDevitt, M. R.; Chattopadhyay, D.; Kappel, B. J.; Jaggi, J. S.; Schiffman, S. R.; Antczak, C. Tumor Targeting with Antibody-Functionalized, Radiolabeled Carbon Nanotubes. *J. Nucl. Med.*, **2007,** *48*, 1180–1189.

Miyako, E.; Hosokawa, C.; Kojima, M.; Yudasaka, M.; Funahashi, R.; Oishi, I. A Photo Thermal Electrical Converter Based on Carbon Nanotubes for Bio-electronic Applications. *Angew. Chem. Int. Ed.* **2011,** *50*, 12266–12270.

Mohseni-Dargah, M.; Akbari-Birgani, S.; Madadi, Z.; Saghatchi, F.; Kaboudin, B. Carbon Nanotube-Delivered iC9 Suicide Gene Therapy for Killing Breast Cancer Cells In Vitro. *Nanomedicine* **2019,** *14*, 1033–1047.

Nakamura, M.; Tahara, Y.; Ikehara, Y.; Murakami, T.; Tsuchida, K.; Iijima, S. Single-Walled Carbon Nanohorns as Drug Carriers: Adsorption of Prednisolone and Anti-Inflammatory Effects on Arthritis. *Nanotechnology* **2011,** *22*, 465102.

Oskoueian, A.; Amin Matori, K.; Bayat, S.; Oskoueian, E.; Ostovan, F.; Toozandehjani, M. Fabrication, Characterization, and Functionalization of Single-Walled Carbon Nanotube Conjugated with Tamoxifen and Its Anticancer Potential against Human Breast Cancer Cells. *J. Nanomater.* **2018,** *2018*, 1–13.

Panchapakesan, B.; Lu, S.; Sivakumar, K.; Taker, K.; Cesarone, G.; Wickstrom, E. Single-Wall Carbon Nanotube Nanobomb Agents for Killing Breast Cancer Cells. *Nanobiotechnology* **2005,** *1*, 133–139.

Pastorin, G.; Wu, W.; Wieckowski, S.; Briand, J. P.; Kostarelos, K.; Prato, M.; Bianco, A. Double Functionalisation of Carbon Nanotubes for Multimodal Drug Delivery. *Chem Commum.* **2006,** 1182–1184.

Pinto, A.; Pocard, M. Photodynamic Therapy and Photothermal Therapy for the Treatment of Peritoneal Metastasis: A Systematic Review. *Pleura Peritoneum* **2018,** *3*, 1–22.

Richard, C.; Doan, B. T.; Beloeil, J. C.; Bessodes, M.; Tóth, É.; Scherman, D. Noncovalent Functionalization of Carbon Nanotubes with Amphiphilic Gd3+ Chelates: Toward Powerful T1 and T2 MRI Contrast Agents. *Nano Lett.* **2008,** *8*, 232–236.

Saifuddin, N.; Raziah, A. Z.; Junizah, A. R. Carbon Nanotubes: A Review on Structure and Their Interaction with Proteins. *J. Chem.* **2013,** 1–18.

Saliev, T. The Advances in Biomedical Applications of Carbon Nanotubes. *C—J. Carbon Res.* **2019,** *5*, 29.

Santos, J. C. C.; Mansur, A. A.; Ciminelli, V. S.; Mansur, H. S. Nanocomposites of Poly (Vinyl Alcohol)/Functionalized-Multiwall Carbon Nanotubes Conjugated with Glucose Oxidase for Potential Application as Scaffolds in Skin Wound Healing. *Int J Polym Mater.* **2014,** *63*, 185–196.

Simmons, T. J.; Lee, S. H.; Park, T. J.; Hashim, D. P.; Ajayan, P. M.; Linhardt, R. J. Antiseptic Single Wall Carbon Nanotube Bandages. *Carbon* **2009,** *47*, 1561–1564.

Simon, J.; Flahaut, E.; Golzio, M. Overview of Carbon Nanotubes for Biomedical Applications. *Materials* **2019,** *12*, 624.

Steel, E. M. *Engineering Hyaluronic Acid-carbon Nanotube Nanofibers: A Peripheral Nerve Interface to Electrically Stimulate Regeneration*; Wayne State University, 2018.

Strasinger, C.; Paudel, K. S.; Wu, J.; Hammell, D.; Pinninti, R. R.; Hinds, B. J.; Stinchcomb, A. Programmable Transdermal Clonidine Delivery Through Voltage-Gated Carbon Nanotube Membranes. *J. Pharm. Sci.* **2014,** *103*, 1829–1838.

Tanaka, M.; Aoki, K.; Haniu, H.; Kamanaka, T.; Takizawa, T.; Sobajima, A. Applications of Carbon Nanotubes in Bone Regenerative Medicine. *Nanomaterials* **2020,** *10*, 659.

Tang, Y.; Pei, F.; Lu, X.; Fan, Q.; Huang, W. Recent Advances on Activatable NIR II Fluorescence Probes for Biomedical Imaging. *Adv. Opt. Mater.* **2019,** *7*, 1900917.

Thakare, V. S.; Das, M.; Jain, A. K.; Patil, S.; Jain, S. Carbon Nanotubes in Cancer Theragnosis. *Nanomedicine* **2010,** *5*, 1277–1301.

Torti, S. V.; Byrne, F.; Whelan, O.; Levi, N.; Ucer, B.; Schmid, M. Thermal Ablation Therapeutics Based on CNx Multi-Walled Nanotubes. *Int. J. Nanomed.* **2007,** *2*, 707.

Tran, P. A.; Zhang, L.; Webster, T. J. Carbon Nanofibers and Carbon Nanotubes in Regenerative Medicine. *Adv. Drug Delivery Rev.* **2009,** *61*, 1097–1114.

Varkouhi, A. K.; Foillard, S.; Lammers, T.; Schiffelers, R. M.; Doris, E.; Hennink, W. E.; Storm, G. SiRNA Delivery with Functionalized Carbon Nanotubes. *Int. J. Pharm.* **2011,** *416*, 419–425.

Vittorio, O.; Duce, S. L.; Pietrabissa, A.; Cuschieri, A. Multiwall Carbon Nanotubes as MRI Contrast Agents for Tracking Stem Cells. *Nanotechnology* **2011,** *22*, 095706.

Wang, C. H.; Huang, Y. J.; Chang, C. W.; Hsu, W. M.; Peng, C. A. In Vitro Photothermal Destruction of Neuroblastoma Cells Using Carbon Nanotubes Conjugated with GD2 Monoclonal Antibody. *Nanotechnology* **2009,** *20*, 315101.

Wang, H.; Wang, J.; Deng, X.; Sun, H.; Shi, Z.; Gu, Z. Bio-Distribution of Carbon Single-Wall Carbon Nanotubes in Mice. *J. Nanosci. Nanotechnol.* **2004,** *4*, 1019–1024.

Wu, L.; Cai, X.; Nelson, K.; Xing, W.; Xia, J.; Zhang, R. A Green Synthesis of Carbon Nanoparticles from Honey and Their Use in Real-Time Photoacoustic Imaging. *Nano Res.* **2013,** *6*, 312–325.

Yang, W.; Thordarson, P.; Gooding, J. J.; Ringer, S. P.; Braet, F. Carbon Nanotubes for Biological and Biomedical Applications. *Nanotechnology* **2007,** *18*, 412001.

Yu, J. G.; Jiao, F. P.; Chen, X. Q.; Jiang, X. Y.; Peng, Z. G.; Zeng, D. M.; Huang, D. S. Irradiation-Mediated Carbon Nanotubes' Use in Cancer Therapy. *J Cancer Res Ther.* **2012,** *8*, 348.

Zhou, F.; Da, X.; Ou, Z.; Wu, B.; Resasco, D. E.; Chen, W. R. Cancer Photothermal Therapy in the Near-Infrared Region by Using Single-Walled Carbon Nanotubes. *J. Biomed. Opt.* **2009,** *14*, 021009.

Zhou, F.; Wu, S.; Song, S.; Chen, W. R.; Resasco, D. E.; Xing, D. Antitumor Immunologically Modified Carbon Nanotubes for Photothermal Therapy. *Biomaterials* **2012,** *33*, 3235–3242.

Zhu, M. X.; Wei, B. W. Preparation and Properties of Hybrid Functionalized Carbon Nanotube Polyurethane Composites. *Chinese J. Mater. Res.* **2012,** *26*, 191–19.

# CHAPTER 8

# Mechanistic Insight into the Tuneable Electronic Properties of Chemically Functionalized Graphene Quantum Dots

MIHIR RANJAN SAHOO[1], SATYAJIT RATHA[1], and ANEEYA K. SAMANTARA[2]

*[1]School of Basic Sciences, Indian Institute of Technology Bhubaneswar, India*

*[2]School of Chemical Sciences, National Institute of Science Education and Research, Bhubaneswar, India*

## ABSTRACT

The emerging graphene quantum dots (GQDs), carbon-based zero-dimensional materials have gained significant attention in research community for their promising abilities for optoelectronic, biomedical, and energy-related applications. The combination of quantum confinement and edge effect makes these materials possess exceptional chemical and physical properties. This chapter aims to update regarding effect of size and shape on electronic properties of GQDs studied by various groups in the past few years on theoretical aspects. The modification of electronic properties of GQDs by chemical functionalization and the underlying mechanism behind the changes in HOMO–LUMO energy gaps of functionalized GQDs are systematically summarized. Herein, we discuss a mechanism which is based on the competition and collaboration between charge transfer and frontier orbital hybridization in edge-functionalized GQDs.

New Forms of Carbon: Nanocarbons. Aneeya Kumar Samantara & Satyajit Ratha (Eds.)

This study can be useful for better tuning of electronic properties for practical device applications.

## 8.1 INTRODUCTION

In the era of global digitalization, the huge demand for the miniaturization of electronic and memory storage devices often raises the most common question: "what could be the smallest possible size of transistors, resistors, memory chips, etc., achievable, to acquire both efficiency and compactness?" Miniaturization plays a major role in the technological revolution, since it has several inherent advantages such as low cost, high speed, and large density. Thus, miniaturized devices are fast, energy efficient, have high aesthetical value, and they are both convenient and extremely portable as well. In other words, if the size or dimension of the device is small, then most of its physical properties can be controlled through quantum mechanical effects and are often labeled as nano-electronic devices. Recently, the research on low-dimensional materials has become a topic of great interest, as it could lead the technological revolution of the world. At nanoscale, the properties of the materials not only depend upon the chemical bonds between atoms but also on the size and shape of the materials. From the physics point of view, in the case of bulk systems, the dimensions or the structures are much larger than the de Broglie wavelength of the charge carrier, which allows their free movement along any direction. But, in nanostructures, the confined dimension is approximately in the range of the carrier wavelength. For a typical semiconductor, the range of the length scale for the confinement effect lies between 1 and 25 nm. Here, the movement of the charge carriers is controlled through the quantum confinement effect, where electron–hole pairs are confined spatially in one or more dimensions within the material. These geometrical constraints can act as particle boundaries for electrons, which forces the electrons to adjust their energy to give the response to the change in system dimensions. Here, the periodicity of the system can be introduced artificially in which the Brillouin zones are folded into smaller ones and lead to the modification in the conduction band. Thus, low-dimensional systems show unique electronic, optical, thermal, mechanical, and other physical properties, which make them promising candidates for the designing of nanoscale devices.

Basically, low-dimensional materials can be classified into three categories, depending upon the number of confined dimensions. If the structure is confined in one dimension and extended in the other two dimensions, then it is termed as a two-dimensional (2D) material or a quantum well. Similarly, systems confined in two- and three-dimensions are considered as one-dimensional (1D) or quantum wires, and zero-dimensional (0D) or quantum dots (QDs) materials, respectively. By increasing the number of confined dimensions or reducing the number of atoms in a material, the discreteness of energy levels can be increased. The narrow potential well, created due to the quantum confinement, can trap the charge carriers, resulting in a change in the energy levels from continuous (in bulk system) to delocalized (in QDs) states.

As per Heisenberg's uncertainty principle, if the position of a particle is known more precisely, then there will be more uncertainty in the calculation of momentum/energy. So, when electrons are more localized in a confined system, then their momentum/energy ranges get broadened, which leads to an increase in the average energy of electrons in the conduction band. As a result, the spacing between the valence band and conduction band will increase, resulting in a bandgap. Since there is no bandgap in the case of metals, therefore, the quantum size effect is not usually observed. The effect is more prominent in the case of semiconductors, due to the presence of a finite bandgap. The quantum-size effect has significant impact on the material properties of semiconductors, when the size is nearly equal to or less than Bohr's exciton radius of the corresponding bulk semiconductor. Bohr's exciton radius of a semiconductor particle is defined as (Yoffe 1993)

$$a_B = \varepsilon \frac{m}{m^*} a_0 \tag{8.1}$$

where $\varepsilon$, $m$, $m^*$, and $a_0$ represent dielectric constant of the material, rest mass of the electron, mass of the particle, and Bohr's radius of the hydrogen atom, respectively. When the size of particles reaches the order of Bohr's exciton radius, then the excitonic transition energy increases, leading to blue shift in the absorption and luminescence spectra. In this chapter, we will be focusing on QDs only.

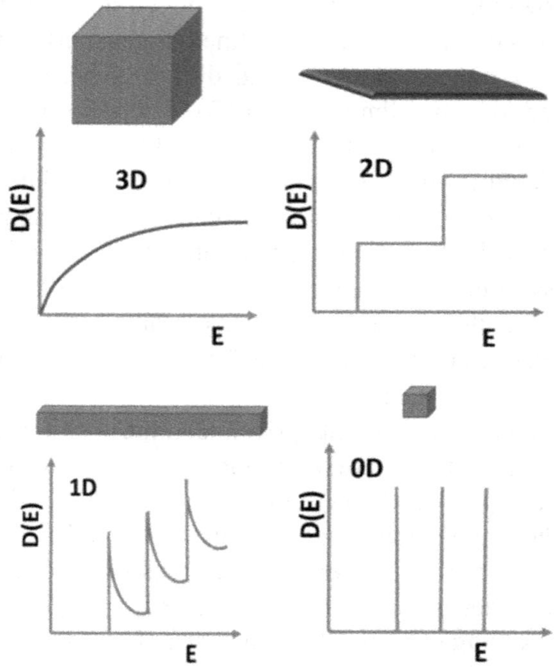


**FIGURE 8.1** Electron density of states (DOS) of bulk and confined nanostructures in different dimensions. The absorption to explain the band gap is roughly proportional to the DOS.

The quantum dots are the type of materials in which quantum confinement effect is purely observed. These are generally semiconductor nanocrystals consisting of hundred to thousand atoms. Thus, electrons and electron–hole pairs (excitons) are tightly confined in all three spatial dimensions. QDs belong to the nanomaterial family which consists of metals, semimetals, semiconductors, insulators, and organic materials. But, in general, we consider semiconductor nanocrystals as quantum dots. From the basic principle of quantum mechanics, it is known that when an electron is confined in space, it acquires kinetic energy and the energy spectrum gets discrete. In bulk semiconductor, the electrons can move freely inside the solid, leading to a continuous energy spectrum, and the density of states (DOS) changes with the square root of energy. However, in a very small piece of synthesized semiconductor, the electrons feel confined and the DOS becomes discrete with the increase in the energy gap (Wise, 2000), which is shown in Figure 8.1. This figure indicates that the optical properties of a semiconductor can be triggered by controlling the size, as DOS changes due to the confinement effect. Out of different parameters like shape, defect, surface directions, and so on, the size effect has a greater influence on the optical properties of semiconductor QDs. The change in surface-to-volume ratio, with change in size, modifies the DOS near the band edge of the QDs and also the optical and electronic properties (Brus, 1991; Choudhury et al., 2011). When a photon is absorbed in 3D bulk semiconductor, then an electron gets excited from the valence band to the conduction band. The excited electron falls to valence band from the conduction band by losing its energy and recombines with the hole, resulting in the formation of light particles of a certain energy. Thus, irrespective of the size of the semiconductor, the light particle always interacts with the exciton (one pair of electron and hole) that leads to the conservation of total or integrated absorption. So, in the ideal scenario of a QD, the DOS becomes a series of delta functions to make the transition, strong enough to conserve the total absorption, which is indicated in Figure 8.1. Since, in a quantum dot, the confined volume is smaller than Bohr's radius, the exciton formed in it requires more energy to be confined. This leads to an increase in the band gap of the QDs. If the energy gap increases, the light particle with high frequency or low wavelength will be created due to electron–hole recombination, resulting in blue shift of light. In other words, decreasing the size of QDs or more confining the spatial volume of the materials leads to blue shift (lower wavelength side) of emission or absorption spectra.

For the past two decades, due to spectroscopic electro-optical properties, the semiconductor quantum dots have gained significant attention in the research community, as they have potential applications in the biomedical industries (Jin and Hildebrandt, 2012; Shao et al., 2011; Yu, 2008; Zhou et al., 2015), bio-sensing (Freeman et al., 2007; Hildebrandt, 2011; Jin and Hildebrandt, 2012; Martynenko et al., 2017; Xing et al., 2009), bio-imaging (Arya et al., 2005; Green, 2004; Jin and Hildebrandt, 2012; Martynenko et al., 2017; Xing et al., 2009), photovoltaic solar cells (Hetsch et al., 2011; Nozik et al., 2010; Raffaelle et al., 2002; Tian and Cao, 2013), quantum information processing (Biolatti et al., 2002; Troiani et al., 2000; Wu et al., 2004), and many optoelectronic devices (Masut et al., 1995; Reithmaier et al., 2006; Skolnick and Mowbray, 2004; Wang et al., 2007). However, high market cost and toxicity associated with inorganic semiconductor QDs hinder their applicability in industry and biomedicine. These limitations have stimulated the researches to find out an alternative for these inorganic QDs, and with the advanced technology, the carbon-based QDs, that is, carbon quantum dots (CQDs) and graphene quantum dots (GQDs), are found to be the best alternatives, due to their cost-effective and nontoxic nature. Furthermore, good solubility, high dispersibility, good surface grafting, chemical inertness, and better photostability make them more compatible in biomedical as well as other optoelectronic and energy device applications. In this chapter, we will mainly focus on the electronic properties of GQDs, through chemical functionalization.

### 8.1.1 GRAPHENE QUANTUM DOTS

Carbon has been considered as an age-old companion of human beings, as it supplies basic energy for life and provides key substances produced from its various allotropes, for the development of modern society. With the help of advanced science and technology, nanoscale carbonaceous materials like buckminsterfullerene C60 (Kroto et al., 1985), carbon nanotubes (Iijima, 1991), and graphene (Geim and Novoselov, 2007; Novoselov et al., 2004) were successively discovered. Excellent mechanical strength, high intrinsic carrier mobility, large surface area, long spin relaxation length, and presence of quantum Hall effect at room temperature make graphene a unique material for electronic, spintronics, and energy device applications (Karpan et al., 2008; Sahoo et al., 2019; Sun et al., 2012, 2018). The interest toward graphene has been mostly powered by its ability to upraise the practical technologies involved in the area of transistors, electronic displays, integrated circuits, sensors, and so on. In spite of various enhanced properties, the practical application of 2D graphene sheet is limited due to having a zero-band gap, forming agglomerations during synthesis, featuring low absorptivity, and poor dispersibility (Zhang et al., 2012). Subsequent investigations show that these limitations can be mitigated through structural modification of graphene. For example, graphene nanoribbon can be created by cutting the graphene sheet in one-dimensional stripe, and the resultant product shows interesting electronic and optical properties due to the addition of an extra confined dimension (Han et al., 2007; Todd et al., 2009). These results stimulated the researchers to check the properties of graphene, after one more dimension is confined, or to investigate, how the electronic and optical properties of graphene get modified if there would be a transition from 2D sheet to 0D or quantum dots (GQDs). GQDs are ultrasmall fragments of graphene sheet with lateral size less than 100 nm. An ideal GQD should have only a single layer of carbon atoms (Chen et al., 2018). However, in practice, the synthesized GQDs contain oxygen and hydrogen atoms and also have few additional atomic layers with size less than 100 nm (Choi, 2017; Liu et al., 2011). Due to quantum size and edge effect, properties like unique fluorescent and good dispersibility, in various solvents, are exhibited in GQDs. Furthermore, enhanced catalytic properties of GQDs are observed, since a high surface-to-volume ratio provides more active sites to carry abundant functional moieties. Typically, the carboxylic acid moieties contained at the edge of the GQDs enhance the water solubility capacity and create easy functionalization with various organic, inorganic, and polymer species (Shen et al., 2012). The GQDs are different from the CQDs, as they possess honeycomb lattice of graphene. But GQDs have nonzero band gap, unlike graphene, which enhances their capability to replace inorganic QDs in optical and electronic devices.

Due to the reduction in the large spatial dimension to a nanoscale range in GQDs, the electron distribution gets modified by the crystal boundary, which leads to the opening of a bandgap in GQDs. Thus, semimetal graphene changes to either semiconductor or insulator GQDs, depending upon the height of the energy gap. As a result, higher energy spectra are created from the broadening of the optical absorption of graphene (Yan et al., 2019). In comparison to other carbon-based nanomaterials like carbon nanotubes, fullerene, carbon dots, and so on, GQDs show different physical and chemical properties due to both the edge and quantum confinement effect.

## 8.2 ELECTRONIC PROPERTIES OF GQDs

As discussed earlier, the unique physical properties possessed by graphene are due to the massless Dirac spectrum in the electronic band structure near the Fermi level and make graphene an important building block for the new generation nano-electronic, photonic, spintronic applications. However, the zero bandgap in graphene's band structure does not allow the conductivity to fall below a minimum value, which acts as a hurdle for manufacturing essential electronic gate circuit due to the absence of "off" state. But this problem can be tackled by confining charge carriers in a low-dimensional structure like 1D graphene nanoribbon or 0D GQD which could produce a finite band gap. The bandgap can be tuned from ultraviolet to infrared through visible spectrum by modifying the sizes of GQDs. Furthermore, for the successful utilization of these QDs in optoelectronic and photonic devices, it is important to understand the electronic excitations in them. Mandal et al. have studied, theoretically, the electronic structure of GQDs through density functional tight binding method (Mandal et al., 2012). They found out that the HOMO–LUMO gap can be modified by controlling size and shape of GQDs. At first, two large-sized quantum dots of different shapes were modeled for theoretical study, which have been synthesized earlier in a controlled manner through solution chemistry (Yan et al., 2010).

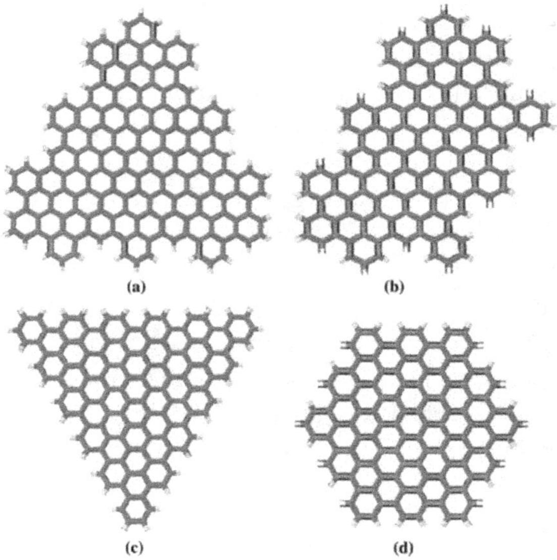


**FIGURE 8.2** Optimized structures of GQDs of four different types: (a) type 1, (b) type 2, (c) type 3, and (d) type 4.

*Source*: Reprinted with permission from Mandal et al., 2012.

The two GQDs contain 168 and 132 conjugated carbon atoms and named type 1 and type 2, respectively (Fig. 8.2a and b). Furthermore, to understand the shape effect on the band gap, they also modeled another pair of GQD structures, where 126 and 114 conjugated carbon atoms are present. As shown in Figure 8.2c and d, these two QD structures, respectively, have triangular shape with armchair edges and hexagonal shapes, named type 3 and type 4 GQDs. The bandgaps between the HOMO and LUMO of these GQDs are shown in Figure 8.3. This figure indicates that the type 1 GQD is having the highest number of C atoms (168 C atoms), a bandgap of energy 1.36 eV, whereas the GQD, having the lowest number of C atoms (type 4), shows the highest bandgap (1.60 eV) among these four GQDs. The sizes of these four GQDs are increased by replicating the unit cells in the *XY* plane, while retaining the original symmetries, and their HOMO–LUMO energy gaps were calculated. Thus, GQDs, up to the size of 10 nm, were studied to observe the change in the bandgap with respect to size, shape, and the number of atoms. It is clear from Figure 8.3 that the bandgap is inversely proportional to the size ($L$) of the quantum dot, and number of atoms: larger the size, lesser the difference between HOMO and LUMO. This is the basic difference between graphene QDs and semiconductor QDs, because in conventional semiconductor or inorganic QDs, the energy gaps vary inversely with the square of sizes ($L^2$) of the QDs. However, the HOMO–LUMO gaps of GQDs are mostly independent of the shapes of QDs. Further studies on the DOS of these GQDs reveal that the contribution of core states on bandgap is more as compared to the edge states.

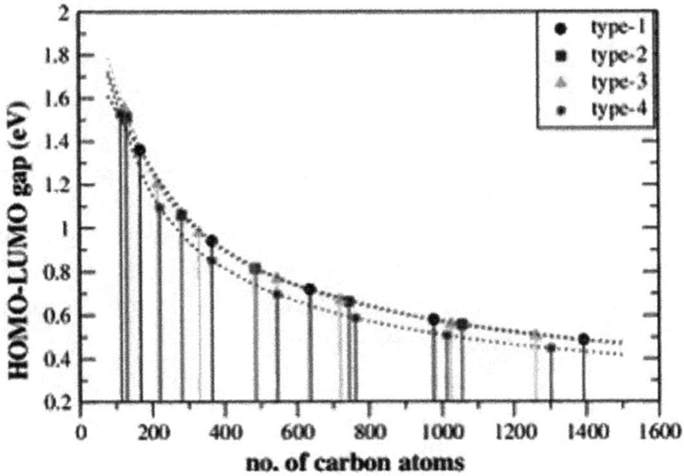


**FIGURE 8.3** The HOMO–LUMO energy gap as a function of total number of conjugated carbon atoms in four different types of GQD structures (Mandal et al., 2012).

*Source*: Reprinted with permission from Mandal et al., 2012.

### *8.2.1 MODULATION OF ELECTRONIC PROPERTIES BY CHEMICAL FUNCTIONALIZATION*

The properties and applications of GQDs can be manipulated through functionalization with organic (Li et al., 2015a; Qian et al., 2013; Qu et al., 2013), inorganic molecules (Mehrdad et al., 2019), biopolymers (Bayoumy et al., 2020), or heavy metal atoms (Abdelsalam et al., 2019), either at the edges or on the surface of GQDs. Abundancy of raw materials, mass-scale production, and easy processing are the main reasons to choose the top-down approach for the synthesis of GQDs. However, most of the oxygen-containing functional groups like hydroxyl (–OH), carboxyl (–COOH), ether (–$OCH_3$), epoxy (–O), and carbonyl (–(C=O)) are retained with GQDs during the synthesis in top-down approach (Li et al., 2013). It is essential to understand the detailed mechanisms of change in the electronic properties of GQDs containing these types of functional groups, which can lead to the synthesis of GQDs, to achieve desired properties for specific applications. In addition to that, the chemical features (type of atoms and bonding), configurations, distributions of functional groups, and optimized system symmetry do also play a major role in determining the optoelectronic features (Cocchi et al., 2012). Feng et al. (2017) studied the effect of functionalization of GQDs, with functional groups containing oxygen, on the electronic properties, through time-dependent density functional theory. The GQD structure, they considered, is made up of 132 carbon atoms and 5 oxygen-containing functional groups (i.e., –OH, –CHO, –COOH, –$OCH_3$, and –COC–), decorated either at the edge sites or on the basal plane of GQD (Figure 8.4a and b). The bandgap of pristine GQD was calculated to be 2.34 eV, whereas the GQDs, functionalized with abovementioned functional groups at the edge site, have a very minor impact on HOMO–LUMO energy gap, that is, only a slight decrease, which is shown in Figure 8.4c. The bandgap of the functionalized GQDs with epoxy, aldehyde, carboxyl, ether, and hydroxyl were calculated to be 2.15, 2.27, 2.30, 2.31, and 2.32 eV, respectively.

On the other hand, the scenario of HOMO–LUMO energy gap is totally changed, when the functional groups are attached to the GQD, on the basal plane (Figure 8.4d). The calculated energy gaps for GQDs, functionalized with –COC–, –CHO, –COOH, –$OCH_3$, and –OH, are 1.86, 0.39, 0.32, 0.37, and 0.36 eV, respectively. These results provide a clear picture that the edge-functionalization of GQD with these functional does not have any significant impact on the modulation of bandgap. However, slight distortion in the charge density, at the edge of GQD, may be the reason for minor change in the bandgap. This study shows that a significant tuning of HOMO–LUMO gap in GQD can be possible by considerable geometry distortion due to functionalization at the surface, which increases the orbital hybridizations.

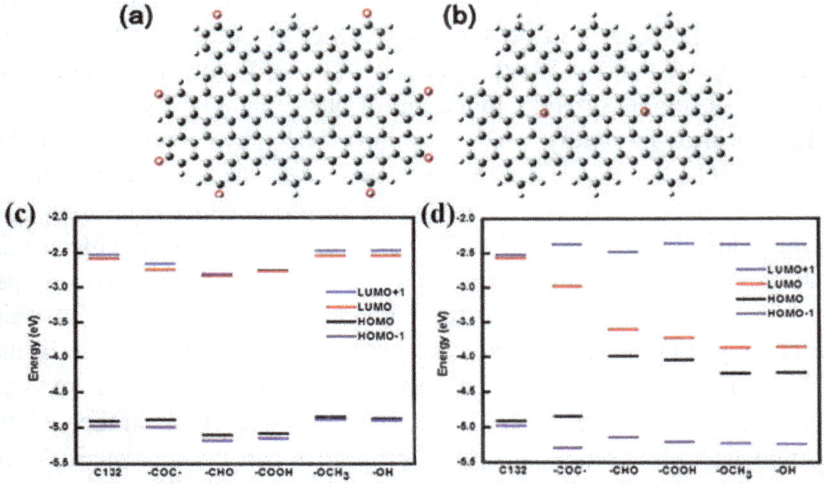


**FIGURE 8.4** (a) Edge functionalization position and (b) surface functionalization positions of GQDs. Energy level diagrams for (c) edge-functionalized and (d) base-functionalized GQDs for different oxygen-containing groups (Feng et al., 2017).

*Source*: Reprinted with permission from Feng et al., 2017.

Larger the hybridization, more reduction in bandgap will be obtained. It is already known that carbon atoms are bonded with each other forming $sp^2$ hybridization in pristine graphene or GQDs, and the optoelectronic properties are mostly determined by these $\pi$ states of $sp^2$ sites. But when the functional group containing oxygen is attached to the basal plane, then carbon in the GQD will change from $sp^2$ to $sp^3$ hybridization by making $\sigma$ bonds with oxygen or carbon of the functional groups. A strong localization occurs due to the presence of $\pi$ and $\pi^*$ electronic levels of $sp^2$ cluster in between $\sigma$ and $\sigma^*$ states of $sp^3$ matrix, leading to the modification of absorption spectra and also the HOMO–LUMO energy gap of the functionalized GQDs. A similar theoretical study has also been performed (Li et al., 2015a), which describes how the HOMO–LUMO energy gap of GQDs can be tuned through edge functionalization. In this work, they compared the results obtained through DFT and the GW method (Hedin 1965). Li and his group considered a hexagonal zigzag-edged GQD as the prototype for the study, which contained 54 carbon atoms, and the edges were saturated with hydrogen atoms. Since the geometry of armchair-edged GQDs can be easily distorted with functionalization, they are not considered as suitable prototypes to study the effect of functional groups on GQDs. Previous studies reported that amino, halogens, hydroxyls, and carbonyl functionalized GQDs can be synthesized through experiments (Feng et al., 2013; Tetsuka et al., 2012; Zhao et al., 2014a, b). Inspired from these experiments, eight different functional groups, namely, hydrogen atom (–H), fluorine (–F), methyl (–$CH_3$), amino (–$NH_2$), hydroxyl (–OH), aldehyde (–CHO), ketone (–$COCH_3$), and carboxyl (–COOH), were chosen to be introduced at the edge sites of the proposed GQD (Fig. 8.5a). To get more insight into the mechanism of the modification of the electronic structures, the functionalized GQDs were categorized into two groups, that is, (i) the GQDs functionalized with carbonyl groups which contain carbon–oxygen double bond (ketone, carboxyl, and aldehyde) and (ii) the GQDs functionalized with the group with the absence of carbon–oxygen double bond. Here, due to very minor structural distortions in the GQDs, its impact on electronic and optical properties can be neglected. When the energy level alignments are compared between –$CH_3$ and –$COCH_3$ or –OH and –COOH group, it is clear from Figure 8.5b that the presence of carbon–oxygen double bond plays a key role for the down-shifting of the energy levels. This may be due to the interactions between π electrons present in C=O and GQD moiety (Cocchi et al., 2012). It was also observed that in the presence of C=O, the LUMO shift more downward than those of HOMO (i.e., LUMO is more hybridized), while an opposite case was observed in the functionalized GQDs without C=O (i.e., HOMO is more hybridized). With the help of frontier orbital hybridization, the mechanism of uneven shifting of energy levels and the subsequent change in the bandgaps can be explained. Amino and aldehyde functionalized GQDs show the lowest HOMO–LUMO gaps, for without and with C=O, respectively. The decrease in the energy gap is a reflection of amount of frontier orbital hybridization. Previous theoretical studies also reported that the GQDs, functionalized at the edges, show reduction in the HOMO–LUMO gaps (Cocchi et al., 2011; Mandal et al., 2015; Sk et al., 2014; Zhao et al., 2014a, b). The distortion induced by the heavy functionalization of GQD can reduce 35–55% of the energy gap (Cocchi et al., 2012). Since the distortions are very small, the shift in the energy levels is mainly due to frontier orbital hybridizations only.

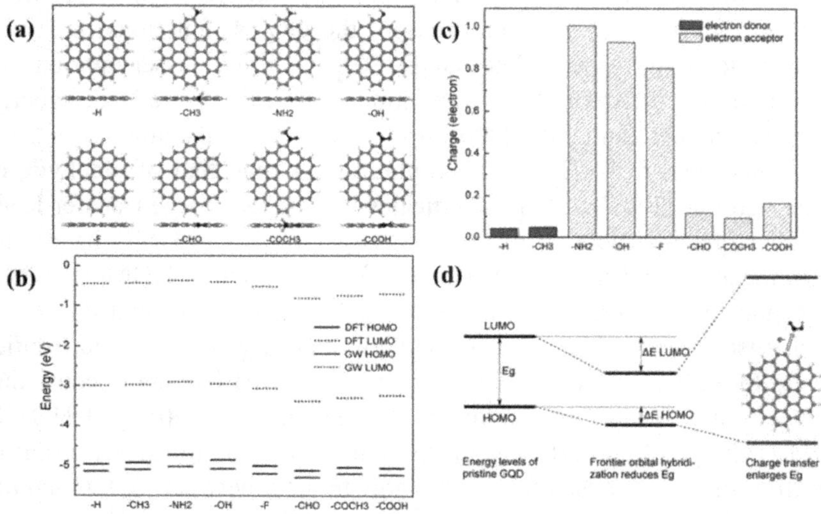


**FIGURE 8.5** (a) Top and side views of different functional groups attached to the ring-shaped GQDs. (b) Energy levels calculated for different functionalized GQDs in DFT and GW approach. (c) Amount of charge transfer between GQDs and functional groups. (d) Competition between frontier orbital hybridization and charge transfer for determining the HOMO–LUMO band gap.

*Source*: Reprinted with permission from Li et al., 2015a. Copyright © 2015, American Chemical Society

The significant increase in the HOMO–LUMO energy gap (approximately 2.7 eV) in GW method than DFT describes the presence of large electron–electron interaction in nanoscale GQD (Li et al., 2015b). Besides these quasi-particle corrections, the energy gap ordering in DFT and GW is quite different. For example, in DFT calculation, the energy gap of GQD–$NH_2$ is close to those of GQD–$COCH_3$ and GQD–COOH. But in GW approach, the calculated energy gap of GQD–$NH_2$ is quite more than those of GQD–$COCH_3$ and GQD–COOH. This deviation is caused due to the charge transfer between GQD and the functional groups. Except –H and –$CH_3$, all the functional groups, are electronegative with respect to pristine GQD (which draws electrons). If the carrier density is increased, then the electron screening effect will be enhanced (Mak et al., 2014; Yang 2011). On the other hand, reduction in the carrier density will lead to a reduction in the screening effect, which enhances the electron–electron interactions. As a result, a large difference between the DFT bandgap and GW bandgap is observed for edge-functionalized GQDs with electron-attracting groups. For the same reason, the GQDs, functionalized with electron-donating groups like GQD–$CH_3$, exhibits lesser amount of quasi-particle correction than the pristine GQDs. From the Bader charge analysis (Henkelman et al., 2006), it is observed that the amount of electron transferred to –$NH_2$ group is quite larger than that of aldehyde or ketone groups (Fig. 8.5c). This indicates that larger charge transfer leads to larger quasi-particle correction and larger HOMO–LUMO band gap. From the above discussions, it is clear that the amount of HOMO–LUMO band gap of the edge-functionalized GQDs can be explained through a mechanism based on the competition between charge transfer and frontier orbital hybridization (Fig. 8.5d). A reduction in the bandgap occurs due to the frontier orbital hybridizations, by asymmetric shifting of HOMO and LUMO levels, whereas bandgap is increased due to the reduction in electron screening caused from the charge transfer between GQDs and the attached functional groups. In the case of GQD–$NH_2$, the simultaneous effect of large charge transfer and strong frontier orbital hybridization makes –$NH_2$ an inefficient functional group for the modulation of bandgap in GQDs. It has also been studied theoretically that the GQDs, passivated with oxygen at edges, show reduced bandgap (Abdelsalam et al., 2018). Hence, changing the oxygen content in a functional group can be considered as an effective way to tune the optical and electronic properties of GQDs.

## 8.3 CONCLUSIONS

In this chapter, an elementary idea about quantum dots has been introduced from quantum mechanics point of view. Then, the advantages of GQDs over conventional semiconductor or inorganic quantum dots are briefly discussed. For a practical nano-electronic device application, tuning of electronic properties of GQDs is quite necessary. The HOMO–LUMO energy gap of GQDs can be tuned by changing the size, whereas it remains almost invariant with change in shape. Furthermore, the functional groups containing carbon–oxygen double bond show better ability in tuning the optoelectronic properties of GQDs, due to strong frontier orbital hybridization, and less charge transfer between GQDs and functional groups. These insightful theoretical mechanisms can provide efficient control on the optoelectronic properties of GQDs to assess their potential applications in various nanodevices.

## KEYWORDS


- **graphene quantum dots**
- **functionalization**
- **quantum confinement**


## REFERENCES


Abdelsalam, H.; Teleb, N. H.; Yahia, I. S.; Zahran, H. Y.; Elhaes, H.; Ibrahim, M. A. First Principles Study of the Adsorption of Hydrated Heavy Metals on Graphene Quantum Dots. *J. Phys. Chem. Solids* **2019,** *130*, 32–40.

Abdelsalam, H.; Elhaes, H.; Ibrahim, M. A. Tuning Electronic Properties in Graphene Quantum Dots by Chemical Functionalization: Density Functional Theory Calculations. *Chem. Phys. Lett.* **2018,** *695*, 138–148.

Arya, H.; Kaul, Z.; Wadhwa, R.; Taira, K.; Hirano, T.; Kaul, S. C. Quantum Dots in Bio-Imaging: Revolution by the Small. *Biochem. Biophys. Res. Commun.* **2005,** *329*, 1173–1177.

Bayoumy, A. M.; Refaat, A.; Yahia, I. S.; Zahran, H. Y.; Elhaes, H.; Ibrahim, M. A.; Shkir, M. Functionalization of Graphene Quantum Dots (GQDs) with Chitosan Biopolymer for Biophysical Applications. *Opt. Quant. Electron.* **2020,** *52*, 16.

Biolatti, E.; D'Amico, I.; Zanardi, P.; Rossi, F. Electro-Optical Properties of Semiconductor Quantum Dots: Application to Quantum Information Processing. *Phys. Rev. B* **2002,** *65*, 1–23.

Brus, L. Quantum Crystallites and Nonlinear Optics. *Appl. Phys. A Solids Surfaces* **1991,** *53*, 465–474.

Chen, W.; Guo, Lv.; Hu, W.; Li, D.; Chen, S.; Dai, Z. Synthesis and Applications of Graphene Quantum Dots: A Review. *Nanotechnol. Rev.* **2018,** *7*, 157–185.

Choi, S. H. Unique Properties of Graphene Quantum Dots and Their Applications in Photonic/Electronic Devices. *J. Phys. D* **2017,** *50*, 103002.

Choudhury, K. R.; So, F.; Kafafi, Z. Colloidal Semiconductor Nanocrystal-Enabled Organic/Inorganic Hybrid Light Emitting Devices. In *Comprehensive Nanoscience and Technology*; Andrews, D. L., Scholes, G. D., G. P. B. T.-C. N.; Wiederrecht, T., Eds., Vols. 1–5; Academic Press: Amsterdam, 2011; pp 183–214.

Cocchi, C.; Prezzi, D.; Ruini, A.; Caldas, M. J.; Molinari, E. Optical Properties and Charge-Transfer Excitations in Edge-Functionalized All-Graphene Nanojunctions. *J. Phys. Chem. Lett.* **2011,** *2*, 1315–1319.

Cocchi, C.; Prezzi, D.; Ruini, A.; Caldas, M. J.; Molinari, E. Electronics and Optics of Graphene Nanoflakes: Edge Functionalization and Structural Distortions. *J. Phys. Chem. C* **2012,** *116*, 17328–17335.

Feng, J.; Dong, H.; Yu, L.; Dong, L. The Optical and Electronic Properties of Graphene Quantum Dots with Oxygen-Containing Groups: A Density Functional Theory Study. *J. Mater. Chem. C* **2017,** *5*, 5984–5993.

Feng, Q.; Cao, Q.; Li, M.; Liu, F.; Tang, N.; Du, Y. Synthesis and Photoluminescence of Fluorinated Graphene Quantum Dots." *Applied Physics Letters* **2013,** *102*, 013111.

Freeman, R.; Gill, R.; Beissenhirtz, M.; Willner, I. Self-Assembly of Semiconductor Quantum-Dots on Electrodes for Photoelectrochemical Biosensing. *Photochem. Photobiol. Sci.* **2007,** *6*, 416–422.

Geim, A. K.; Novoselov, K. S. The Rise of Graphene. *Nat. Mater.* **2007,** *6*, 183–191.

Green, M. Semiconductor Quantum Dots as Biological Imaging Agents. *Angew. Chem. Int. Ed.* **2004,** *43*, 4129–4131.

Han, M. Y.; Özyilmaz, B.; Zhang, Y.; Kim, P. Energy Band-Gap Engineering of Graphene Nanoribbons. *Phys. Rev. Lett.* **2007,** *98*, 206805.

Hedin, L. New Method for Calculating the One-Particle Green's Function with Application to the Electron-Gas Problem. *Phys. Rev.* **1965,** *139*, A796.

Henkelman, G.; Arnaldsson, A.; Jónsson, H. A Fast and Robust Algorithm for Bader Decomposition of Charge Density. *Comput. Mater. Sci.* **2006,** *36*, 354–360.

Hetsch, F.; Xu, X.; Wang, H.; Kershaw, S. V.; Rogach, A. L. Semiconductor Nanocrystal Quantum Dots as Solar Cell Components and Photosensitizers: Material, Charge Transfer, and Separation Aspects of Some Device Topologies. *J. Phys. Chem. Lett.* **2011,** *2*, 1879–1887.

Hildebrandt, N. Biofunctional Quantum Dots: Controlled Conjugation for Multiplexed Biosensors. *ACS Nano* **2011,** *5*, 5286–5290.

Iijima, S. Helical Microtubules of Graphitic Carbon. *Nature* **1991,** *354*, 56–58.

Jin, Z.; Hildebrandt, N. Semiconductor Quantum Dots for in Vitro Diagnostics and Cellular Imaging. *Trends Biotechnol.* **2012,** *30*, 394–403.

Karpan, V. M.; Khomyakov, P. A.; Starikov, A. A.; Giovannetti, G.; Zwierzycki, M.; Talanana, M.; Brocks, G.; Brink, J.; Kelly, P. J. Theoretical Prediction of Perfect Spin Filtering at Interfaces between Close-Packed Surfaces of Ni or Co and Graphite or Graphene. *Phys. Rev. B* **2008,** *78*, 195419.

Kroto, H. W.; Heath, J. R.; O'Brien, S. C.; Curl, R. F.; Smalley, R. E. $C_{60}$: Buckminsterfullerene. *Nature* **1985,** *318*, 162–163.

Li, L.; Wu, G.; Yang, G.; Peng, J.; Zhao, J.; Zhu, J. J. Focusing on Luminescent Graphene Quantum Dots: Current Status and Future Perspectives. *Nanoscale* **2013,** *5*, 4015–4039.

Li, Y.; Shu, H.; Niu, X.; Wang, J. Electronic and Optical Properties of Edge-Functionalized Graphene Quantum Dots and the Underlying Mechanism. *J. Phys. Chem. C* **2015a,** *119*, 24950–24957.

Li, Y.; Shu, H.; Wang, S.; Wang, J. Electronic and Optical Properties of Graphene Quantum Dots: The Role of Many-Body Effects. *J. Phys. Chem. C* **2015b,** *119*, 4983–4989.

Liu, R.; Wu, D.; Feng, X.; Müllen, K. Bottom-up Fabrication of Photoluminescent Graphene Quantum Dots with Uniform Morphology. *J. Am. Chem. Soc.* **2011,** *133*, 15221–15223.

Mak, K, F.; Jornada, F. H. D.; He, K.; Deslippe, J.; Petrone, N.; Hone, J.; Shan, J.; Louie, S. G.; Heinz, T. F. Tuning Many-Body Interactions in Graphene: The Effects of Doping on Excitons and Carrier Lifetimes. *Phys. Rev. Lett.* **2014,** *112*, 207401.

Mandal, B.; Sarkar, S.; Sarkar, P. Exploring the Electronic Structure of Graphene Quantum Dots. *J. Nanoparticle Res.* **2012,** *14*, 1–8.

Mandal, B.; Sarkar, S.; Sarkar, P. Theoretical Studies on Understanding the Feasibility of Porphyrin-Sensitized Graphene Quantum Dot Solar Cell. *J. Phys. Chem. C* **2015,** *119*, 3400–3407.

Martynenko, I. V.; Litvin, A. P.; Purcell-Milton, F.; Baranov, A. V.; Fedorov, A. V.; Gun'ko. Y. K. Application of Semiconductor Quantum Dots in Bioimaging and Biosensing. *J. Mater. Chem. B* **2017,** *5*, 6701–6727.

Masut, R. A.; Tran, C. A.; Beaudoin, M.; Leonelli, R. Strained InAs/InP Quantum Wells and Quantum Dots for Optoelectronic Device Applications. In *Circular-Grating Light-Emitting Sources*; Najafi, S. I., Fallahi, M., Peyghambarian, N., Eds., Vol. 2398; SPIE, 1995; pp 116–24.

Mehrdad, V. B.; Pourhashem, S.; Sedghi, M.; Vaezi, Z.; Givi, B. S.; Rashidi, A.; Manesh, H. N. A Novel Aspect of Functionalized Graphene Quantum Dots in Cytotoxicity Studies. *Toxicol. In Vitro* **2019,** *61*, 104649.

Novoselov, K. S.; Geim, A. K.; Morozov, S. V.; Jiang, D.; Zhang, Y.; Dubonos, S. V.; Grigorieva, I. V.; Firsov, A. A. Electric Field in Atomically Thin Carbon Films. *Science* **2004,** *306*, 666–669.

Nozik, A. J.; Beard, M. C.; Luther, J. M.; Law, M.; Ellingson, R. J.; Johnson, J. C. Semiconductor Quantum Dots and Quantum Dot Arrays and Applications of Multiple Exciton Generation to Third-Generation Photovoltaic Solar Cells. *Chem. Rev.* **2010,** *110*, 6873–6890.

Qian, Z.; Ma, J.; Shan, X.; Shao, L.; Zhou, J.; Chen, J.; Feng, H. Surface Functionalization of Graphene Quantum Dots with Small Organic Molecules from Photoluminescence Modulation to Bioimaging Applications: An Experimental and Theoretical Investigation. *RSC Adv.* **2013,** *3*, 14571–14579.

Qu, Z. B.; Zhou, X.; Gu, L.; Lan, R.; Yu, D.; Shi, G. Boronic Acid Functionalized Graphene Quantum Dots as a Fluorescent Probe for Selective and Sensitive Glucose Determination in Microdialysate. *Chem. Commun.* **2013,** *49*, 9830–9832.

Raffaelle, R. P.; Castro, S. L.; Hepp, A. F.; Bailey, S. G. Quantum Dot Solar Cells. *Progress Photovoltaics Res. App.* **2002,** *10*, 433–439.

Reithmaier, J. P.; Somers, A.; Kaiser, W.; Deubert, S.; Gerschütz, F.; Forchel, A.; Parillaud, O.; Krakowski, M.; Alizon, R.; Hadass, D.; Bilenca, A.; Dery, H.; Mikhelashvili, V.; Eisenstein, G.; Gioannini, M.; Montrosset, I.; Berg, T. W.; Poel, M.; Mørk, J.; Tromborg, B. Semiconductor Quantum Dots Devices: Recent Advances and Application Prospects. *Phys. Status Solidi (B)* **2006,** *243*, 3981–3987.

Sahoo, M. R.; Kushwaha, A. K.; Pati, R.; Ajayan, P. M.; Nayak, S. K. First-Principles Study of a Vertical Spin Switch in Atomic Scale Two-Dimensional Platform. *J. Magnet. Magnet. Mater.* **2019,** *484*, 462–471.

Shao, L.; Gao, Y.; Yan, F. Semiconductor Quantum Dots for Biomedicial Applications. *Sensors* **2011,** *11*, 11736–11751.

Shen, J.; Zhu, Y.; Yang, X.; Li, C. Graphene Quantum Dots: Emergent Nanolights for Bioimaging, Sensors, Catalysis and Photovoltaic Devices. *Chem. Commun* **2012,** *48*, 3686–3699.

Sk Alam, M.; Ananthanarayanan, A.; Huang, L.; Lim, K. H.; Chen, P. Revealing the Tunable Photoluminescence Properties of Graphene Quantum Dots. *J. Mater. Chem. C* **2014,** *2*, 6954–6960.

Skolnick, M. S.; Mowbray, D. J. Self-Assembled Semiconductor Quantum Dots: Fundamental Physics and Device Applications. *Annu. Rev. Mater. Res.* **2004,** *34*, 181–218.

Sun, J.; Lowe, S. E.; Zhang, L.; Wang, Y.; Pang, K.; Wang, Y.; Zhong, Y.; Liu, P.; Zhao, K.; Tang, Z.; Zhao, H. Ultrathin Nitrogen-Doped Holey Carbon@Graphene Bifunctional Electrocatalyst for Oxygen Reduction and Evolution Reactions in Alkaline and Acidic Media. *Angew. Chem. Int. Ed.* **2018,** *57*, 16511–16515.

Sun, Y. Li, C.; Shi, G. Nanoporous Nitrogen Doped Carbon Modified Graphene as Electrocatalyst for Oxygen Reduction Reaction. *J. Mater. Chem.* **2012,** *22*, 12810–12816.

Tetsuka, H.; Asahi, R.; Nagoya, A.; Okamoto, K.; Tajima, I.; Ohta, R.; Okamoto, A. Optically Tunable Amino-Functionalized Graphene Quantum Dots. *Adv. Mater.* **2012,** *24*, 5333–5338.

Tian, J.; Cao, G. Semiconductor Quantum Dot-Sensitized Solar Cells. *Nano Rev.* **2013,** *4*, 22578.

Todd, K.; Chou, H. T.; Amasha, S.; Gordon, D. G. Quantum Dot Behavior in Graphene Nanoconstrictions. *Nano Lett.* **2009,** *9*, 416–421.

Troiani, F.; Hohenester, U.; Molinari, E. Exploiting Exciton-Exciton Interactions in Semiconductor Quantum Dots for Quantum-Information Processing. *Phys. Rev. B* **2000,** *62*, R2263–R2266.

Wang, K. L.; Cha, D.; Liu, J.; Chen, C. Ge/Si Self-Assembled Quantum Dots and Their Optoelectronic Device Applications. *Proc. IEEE* **2007,** *95*, 1866–1883.

Wise, F. W. Lead Salt Quantum Dots: the Limit of Strong Quantum Confinement. *Acc. Chem. Res.* **2000,** *33*, 773–780.

Wu, Y.; Li, X.; Steel, D.; Gammon, D.; Sham, L. J. Coherent Optical Control of Semiconductor Quantum Dots for Quantum Information Processing. *Phys. E* **2004,** *25*, 242–248.

Xing, Y.; Xia, Z.; Rao, J. Semiconductor Quantum Dots for Biosensing and in Vivo Imaging. *IEEE Trans. Nanobiosci.* **2009,** *8*, 4–12.

Yan, X.; Cui, X.; Shi L. Synthesis of Large, Stable Colloidal Graphene Quantum Dots with Tunable Size. *J. Am. Chem. Soc.* **2010,** *132*, 5944–5945.

Yan, Y.; Gong, J.; Chen, J.; Zeng, Z.; Huang, W.; Pu, K.; Liu, J.; Chen, P. Recent Advances on Graphene Quantum Dots: From Chemistry and Physics to Applications. *Adv. Mater.* **2019,** *31*, 1808283 (1–22).

Yang, Li. Excitonic Effects on Optical Absorption Spectra of Doped Graphene. *Nano Lett.* **2011,** *11*, 3844–3847.

Yoffe, A. D. Low-Dimensional Systems: Quantum Size Effects and Electronic Properties of Semiconductor Microcrystallites (Zero-Dimensional Systems) and Some Quasi-Two-Dimensional Systems. *Adv. Phys.* **1993,** *42*, 173–262.

Yu, W. W. Semiconductor Quantum Dots: Synthesis and Water-Solubilization for Biomedical Applications. *Expert Opin. Biol. Ther.* **2008,** *8*, 1571–1581.

Zhang, Z.; Zhang, J.; Chen, N.; Qu, L. Graphene Quantum Dots: An Emerging Material for Energy-Related Applications and Beyond. *Energy Environ. Sci.* **2012,** *5*, 8869–8890.

Zhao, J.; Tang, L.; Xiang, J.; Ji, R.; Yuan, J.; Zhao, J.; Yu, R.; Tai, Y.; Song, L. Chlorine Doped Graphene Quantum Dots: Preparation, Properties, and Photovoltaic Detectors. *Appl. Phys. Lett.* **2014a,** *105*, 111116.

Zhao, M.; Yang, F.; Xue, Y.; Xiao, D.; Guo, Y. Effects of Edge Oxidation on the Stability and Half-Metallicity of Graphene Quantum Dots. *ChemPhysChem* **2014b,** *15*, 157–164.

Zhou, J.; Yang, Y.; Zhang, C. Y. Toward Biocompatible Semiconductor Quantum Dots: From Biosynthesis and Bioconjugation to Biomedical Application. *Chem. Rev.* **2015,** *115*, 11669–11717.

**CHAPTER 9**

# Carborane Clusters for Promoting Medicinal Applications

BIBHUTI BHUSAN JENA[1] and MANAS R. PATTANAYAK[2]

[1]*Department of Chemistry, Ravenshaw University, Cuttack, Odisha, India*

[2]*Syngene International LTD, Medchal, Hyderabad, India*

## ABSTRACT

Carboranes belong to a specific group of electron delocalized clusters, which constitute both boron and carbon atoms. Typically, carboranes with icosahedral structures, have good chemical and thermal stability. The lipophilicity or amphiphilicity properties make carborane particularly suitable to act as a hydrophobic pharmacophore. Carborane clusters can find applications as enhancers of hydrophobic interactions between pharmaceuticals and their receptors, which imparts improved *in vivo* stability and increases the bioavailability of the relevant compounds. The novel carboranyl cholesterol mimics are excellent building blocks, for the construction of various boronated liposomes (including both nontargeted and receptor-targeted types), which are crucial for high-end therapeutics, for example, boron neutron capture therapy (BNCT) (for the treatment of cancer) and for other applications as well. In the area of anticancer drugs, inhibitors of platelet aggregation, and modulators of important hormone receptors, carboranes are attracting fast-growing research interest in the quest for novel drugs that could overcome the drawbacks, associated with the currently available products. Carboranes are very useful pharmacophores

New Forms of Carbon: Nanocarbons. Aneeya Kumar Samantara & Satyajit Ratha (Eds.)

for the design of transthyretin amyloidosis and $\alpha$-human thrombin inhibitors. Modification of adenosine with a carborane cluster had the capacity to open the way for biological screening, evaluation of the new class of bioorganic–inorganic conjugates, modulation of human blood platelet activities, and boron carrier for BNCT.

## 9.1 INTRODUCTION

The term, "chemotherapy," was made familiar to the scientific fraternity, by Paul Ehrlich, in the early 20th century. He had worked extensively on therapeutic treatments, based on chemical substances ("magic bullets," in German Zauberkugel) that are characterized by properties such as selective affinity or toxicity, targeting specific biological entities, namely, pathogens. In 1908, he was conferred with the Nobel Prize in Physiology and Medicine for his discoveries and was considered the pioneer/father of chemotherapy. Chemical-based therapies are worldwide approved for the treatment of tumors, viral, parasitic, inflammation, pain, and a combination thereof over the last 100 years. The commercially available drugs, for example, organic compounds like doxorubicin, are structurally very small molecules (MWo 1000 Da), a topoisomerase II inhibitor, which can treat a wide range of cancerous conditions that include leukemia, Hodgkin's lymphoma, breast cancer, etc. Further belonging to this category, are chloroquine, which is a nonsteroidal anti-malarial drug, and the nonsteroidal anti-inflammatory drug (NSAIDs), for example, ibuprofen (Imming, 2015). In addition to this, metal-based drugs and a few platinum-containing complexes have been approved for the treatment of a range of cancerous conditions (i.e., ovarian, testicular, cervical, breast, etc.). Cisplatin, oxaliplatin, and carboplatin are the frontline examples of this (Wilson et al., 2011).

In the last 20 years, hybrid organic–inorganic compounds, as novel small molecule-based drugs, which contain boron–carbon–hydrogen clusters, for example, dicarba-*closo*-dodecaboranes (otherwise known as carbaboranes or carboranes), have drawn great interest. Carboranes are electronic delocalized clusters ranging from 4-vertex (sub-icosahedral) to 12-vertex (icosahedral) and 14-vertex (supra-icosahedral) structures (Greme, 2016). In the present book chapter, we largely emphasize the 12-vertex clusters, which are very robust in nature and are characterized by a three-dimensional

(3D) structure (Poater et al., 2014), high stability (chemical, thermal, and photochemical) (Cabrera-Gonzalez et al., 2017; Nunez et al., 2016), good electron-withdrawing property (Nunez et al., 2006), high hydrophobicity (Goszczynski et al., 2017), and low toxicity (Wu et al., 2019; Scholz et al., 2011). They act in a different manner as compared to the phenyl ring structure (Teixidor et al., 2005). They primarily exist in three isomeric forms: *closo*-1,2-$C_2B_{10}H_{12}$ (*ortho*-carborane **1**), *closo*-1,7-$C_2B_{10}H_{12}$ (*meta*-carborane **2**), and *closo*-1,12-$C_2B_{10}H_{12}$ (*para*-carborane **3**) and all are very stable (Fig. 9.1). Dicarba-*closo*-dodecacarboranes ($C_2B_{10}H_{12}$) (**1–3**) were well characterized by an exceptional hydrophobic character. The high hydrophobicity of many carborane clusters and their derivatives can be explained because of the slightly polarized hydrogen atoms on the boron atoms and their "hydride-like" character. This prevents them to form hydrogen bonds and makes the carborane clusters hydrophobic in nature. Unconventional hydrogen bondings are usually observed in the case of boranes, due primarily to the electronegative nature of the constituent hydrogen atoms. These bonds are referred to as the dihydrogen bonds and are weaker when compared to the conventional/classical hydrogen bonds, and a net repulsive effect toward surrounding water molecules prevails making the carborane clusters hydrophobic in nature. Again, depending on the position of the carbon atoms in the carborane cage the hydrophobicity nature of carboranes also varies with the dipole moment. However, the functionalizations of hydrophobic *ortho*-, *meta*-, and *para*-carboranes (**1–3**) are much easier than the hydrophilic dodecaborate ion which prevents its wider application in synthetic and medicinal chemistry (Nishiyama et al., 2005; Sivaev and Bregadze, 2009; Lesnikowski, 2007a, b).

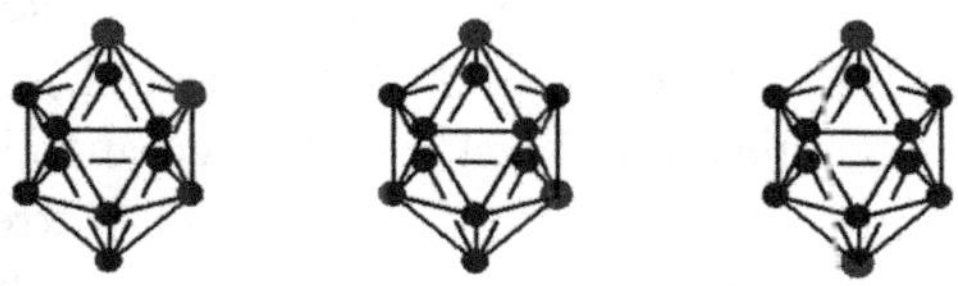

*Ortho*-carborane 1,2-$C_2B_{10}H_{12}$ **1**  *Meta*-carborane 1,7-$C_2B_{10}H_{12}$ **2**  *Para*-carborane 1,12-$C_2B_{10}H_{12}$ **3**

**FIGURE 9.1** Structure of carborane clusters.

While the carborane-based drugs are yet to get approval, most of them have shown a strong candidature, for therapeutic application, for example, boron neutron capture therapy (BNCT) (Barth et al., 2018), enzyme inhibitors for potent (Lesnikowski, 2016) and selective inhibition of HIV-1 protease, similar to the case of carborane-based epidermal growth factor receptor (EGFR) inhibitors (Couto et al., 2018). The high boron content with thermally stability makes carborane a boron carrier for BNCT. The basic knowledge about toxicity and pharmacokinetics of particular classes of carborane-containing compounds studied as boron carriers for BNCT can help in the development of the other biological applications of boron compounds. Moreover, the medicinal chemistry of carborane-based drugs is quite similar to the organic/organometallic drugs, both of which are based on the target-vector recognition approaches (Scholz et al., 2011). The unique medicinal application of carboranes in the field of medicine is because of the properties characterized by (i) highly hydrophobic and low polarity hydridic B–H groups, which is effective for improving the transportation across the cell membranes and the blood–brain barrier (BBB) (Issa et al., 2011), (ii) inorganic nature, which allows them to remain unaffected by the conventional enzymatic degradation processes, thus promoting drug stability under various biological conditions, (iii) the added feature of 3D orthogonal structure provides the cluster with much better structural flexibility than the benzene systems, which are limited to two dimensions, and (iv) the multitude of possible noncovalent interactions with biological targets: B-H···H–X (X = N, C, and S) dihydrogen bonds, B–H···$Na^+$ bridging interactions, and C–H···X (X = O, N, S, F, π system) hydrogen bonds (Scholz et al., 2011). The self-assembly process, in an aqueous solution, spontaneously forms nano- to micrometer-sized particles, which is an added advantage with the carborane clusters (Fernandez-Alvarez et al., 2018). The nanometer-sized particles potentially represent a very efficient way for the selective delivery of chemotherapeutics to cancer cells (Nakamura et al., 2016). Till-now, the self-assembly properties have been investigated and characterized for only the anionic boron clusters (COSAN species) (Tarres et al., 2015) and neutral carborane-containing compounds (He et al., 2017). However, studies on the spontaneous behavior along with the biological activity of carborane-containing drugs are quite limited.

Generally, carboranes tend to replace the hydrophobic components, such as a phenyl ring or adamantane, when introduced into the structures of bioactive compounds, subsequently enhancing the interactions with the receptors of those compounds (Julius et al., 2007; Nakamura et al., 2013).

Carborane clusters play a key role in the modification of biomolecules because they can be functionalized easily, and these biomolecules offer high resistance toward degradation and have improved stability as well as bioavailability of bioactive molecules (Issa et al., 2011). These applications of carboranes make it a best candidate for pharmacophores in nonsteroidal anti-inflammatory drugs (Scholz et al., 2011), anti-folates (Reynolds et al., 2007), carbonic anhydrase inhibitors (Brynda et al., 2013), thrombin inhibitors (Page et al., 2008), hypoxia-inducible factor (HIF) inhibitors (Nakamura et al., 2013; Ban et al., 2010), purinergic receptor ligands (Bednarska et al., 2010a, b; Olejniczak et al., 2007a, b; Wilkinson et al., 2014), lidocaine (Kracke et al., 2015), antiviral drugs (Olejniczak et al., 2013; Adamska et al., 2012) and other bioactive molecules.

## 9.2 STRUCTURE OF CARBORANE

The incorporation of one or more carbon atom(s) into the structural framework of polyhedral borane is called a carborane. A borane can be converted into *ortho*-carborane **1** (Fig. 9.2) by replacing two (B–H) fragments with the help of two (C–H) fragments because of the same isoelectronic structure and solubility. Here both fragment possesses similar frontier molecular orbital (FMO) properties including energy, symmetry, extent in space, and electronic occupation. Finally, replacing two (B–H) units with two (C–H) units produces a neutral compound with the general formula $C_2B_{n-2}H_n$. The nucleophilic attack occurs in a polyhedral carborane at the boron closest to the carbon, while the electrophilic attack is preferred at the boron furthest away from the carbon.

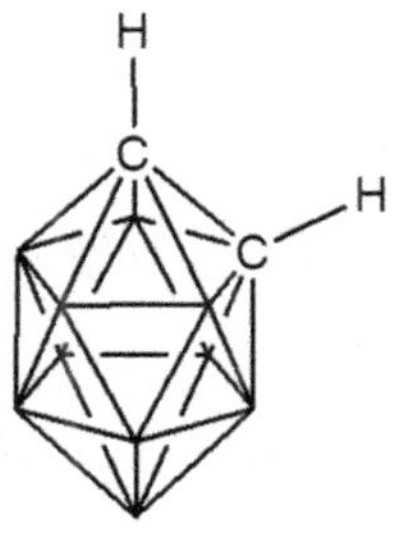

**1**

**FIGURE 9.2** The structure of 1,2-*closo*-carborane or *ortho*-carborane.

## 9.3 NOMENCLATURE OF CARBORANE

The term carborane originates from the IUPAC-approved designation of carbaborane. It was used to describe the molecules composed as clusters of carbon and boron atoms. According to Wade"s Rules, the prefix "*closo*" is used in the polyhedra that contain all of the expected vertices. Numbering of the cage vertices in *closo*-carborane begins at the apex atom and in a clockwise direction the successive atoms are listed in "rings". Here the carbon atoms are given the lowest numbers and are designated in the formula front (Dilthey, 1921). So, the proper name for *ortho*-carborane **1** is 1, 2-dicarba-*closo*-dodecaborane. Furthermore, the substitution of the cage is designated in the name by a numerical suffix (i.e., 1-(1,2-dicarba-*closo*-dodecaborane)-(pendant group name) for C-substituted *ortho*-carborane) (Wu et al., 1997; Nabakka et al., 1998). Alternatively, in a suffixal manner, the point of cage substitution can be indicated at the end of a polyhedral name (i.e., (1, 2-dicarba-*closo*-dodecaboran-1-yl)-(pendant group name). Similarly, the corresponding nomenclature for *nido*-carborane **4** would be 7,8-dicarba-*nido*-dodecaborane since the numbering of the cage atom begins at the vertex opposite to the open face and the carbon atoms are given the lowest possible numerical designation. The total discussions are outlined in Figure 9.3.

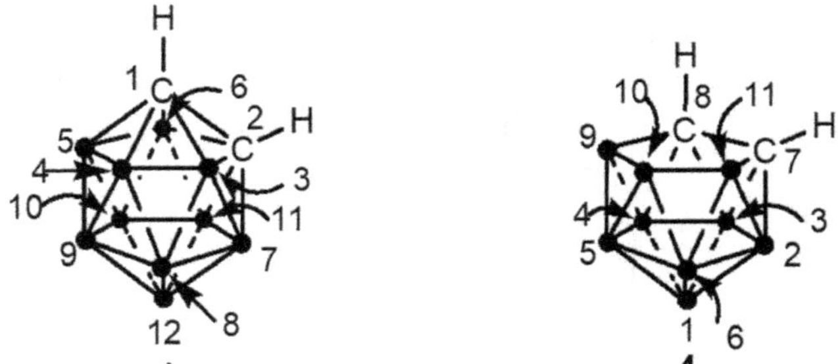


**1** **4**

**FIGURE 9.3** Numbering system for *closo* and *nido* cage atoms.

## 9.4 PREPARATION OF 1,2-*CLOSO*-$C_2B_{10}H_{12}$

Dicarba-*closo*-dodecaboranes are characterized by the presence of boron, carbon, and hydrogen atoms, with the general formula, $C_2B_nH_{n+2}$. The first example of an icosahedral carborane ($C_2B_{10}H_{12}$) was reported in 1963 (Heying et al., 1963). Decaborane **5** is the main starting material for the synthesis of icosahedral carborane. The synthesis of *ortho*-carborane **1** was reported first in 1957, prepared by insertion of acetylenes **7** into decaborane **5** mediated by Lewis bases. This is a two process involves two steps reaction (Scheme 9.1). At first activation of decaborane occurs by reacting with a weak base ($CH_3CN$, $R_3N$, and $R_2S$) to generate $B_{10}H_{12}L_2$, and then in the second step insertion of an alkyne into the open face of the borane happens which replaces both Lewis bases and the two remaining bridging hydrogen atoms.

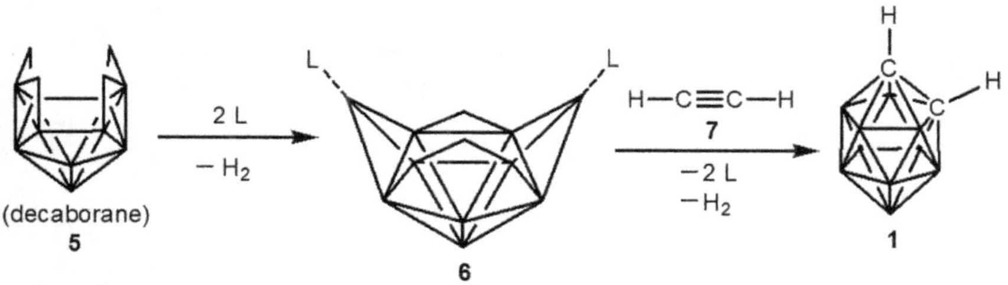


**SCHEME 9.1** Synthesis of 1,2-*closo*-$C_2B_{10}H_{12}$ via acetylene insertion.

Formation of *closo*-carboranes proceeds via three types of reactions:

1. Removal of hydrogen and introduction of various functional groups at the cage carbon atoms and the boron atoms.
2. Bases deboronate the cage structure and followed metalation with suitable metal fragments.
3. Lastly, polyhedral expansion with Na/Li metal in the presence of naphthalene followed by recapitulation with suitable metal fragments or boron fragments.

Furthermore, functional group can be added into the carbon atoms by using the appropriately substituted alkyne (internal and external) derivatives with $B_{10}H_{12}L_2$ (Wu et al., 1998). The decision as to which procedure is best depends upon the nature of the functional groups presents in the synthetic target compounds (Wiesboeck and Hawthorne, 1964; Lee et al. 1999) (see Scheme 9.2).

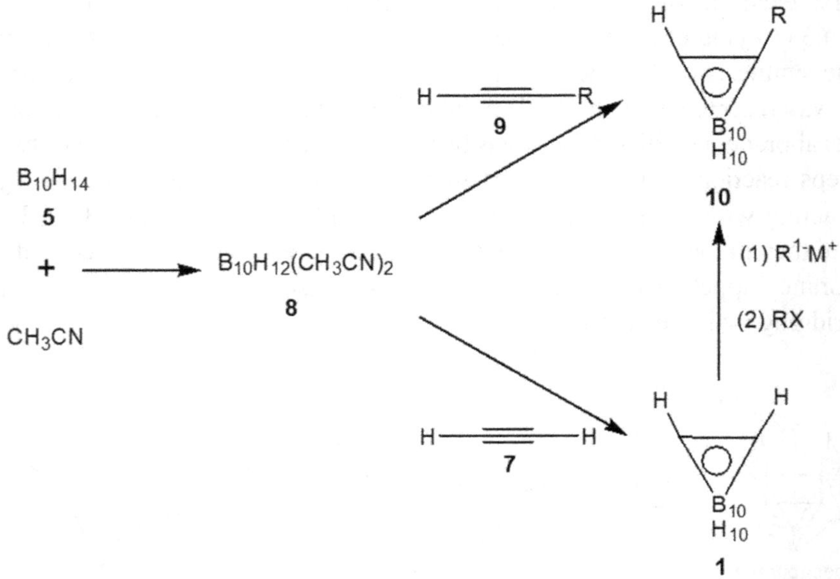

**SCHEME 9.2** Synthetic route to substituted carborane.

## 9.5 CHARACTERIZATION OF CARBORANE

Carboranes can be characterized by using different techniques such as X-ray crystallography, nuclear magnetic resonance, infrared spectroscopy, etc. (Leites, 1992; Hermanek, 1992). Infrared spectroscopies providing useful information during the characterization of carborane include B–H stretching around 2600 and 2520 $cm^{-1}$ for *closo*- and *nido*-carborane, respectively. Moreover, C–H stretching appears at 3079, 3070, and 3065 $cm^{-1}$ for *ortho*- (**1**), *meta*- (**2**), and *para*-carborane (**3**), respectively. Regarding nuclear magnetic resonance (NMR), the protons directly attached to the boron atoms of the carborane cage show a broad signal between 3.00 and –0.75 ppm in the proton spectra of carborane. Again, protons attached to the carbon atoms appear as well-defined singlets between 2.0 and 3.5 ppm. Doublets between –2.5 and–3.0 ppm appear for carborane in the *nido* form. For boron NMR, the most abundant isotope of boron ($^{11}B$, 80.2% abundancy) is used only (Hermanek, 1992). The boron spectrum consists of a series of signals which appear as doublets due to coupling with hydrogen atoms ($J$ = 125−205 Hz). These doublets become singlets when the spectrum is decoupled from the proton. Hence, NMR is extremely useful in identifying carborane cages clusters.

## 9.6 CARBORANE ISOMERIZATION

1, 2-*closo*-$C_2B_{10}H_{12}$ (*ortho*-carborane **1**) quantitatively isomerized to 1,7-*closo*-$C_2B_{10}H_{12}$ (*meta*-carborane **2**) at very high temperatures (Grafstein and Dvorak, 1963). One year later, the 1,7-*closo*-$C_2B_{10}H_{12}$ (*meta*-carborane **2**) to 1,12-*closo*-$C_2B_{10}H_{12}$ (*para*-carborane **3**) rearrangement was also observed by Albeit in lower yields, resulting from decomposition due to the increased temperature needed (Papetti and Heying, 1964). The driving force behind these rearrangements is believed to be the separation of the relatively negative carbon atoms. All three isomers are similar to each other in appearance (white color solids), odor, and solubility but significant differences in IR and NMR spectroscopic spectra. The reverse isomerization was also observed soon after via successive two-electron reductions and oxidations through high-temperature thermal rearrangement under an inert atmosphere (Scheme 9.3). At 400 °C, *ortho*-carborane **1** converts to *meta*-carborane **2** which is in rearrange to *para*-carborane **3** at 600 °C. Again, a two-electron reduction of the 1,12-isomer followed by reduction results in the 1,7-isomer formed. Likewise, the same redox process conversion of 1,7-isomer to 1,2-isomer happens (Zakharkin et al., 1970). To date, the actual mechanism of these thermal isomerization processes is still not fully established but many theories have been postulated and have been studied experimentally and computationally.

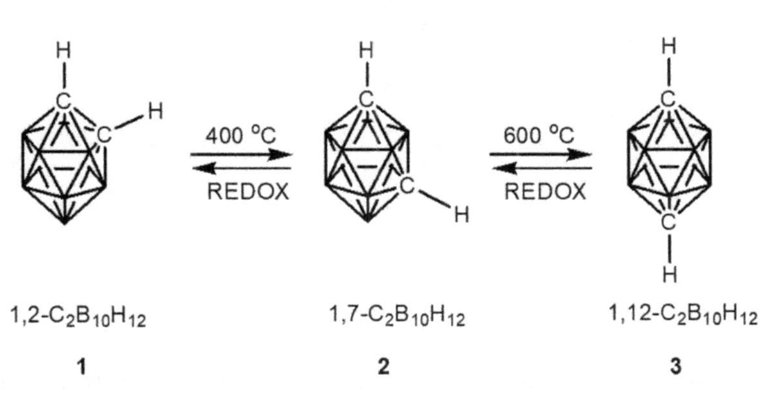


**SCHEME 9.3** Thermal and electrochemical carborane rearrangement.

## 9.7 CARBORANE CHEMISTRY FOR MEDICINAL APPLICATION

Carboranes such as icosahedral dicarba-*closo*-dodecacarboranes ($C_2B_{10}H_{12}$) (**1–3**) were characterized by exceptional lipophilicity or amphiphilicity character. The lipophilicity or amphiphilicity properties make them particularly suitable as a hydrophobic component in biologically active molecules, particularly those which interact hydrophobically with molecular targets in the cellular environment. The high hydrophobicity nature of many carborane clusters and their derivatives can be explained by the presence of partial negative charge (density differs for different hydrogens and depends upon the type of cluster) located on boron-bound hydrogen atoms in B–H groups. Such "hydride-like" character prevents them from forming classical hydrogen bonds, which causes a lipophilic/hydrophobic character of the carborane clusters (Fig. 9.4).

Consequently, the hydrophobicity of dicarbadodecaborane ($C_2B_{10}H_{12}$) (**1–3**) isomers increases in the order: 1,2-dicarba-*closo*-dodecaborane (*ortho*-carborane) **1** < 1,7-dicarba-*closo*-dodecaborane (*meta*-carborane) **2**, 1,12-dicarba-*closo*-dodecaborane (*para*-carborane) **3** which is showed in Figure 9.5. Moreover, the removal of the most electrophilic boron atom from the hydrophobic component of neutral *closo*-carborane produces more hydrophilic anionic *nido*-carboranes ($C_2B_9H^-_{12}$).

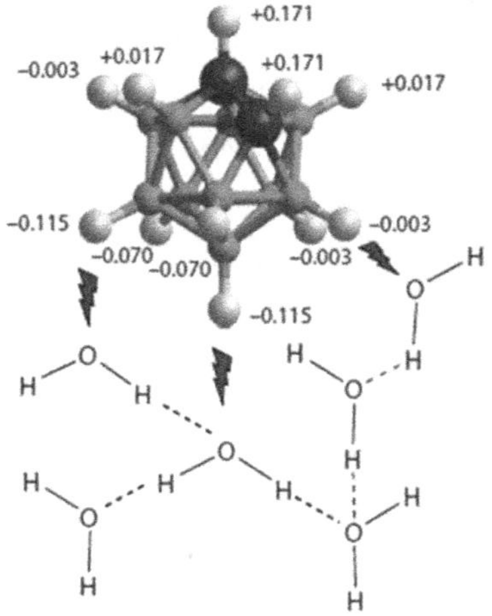


**FIGURE 9.4** Hydridic character of boron cage hydrogens cause inability to contribute in hydrogen bonding with water and participates in hydrophobic character of carboranes.

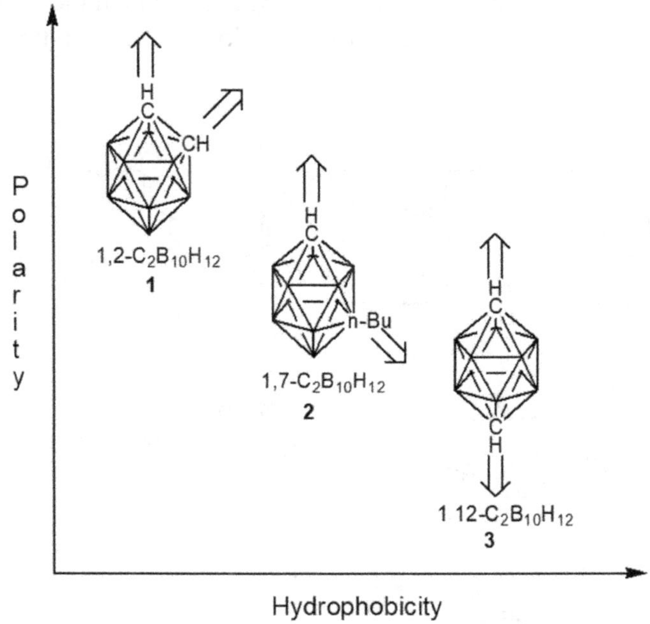


**FIGURE 9.5** An effect of dipole moment on hydrophobicity of dicarbadodecaboranes

## 9.7.1 MEDICAL APPLICATIONS OF CARBORANE CLUSTERS

The cage-like with ball-shaped dicarbadodecaborane structure mimics well the dodecahedral volume created by the rotation of the planar phenyl ring over 360°. However, it is bigger and has a much more hydrophobic moiety as compared to the phenyl ring. Its higher volume and surface area may explain the high efficacy interactions of carborane-containing biomolecules with hydrophobic domains of proteins such as receptors. These advantages first help with the modification of amino acids and peptides.

### 9.7.1.1 RETINOID RECEPTOR LIGANDS HAVING A DICARBA-CLOSO-DODECABORANE AS A HYDROPHOBIC MOIETY

Retinoids and their analogs were used as chemopreventive and therapeutic agents in the field of dermatology and oncology. The biological activities of retinoids were mediated by binding to and activation of the retinoic acid receptors (RARs), with subsequent modulation of the gene transcription by the complex.

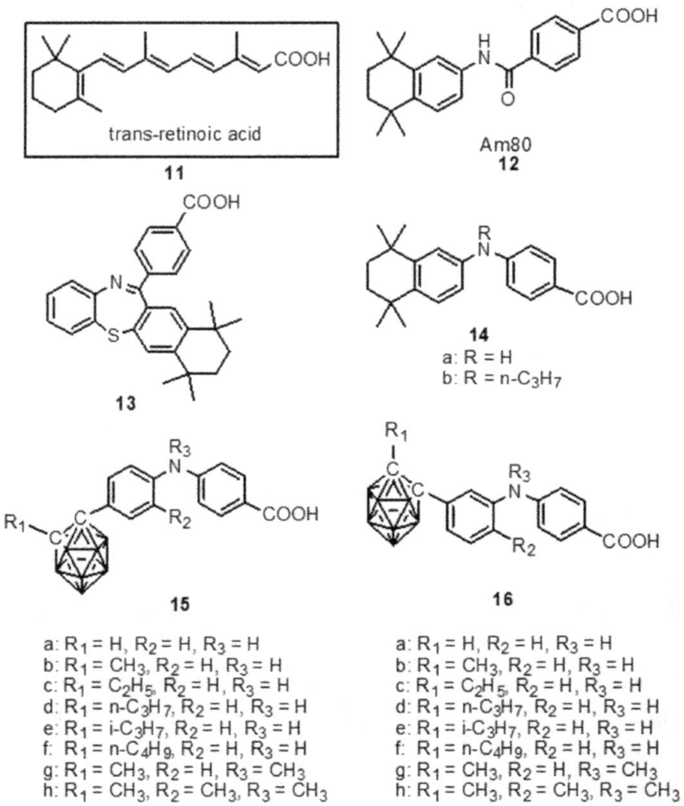


**FIGURE 9.6** Selected examples of two types of specific nuclear receptors ligands (retinoic acid receptors [RARs] and retinoid X receptors [RXRs]), unmodified and modified with boron cluster pharmacophore.

The carboxylic acid moiety with an appropriate hydrophobic group which interacts with the hydrophobic cavity of the RAR ligand binding domain (LBD) is necessary for high-binding activity. A variety of diphenylamines bearing a 1,2-dicarba-*closo*-dodecaborane **1** moiety have been synthesized (Fig. 9.6), and their retinoidal activity was evaluated through the differentiation-inducing ability toward human promyelocytic leukemia

HL-60 cells. The potent agonistic activity of 4-[4-(1,2-dicarba-*closo*-dodecaboran-1-yl)phenylamino]benzoic acids **15** and 4-[3-(1,2-dicarba-*closo*-dodecaboran-1-yl)phenylamino]benzoic acids **16** showed at concentrations of $10^{-8}$ to $10^{-9}$ M. Among all the synthesized compounds 4-[4-(2-propyl-1, 2-dicarba-*closo* dodecaboran-1-yl) phenylamino]benzoic acid (**15d**: $R_1$ = $n$-$C_3H_7$, $R_2$ = H, $R_3$ = H) exhibited biological activity almost equal to that of the natural ligand. It was also clear that by the introduction of a methyl group on the aromatic ring or alkyl groups on the nitrogen atom, the RXR antagonistic activity could be separated from retinoid agonistic activity. The field of carborane analogs of RAR ligands and modulators is still expanding, and provides excellent routes for a new class of therapeutic agents (Endo et al., 1999; Ohta et al., 2004).

### 9.7.2 STEROID ANALOGS BEARING CARBORANE CLUSTER MODIFICATION

Steroids are a class of lipids, characterized by a sterane-core and additional functional groups. A large number of distinct steroids are usually found in plants, animals, and fungi. Steroids were also classified by their chemical composition into cholestanes (e.g., cholesterol), cholanes (e.g., cholic acid), pregnanes (e.g., progesterone), and androstanes (e.g., testosterone and androgen). The hydrophobic structure of steroids is important for their biological activity and role as a scaffold in fixing the spatial positions of hydrogen-bonding functional groups. These open attractive opportunities produce a suitable way to modification of biologically important steroids with carborane cluster lipophilic pharmacophore for the modulation of essential hydrophobic interaction in receptor–ligand complexation.

#### 9.7.2.1 ESTROGEN ANALOGS HAVING A DICARBA-CLOSO-DODECABORANE AS A HYDROPHOBIC MOIETY

Estrogen enhanced the growth, differentiation, and function of many target tissues. In addition to this, estrogen plays a major role in both female and male reproductive system. It has also significant role in bone maintenance, central nervous system, and in the cardiovascular system. The first step in the appearance of these activities was mediated by the binding of hormonal ligands to the α and β estrogen receptor (ER) monomers which resulting the formation of ER dimer. In this case, hydrophobic interactions play a major and important role. Endo and his coworker first time introduced the effective modulation of binding affinity of receptor-ligand complexes, through the implementation of hydrophobic *closo*-carborane clusters (Endo, 2018). They designed the major female sex hormone, 17β-estradiol, with carborane clusters **18**, which plays a significant role in the growth, differentiation, and physiology of both male and female reproductive systems, and in bone marrow maintenance as well. The same group extended their work, more recently, on the synthesis and biological aspects of carboranyl-based compounds, having an affinity for specific ER subtypes (Ohta et al., 2014). A variety of nuclear responses have been mediated through the binding of estrogens (17β-estradiol, estrone, estriol, and estetrol), or estrogen analogs, to the LBD, which produces gene expression upon ligand binding. It is to be noted that the LBDs of ERα and ERβ are rather similar, which allows many ligands to get bound in an unselective manner to both receptor subtypes. Nevertheless, Erβ has drawn greater attention as a novel ER modulator, due to its promising therapeutic ability for the regulation of immune responses toward chronic inflammation. In this context, a small group of tetrahydrofluorenone analogs has been prepared, by varying the length and the position of alkyl substituents, at the cluster vertices, with the n-propyl derivative (Endo, 2018) possessing about seven-fold higher affinity toward ERβ (Fig. 9.7). However, the binding between the aliphatic group and Erα should not be favored, due to steric hindrance with the Met421 residue in the LBD, which is replaced in the ERβ subtype with Ile373. Moreover, the Met336 in ERβ unit replaced the ERα Leu384, which results in a wide range of ERβ-selective ligands, obtaining their ER subtype selectivity through steric, electronic, or hydrophobic interactions with these essential amino acid residues.

In this context, various ERβ-selective derivatives have been reported to reduce the affinity for Erα, through ligand substituents that trigger steric interference with the Met421 content of ERα. It has been found that the structural properties of noncarborane receptor sub-type-selective agents, such as androstane (a C19 steroid with a gonane core) skeleton analogs **19**, or the halogenated version of the carborane **20** (Fig. 9.7), have reduced binding affinities to ERα while showing about 60-fold higher affinity toward ERβ (Fujii, 2016). Although the selectivity of the cyclohexanol derivative **19** is significantly higher than **20**, low binding affinity toward

Erβ significantly limits its application as a potential drug. However, carborane cages, modified with the iodination process, could result in ERβ-selective ligands, such as the 9,10-diiodo-*m*-carborane **21**, with about 14-fold higher affinity toward ERβ than ERα (Ohta et al., 2017).

The first successful estradiol analog **22** (Fig. 9.7), bearing a *para*-carborane, for mimicking the C and D ring of the steroid skeleton, exhibited a higher binding affinity ($K_i$ = 0.1 nM) than estradiol ($K_D$ = 0.4 nM), and acted as an ERα superagonist (Endo, 2018). Recently, different derivatives of 1-(4 methoxyphenyl)-12-hydroxy-methyl-*p*-carborane **23** by following the carborane cluster derivative **22** have been developed. It was reported that there is a direct correlation between enhanced cell growth inhibition and the increased number of phenolic methoxy groups. Moreover, the trimethoxy derivative of carborane clusters **24**, showed the most potent cell growth inhibition, within an assay of 39 human cancer cell lines (the concentration for 50% of maximal inhibition of cell proliferation—$GI_{50}$ value: 5.8 μM). In addition to this, the observation of apoptosis, for a certain type of breast cancer cell line, draws significant research interest (Kaise et al., 2016).

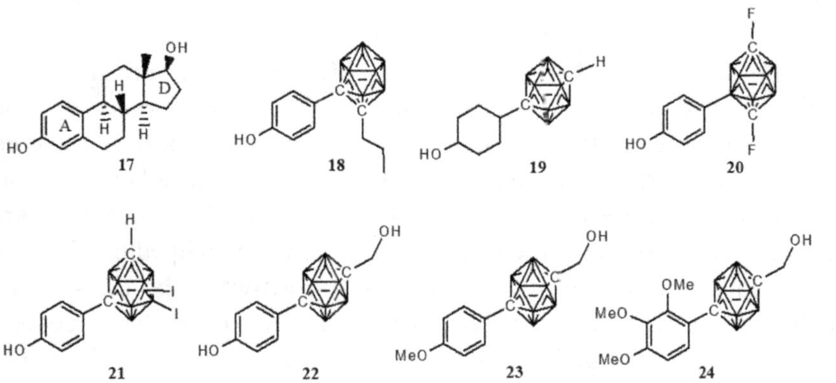

**FIGURE 9.7** Structure of 17β-estradiol **17** and ER ligands **18–24.**

### 9.7.2.2 *ANDROGEN ANALOGS BASED ON CARBORANE CLUSTER STRUCTURE*

The androgen receptor (AR) is essential for the development and maintenance of the male reproductive system. Its biological functions were started by the binding of the steroid hormones, testosterone, and/or 5α-dihydrotestosterone to the AR, where an intricate machinery involving translocation of AR into the nucleus, binding to specific DNA sites, formation of transcriptional complex, and also activation of the expression of specific genes occurs. AR ligands have been clinically used for the treatment of aplastic anemia and prostate cancer. Compounds **26a**, **26b**, and **27**, novel AR-antagonistic *p*- and *m*-carborane derivatives bearing a cyano or nitrobenzyl group (Fig. 9.8) performed anti-androgenic activity similar to that of the well-known anti-androgen flutamide in reporter gene assay using NIH3T3 cells transfected with a human AR expression plasmid (Goto et al., 2005). Here the carborane cage acts as a hydrophobic pharmacophore for the expression of AR-antagonistic activity. After that, the structural analyses of the bicalutamide complex (bound to the mutant AR-LBD) were analyzed through crystallographic characterization, and glycerol and amino glycerol groups were introduced accordingly, at the phenolic para position of the *para*-carborane derivative. The novel compounds, **28** and **29** (Fig. 9.8), showed anti-androgenic activity by inhibiting the mutant AR in LNCaP cell lines. It has been found that, both **28** and **29** exhibit antagonistic activities in LNCaP-proliferation assay ($IC_{50}$ = 0.39 and 0.42 μM, respectively), which is similar to that of the (R)-bicalutamide, the active enantiomer ($IC_{50}$ = 0.43 μM). Furthermore, no suppression in the cell proliferation, in the PC3 cell lines (AR-independent prostate cancer cell line), and activity of the carborane derivatives was observed, which can be attributed to the fact that the inhibition triggered, during the cell proliferation in AR-dependent cell lines, is actually induced by the inhibition of AR activation (Kaise et al., 2018). Molecular docking also cleared the simulations of the two enantiomers of **28**, which helped elucidate the observed isomer-dependent AR antagonistic activity. The marked anti-androgenic activity could be correlated directly to the steric repulsion between (R)-17 and helix-12 of the AR-LBD (Kaise et al., 2018). On the other hand, a series of AR)ligands were introduced as the terminal polar group carborane derivatives. These receptors came with an acidic heterocycle and have the ability to form hydrogen bonds as well (Fujii et al., 2009). The 1,2,4- oxadiazole-5-thione derivatives, **30a** and **30b**, possessed remarkably high affinities for hAR, exceeding that of hydroxyflutamide. This heterocyclic functionality is a potential bioisostere of the nitro and cyano groups which form the polar pharmacophores of known nonsteroidal AR ligands.

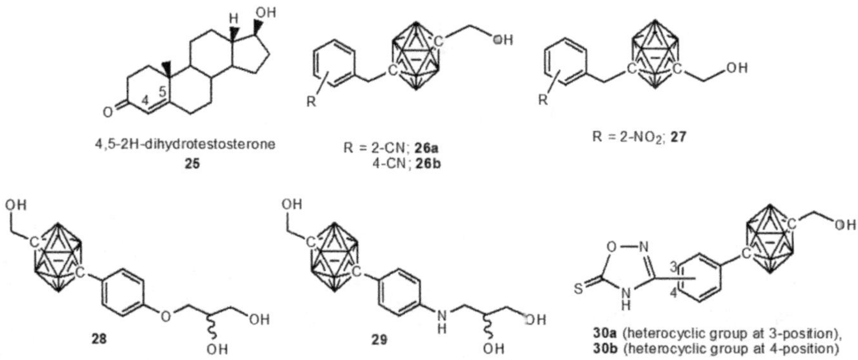


**FIGURE 9.8** Selected examples of AR ligands based on carborane structure.

### *9.7.2.3 CARBORANE CLUSTER BEARING CHOLESTEROL MIMICS*

There are two types of cholesterol containing carborane cluster modification can be defined. The first one includes cholesterols conjugated with carborane clusters **31** and **33** (Feakes et al., 1999; Nakamura et al., 2007), whereas the second choice is cholesterol mimic **34–36** (Zhao et al., 2006) (Fig. 9.9). Recently, novel cholesterol derivatives containing carborane cluster as a pharmacophore **34–36** were proposed, which are obtained by replacing the B and the C rings of cholesterol with carborane cluster, which is analogous to the estradiol modification, proposed earlier (Zhao et al., 2006). Computational analysis indicates the structural features and physicochemical properties of all three carborane-based compounds have very similar to those of cholesterol. One of the synthesized carborane-based cholesterol mimics compound **36** was stably incorporated into nontargeted, folate receptor (FR)-targeted and vascular endothelial growth factor receptor-2 (VEGFR-2)-targeted liposomes. But no major differences were obtained in appearance, size distribution, and lamellarity between conventional DPPC/cholesterol liposomes, nontargeted, and FR-targeted liposomal formulations of this carboranyl derivative compound **36**. These valuable results demonstrate the novel carboranyl cholesterol mimics are excellent lipid bilayer components for the construction of nontargeted and receptor-targeted boronated liposomes for BNCT of cancer and for other applications.

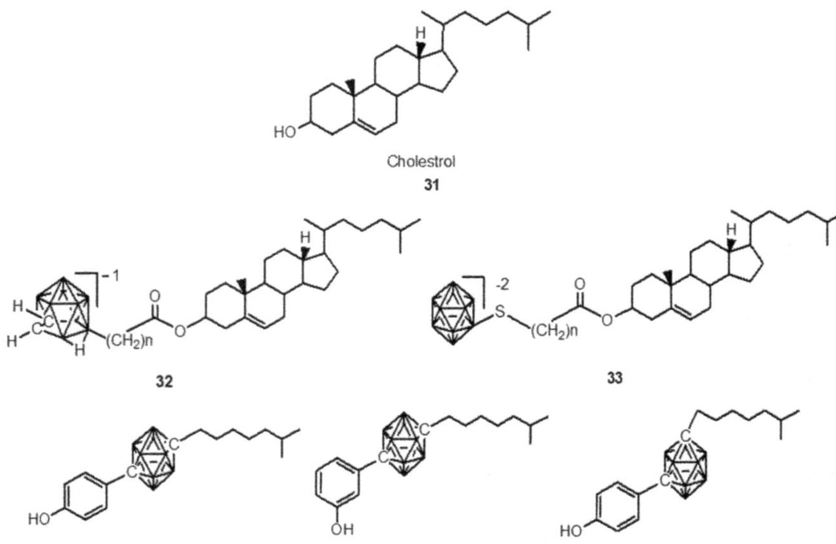


**FIGURE 9.9** Boronated cholesterol conjugate **32** and **33**, and cholesterol mimics **34–36.**

### *9.7.3 TRANSTHYRETIN AMYLOIDOSIS INHIBITORS CONTAINING CARBORANE PHARMACOPHORES*

Transthyretin (TTR) belongs to the class of thyroxine-transport protein, usually found in the blood. It has already been implemented in a variety of amyloid related diseases. Previous investigations have identified a variety of NSAIDs such as flufenamic acid or diflunisal, and structurally related derivatives that introduce kinetic stabilization to TTR. Thus, by inhibiting its dissociative fragmentation and subsequent aggregation, putative toxic amyloid fibrils formed. These pharmaceuticals' ability associated with cyclooxygenase (COX) activity may limit their potential power as long-term therapeutic agents for TTR amyloid diseases.

To solve this problem, several carborane analogs of NSAIADs have been synthesized (Fig. 9.10) and evaluated for inhibition of amyloid fibril formation. Inhibitors were studied by using a 72-h stagnant fibril-formation assay and during this assay time, physiological concentrations of TTR (3.6 μM) were subjected to acid-mediated partial denaturation at pH 4.4, either in the presence or absence of an inhibitor. The formation of fibril was measured by optical density at 400 nm. The obtained results were reported as percent fibril formation (% ff), with TTR in the absence of inhibitor defined as 100% ff. The most valuable promising of these compounds is 1-carboxylic acid-7-[3 fluorophenyl]-1,7-dicarba-*closo*-dodecborane **45** with ff 15 ± 5 at 3.6-μM inhibitor concentration. The significant results shown that along with hydrophobicity nature, the steric bulk and lack of π interactions make the carborane functionality ideal scaffolding for use in the binding channel of TTR. Moreover, when applied to COX inhibitors these same properties are also detrimental (Julius et al., 2007). Consequently, the modified NSAIDs (with known TTR activity), where a carborane moiety substitutes a constituent phenyl ring, have been found to retain the TTR potency, while there was a drastic reduction in the detrimental COX activity. The above-noted unique properties and their biological (and chemical) stability make carboranes very prominent pharmacophores in the design and synthesis of potent and selective pharmaceuticals.

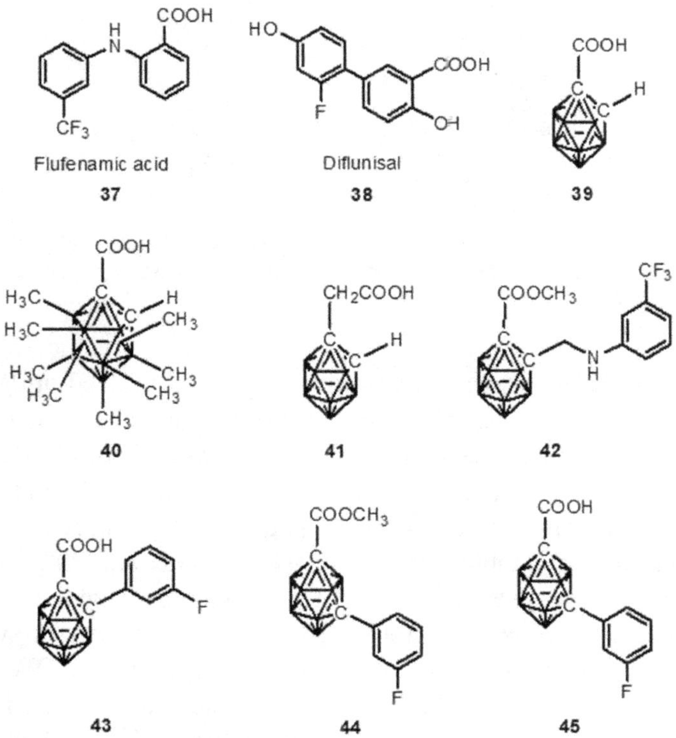


**FIGURE 9.10** Transthyretin (TTR) amyloidosis inhibitors, analogs of NSAIDs flufenamic acid and diflunisal, containing carborane pharmacophors.

### 9.7.4 A-HUMAN THROMBIN INHIBITOR CONTAINING A CARBORANE PHARMACOPHORE

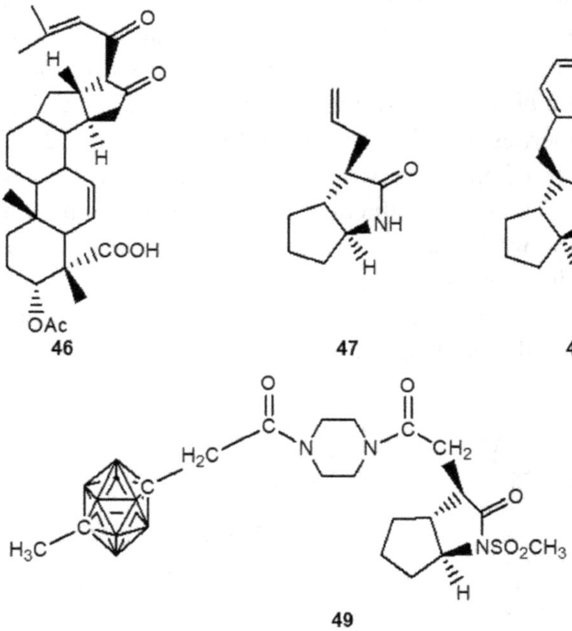


**FIGURE 9.11** Naturally occurring [5,5]-*trans*-lactone thrombin inhibitor **46**, its synthetic analogs **47** and **48**, and lead thrombin inhibitor containing hydrophobic carborane pharmacophore **49.**

$\alpha$-Human thrombin is a potent platelet agonist, involved in the blood coagulation cascade. It has also significant role as an anticoagulant agent, due to its involvement in several debilitating diseases. Recently, the development of thrombin and prothrombin activating factor Xa inhibitors such as synthetic small-molecules has led to a number of potent anti-coagulant compounds arising (Stubbs and Bode, 1993; Bode et al., 1989; Jorg and Jorg, 1999; Srivastava et al., 2005). These include bicyclic *trans*-lactone **46** which is a thrombin inhibitor (IC50 = 4 nM) and also its synthetic analogs **47** and **48** are more resistant to hydrolysis in plasma (Weir et al., 1998; O'Neill et al., 1998). In this direction, a new architecture for size-selective serine protease inhibitors compound **49** was designed by using a computational docking program, that is, flexX (Fig. 9.11). Compound **49** utilize a fully methylated icosahedral *p*-carborane as a dominating hydrophobic pharmacophore. Computationally while positioning an acylating group for the facile irreversible attack at the Ser$^{195}$ hydroxyl group, this compound shown the ability to provide ligand–protein binding interactions throughout the thrombin's main active site (S1–S3). The worthy note displays the importance of fully methylated *para*-carborane derivative over dicarbadodecaborane structure in the design of the $\alpha$-human thrombin inhibitor (Brynda et al., 2013).

### 9.7.5 CARBORANE–NUCLEOSIDE CONJUGATES AS A NEW HUMAN BLOOD PLATELET FUNCTION INHIBITOR

The conjugates of carborane clusters with natural and nonnatural nucleosides provide several analytical and therapeutic applications. These nucleosides, their derivatives, and analogs were widely used as chemotherapeutics. These are mainly utilized as antiviral or anticancer agents because of the crucial role of nucleosides in many metabolic pathways as well as interactions with other biomolecules. Carborane clusters were generally used for lipophilic pharmacophores and modulators of the nucleoside's physicochemical, biological, and pharmacological properties (Lesnikowski, 2003, 2007a, b). The carborane-containing nucleosides could be incorporated directly into DNA and target the genetic material of tumor cells easily. Therefore, nucleosides and their derivatives were used as boron carriers for BNCT of tumors (Lesnikowski et al., 2005). The first example of the carborane–nucleoside conjugate, 5-(1-o-carboranyl)-2′-deoxyuridine was prepared by Yamamoto et al. (1992). After that, a series of carborane derivative conjugates **50–55** with nucleoside were developed by nucleophilic substitution reactions between carborane clusters and nucleoside analog (Fig. 9.12). The nature of the carborane and its suitable chemical properties play a key role in the synthesis of these compounds.

In addition to nucleophilic substitution reaction, cross-coupling and [3 + 2] dipolar cycloaddition reactions were used far less frequently at the step of addition of the carborane clusters to the nucleoside (Fig. 9.13). Recently, the establish of [3 + 2] dipolar cycloaddition of carboranyl-alkyl azides to alkynylated nucleosides makes it possible to prepare some conjugates in one or two steps virtually avoiding the use of protecting groups (Matuszewski et al., 2015) which was surely less laborious than the earlier multistep approaches to the preparation of nucleoside–carborane conjugates. Moreover, several general methods for the modification of pyrimidine as well as purine nucleosides with a variety of different carborane cluster was proposed opening the way for biological screening and evaluation of this new class of bioorganic–inorganic conjugates (Olejniczak et al., 2007a, b; Wojtczak et al., 2008).

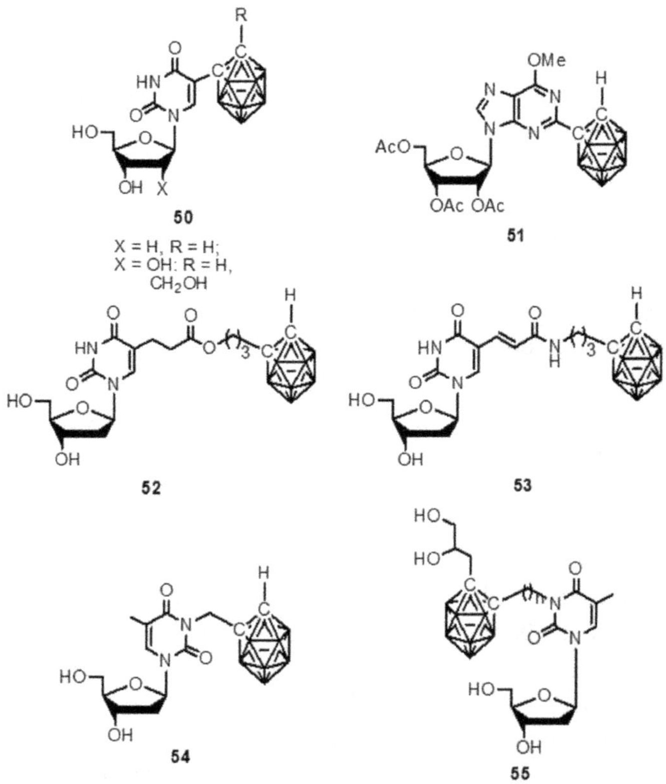


**FIGURE 9.12** A series of polyhedral boron compound conjugate with nucleoside were developed by nucleophilic substitution reactions between boron clusters and nucleoside.

A variety of adenosine and 2′-deoxyadenosine conjugates with carborane clusters (Fig. 9.14) were compared their effects on the responses of platelets (aggregation, protein release, and P-selectin expression) to stimulation by the agonists, ADP, and thrombin. When the modification of adenosine at the 2′-C position with a *para*-carborane **3** cluster occurs, the obtained **60** results an efficient inhibition of platelet function, including aggregation. This preliminary finding addition with the new chemistry proposed to form the basis of the development of a new class of adenosine analogs have capacity to modulate human blood platelet activities (Bednarska et al., 2010a, b).

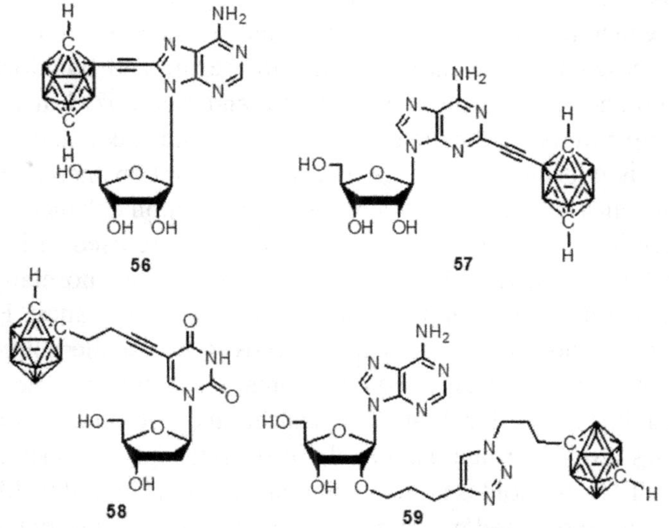

**FIGURE 9.13** A series of polyhedral boron compound formed by cross-coupling and [3 + 2] dipolar cycloaddition reactions.

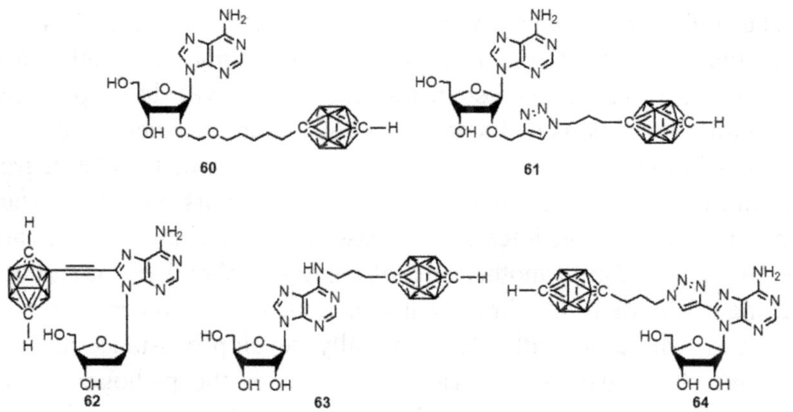

**FIGURE 9.14** Selected examples of adenosine and 2′-deoxyadenosine conjugates with *p*-carborane clusters.

## 9.8 SUMMARY

All of the life forms have been derived (either directly or indirectly) from the element, carbon. Since boron lies next to carbon, in the periodic table of elements, it possesses few characteristics that are similar to those of carbon, but also possesses a number of unique properties (and differences). Carboranes produce from the combination of boron, carbon and hydrogen in a 3c–2e manner, and can participate in biochemical reactions and interactions. These carboranes are very easy to prepare and functionalization too. Carboranes are thermally very stable and isomerized at the appropriate temperature. In the field of bioorganic and medicinal chemistry, carborane is expected to emerge as a next-generation therapeutic drug. Carborane clusters can act as hydrophobic interaction enhancers between pharmaceuticals, and their receptors and entities are used to increase *in vivo* stability and bioavailability of compounds that were normally rapidly metabolized due to their lipophilicity or amphiphilicity nature. However, the carborane–cholesterol conjugate derivatives are excellent lipid bilayer components for the construction of nontargeted and receptor-targeted boronated liposomes for BNCT of cancer and for other applications. Carboranes are performed useful role in pharmacophores, especially for design of TTR amyloidosis and $\alpha$-human thrombin inhibitor. Moreover, carborane clusters-based modification of adenosine are ability to opening the way for biological screening, evaluation of this new class of bioorganic–inorganic conjugates, modulation of human blood platelet activities as well as boron carrier for BNCT.

The high boron content and stability of carborane make it suitable for medicinal applications. They are also providing information about fast-growing research interest in the quest for novel drugs potentially overcoming limitations and side effects of current products, and several examples in the area of anticancer drugs, inhibitors of platelet aggregation, and modulators of important hormone receptors have been shown in this discussion. Anti-infectious disease drugs bearing essential carborane component forms another area of medicinal chemistry of carborane awaiting exploration. It is important to note that unlike other commercial antibiotics, where the pathogens gradually develop resistance/immunity with time, there would be less/no evolution of the pathogens toward carborane-based antibiotic drugs. This makes carborane one of the best choices for the potential anti-microbial drug. Pathogens usually show rapid evolution against any type of commercially available drugs that attack them. However, it is expected that the process of evolution will either be much slower (or entirely absent), in the case of carborane-based compounds. This would not only revolutionize the pharmaceutical entities but also will put us one step ahead in terms of advanced drug delivery and therapeutics.

## KEYWORDS


- **carborane**
- **BNCT**
- **hydrophobic**
- **RARs**
- **pharmacophore**


## REFERENCES


Adamska, A.; Olejniczak, A. B.; Zwolinski, K.; Szczepek, W. J.; Krol, E.; Szewczyk, B.; Grynkiewicz, G.; Lesnikowski, Z. J. Oseltamivir Analog with Boron Cluster Modulator. *Acta Pol. Pharm.* **2012,** *69*, 1218–1223.

Ban, H. S.; Shimizu, K.; Minegishi, H.; Nakamura, H. Identification of HSP60 as a Primary Target of O-Carboranylphenoxyacetanilide, an HIF-1α Inhibitor. *J. Am. Chem. Soc.* **2010,** *132*, 11870–11871.

Barth, R. F.; Mi, P.; Yang, W. Boron Delivery Agents for Neutron Capture Therapy of Cancer. *Cancer Commun.* **2018,** *38*, 35.

Bednarska, K.; Olejniczak, A. B.; Wojtczak, B. A.; Piskała, A.; Sułowska, Z.; Les′nikowski, Z. J. Adenosine and 2′-Deoxyadenosine Modified with Boron Clusters (C2B10H12) as New Classes of Human Blood Platelet and Neutrophil Function Modulators, "Purines 2010." *An International Meeting on Adenine Nucleoside and Nucleotides in Biomedicine*; Tarraco, Spain, May 30–June 2, 2010a.

Bednarska, K.; Olejniczak, A. B.; Wojtczak, B. A.; Sułowska, Z.; Lesnikowski, Z. J. Adenosine and 2′-Deoxyadenosine Modified with Boron Cluster Pharmacophores as New Classes of Human Blood Platelet Function Modulators. *Chem. Med. Chem.* **2010b,** *5*, 749–756.

Bode, W.; Mayr, I.; Baumann, U.; Huber, R.; Stone, S. R.; Hofsteenge, J. The Refined 1.9 A Crystal Structure of Human Alpha-Thrombin: Interaction with D-Phe-Pro-Arg Chloromethylketone and Significance of the Tyr-Pro-Pro-Trp Insertion Segment. *EMBO J.* **1989,** *8*, 3467–3475.

Brynda, J.; Mader, P.; Sicha, V.; Fabry, M.; Poncova, K.; Bakardiev, M.; Gruner, B.; Cigler, P.; Rezacova, P. Carborane-Based Carbonic Anhydrase Inhibitors. *Angew. Chemie Int. Ed.* **2013,** *52*, 13760–13763.

Cabrera-Gonzalez, J.; Ferrer-Ugalde, A.; Bhattacharyya, S.; Chaari, M.; Teixidor, F.; Gierschner, J.; Nunez, R. Fluorescent Carborane–Vinylstilbene Functionalised Octasilsesquioxanes: Synthesis, Structural, Thermal and Photophysical Properties. *J. Mater. Chem. C.* **2017,** *5*, 10211–10219.

Couto, M.; Garcia, M. F.; Alamón, C.; Cabrera, M.; Cabral, P.; Merlino, A.; Teixidor, F.; Cerecetto, H.; Vinas, C. Discovery of Potent EGFR Inhibitors through the Incorporation of a 3D-Aromatic-Boron-Rich-Cluster into the 4-Anilinoquinazoline Scaffold: Potential Drugs for Glioma Treatment. *Chem.—A Eur. J.* **2018,** *24*, 3122–3126.

Dilthey, W. Z. Personal-und Hochschulnachrichten. *Angew. Chem.* **1921,** *34*, 596.

Endo, Y. In *Boron-Based Compounds*; Hey-Hawkins, E., Vin˜as Teixidor, C.; John Wiley & Sons, Ltd: Chichester, 2018; pp 1–19.

Endo, Y.; Yoshimi, T.; Iijima, T.; Yamakoshi, Y. Estrogenic Antagonists Bearing Dicarba-Closo-Dodecaborane as a Hydrophobic Pharmacophore. *Bioorg. Med. Chem. Lett.* **1999,** *9*, 3387–3392.

Feakes, D. A.; Spinler, J. K.; Harris, F. R. Synthesis of Boron-Containing Cholesterol Derivatives for Incorporation into Unilamellar Liposomes and Evaluation as Potential Agents for BNCT. *Tetrahedron* **1999,** *55*, 11177–11186.

Fernandez-Alvarez, R.; Dorďovic, V.; Uchman, M.; Matejicek, P. Amphiphiles without Head-and-Tail Design: Nanostructures Based on the Self-Assembly of Anionic Boron Cluster Compounds. *Langmuir* **2018,** *34*, 3541–3554.

Fujii, S. Expanding the Chemical Space of Hydrophobic Pharmacophores: The Role of Hydrophobic Substructures in the Development of Novel Transcription Modulators. *Medchemcomm* **2016,** *7*, 1082–1092.

Fujii, S.; Ohta, K.; Goto, T.; Kagechika, H.; Endo, Y. Acidic Heterocycles as Novel Hydrophilic Pharmacophore of Androgen Receptor Ligands with a Carborane Core Structure. *Bioorg. Med. Chem.* **2009,** *17*, 344–350.

Goszczynski, T. M.; Fink, K.; Kowalski, K.; Leśnikowski, Z. J.; Boratynski, J. Interactions of Boron Clusters and Their Derivatives with Serum Albumin. *Sci. Rep.* **2017,** *7*, 9800.

Goto, T.; Ohta, K.; Suzuki, T.; Ohta, S.; Endo, Y. Design and Synthesis of Novel Androgen Receptor Antagonists with Sterically Bulky Icosahedral Carboranes. *Bioorg. Med. Chem.* **2005,** *13*, 6414–6424.

Grafstein, D.; Dvorak, J. Neocarboranes, a New Family of Stable Organoboranes Isomeric with the Carboranes. *Inorg. Chem.* **1963,** *2*, 1128–1133.

Greme, R. N. *Carborane*; Academic Press: Amsterdam, 2016.

He, T.; Misuraca, J. C.; Musah, R. A. "Carboranyl-Cysteine"—Synthesis, Structure and Self-Assembly Behavior of a Novel α-Amino Acid. *Sci. Rep.* **2017,** *7*, 16995.

Hermanek, S. Boron-11 NMR Spectra of Boranes, Main-Group Heteroboranes, and Substituted Derivatives. Factors Influencing Chemical Shifts of Skeletal Atoms. *Chem. Rev.* **1992,** *92*, 325–362.

Heying, T. L.; Ager, J. W.; Clark, S. L.; Mangold, D. J.; Goldstein, H. L.; Hillman, M.; Polak, R. J.; Szymanski, J. W. A New Series of Organoboranes. I. Carboranes from the Reaction of Decaborane with Acetylenic Compounds. *Inorg. Chem.* **1963,** *2*, 1089–1092.

Imming, P. *The Practice of Medicinal Chemistry*; Elsevier, 2015; pp 3–13.

Issa, F.; Kassiou, M.; Rendina, L. M. Boron in Drug Discovery: Carboranes as Unique Pharmacophores in Biologically Active Compounds. *Chem. Rev.* **2011,** *111*, 5701–5722.
Jorg, H.; Jorg, S. Synthetic Inhibitors of Thrombin and Factor Xa: From Bench to Bedside. *Thrombosis Res.* **1999,** *93*, 203–241.
Julius, R. L.; Farha, O. K.; Chiang, J.; Perry, L. J.; Hawthorne, M. F. Synthesis and Evaluation of Transthyretin Amyloidosis Inhibitors Containing Carborane Pharmacophores. *Proc. Natl. Acad. Sci.* **2007,** *104*, 4808– 4813.
Kaise, A.; Ohta, K.; Endo, Y. Design, Synthesis, and Anti-Proliferative Activity of 1-(4-Methoxyphenyl)-12-Hydroxymethyl-p-Carborane Derivatives. *Eur. J. Med. Chem.* **2016,** *122*, 257–263.
Kaise, A.; Ohta, K.; Fujii, S.; Oda, A.; Goto, T.; Endo, Y. Novel Androgen Receptor Full Antagonists: Design, Synthesis, and a Docking Study of Glycerol and Aminoglycerol Derivatives That Contain *p*-Carborane Cages. *Bioorg. Med. Chem.* **2018,** *26*, 3805–3811.
Kracke, G. R.; VanGordon, M. R.; Sevryugina, Y. V; Kueffer, P. J.; Kabytaev, K.; Jalisatgi, S. S.; Hawthorne, M. F. Carborane-Derived Local Anesthetics Are Isomer Dependent. *Chem Med Chem* **2015,** *10*, 62–67.
Lee, S. H.; Park, Y. J.; Yoon, C. M. Reductive Etherification of Aromatic Aldehydes with Decaborane. *Tetrahedron Lett.* **1999,** *40*, 6049–6050.
Leites, L. A. Vibrational Spectroscopy of Carboranes and Parent Boranes and Its Capabilities in Carborane Chemistry. *Chem. Rev.* **1992,** *92*, 279–323.
Lesnikowski, Z. J. Boron Clusters – A New Entity for DNA-Oligonucleotide Modification. *Eur. J. Org. Chem.* **2003,** *2003*, 4489–4500.
Lesnikowski, Z. J. Boron Units as Pharmacophores—New Applications and Opportunities of Boron Cluster Chemistry. *Collect. Czechoslov. Chem. Commun.* **2007a,** *72*, 1646–1658.
Lesnikowski, Z. J. DNA as Platform for New Biomaterials. Metal-Containing Nucleic Acids. *Curr. Org. Chem.* **2007b,** *11*, 355–381.
Lesnikowski, Z. J. Challenges and Opportunities for the Application of Boron Clusters in Drug Design. *J. Med. Chem.* **2016,** *59*, 7738–7758.
Lesnikowski, Z. J.; Paradowska, E.; Olejniczak, A. B.; Studzinska, M.; Seekamp, P.; Schuller, U.; Gabel, D.; Schinazi, R. F.; Plesek, J. Towards New Boron Carriers for Boron Neutron Capture Therapy: Metallacarboranes and Their Nucleoside Conjugates. *Bioorg. Med. Chem.* **2005,** *13*, 4168–4175.
Matuszewski, M.; Kiliszek, A.; Rypniewski, W.; Lesnikowski, Z. J.; Olejniczak, A. B. Nucleoside Bearing Boron Clusters and Their Phosphoramidites—Building Blocks for Modified Oligonucleotide Synthesis. *New J. Chem.* **2015,** *39*, 1202–1221.
Nabakka, J. M.; Harwell, D. E.; Knobler, C. B.; Hawthorne, M. F. Synthesis and Structural Characterization of a Thioether-Bridged Commo-Metallabis(Dicarbollide) Species: A Model System for Venus Flytrap Cluster Reagents1Dedicated to Professor Ken Wade on the Occasion of His 65th Birthday and in Recognition of His Outstanding Contributions to the Fields of Organometallic and Inorganic Chemistry.1. *J. Organomet. Chem.* **1998,** *550*, 423–429.
Nakamura, H.; Ueno, M.; Lee, J.-D.; Ban, H. S.; Justus, E.; Fan, P.; Gabel, D. Synthesis of Dodecaborate-Conjugated Cholesterols for Efficient Boron Delivery in Neutron Capture Therapy. *Tetrahedron Lett.* **2007,** *48*, 3151–3154.
Nakamura, H.; Yasui, Y.; Maruyama, M.; Minegishi, H.; Ban, H. S.; Sato, S. Development of Hypoxia-Inducible Factor (HIF)-1α Inhibitors: Effect of Ortho-Carborane Substituents

on HIF Transcriptional Activity under Hypoxia. *Bioorg. Med. Chem. Lett.* **2013,** *23,* 806–810.

Nakamura, Y.; Mochida, A.; Choyke, P. L.; Kobayashi, H. Nanodrug Delivery: Is the Enhanced Permeability and Retention Effect Sufficient for Curing Cancer? *Bioconjug. Chem.* **2016,** *27,* 2225–2238.

Nishiyama, N.; Iriyama, A.; Jang, W.-D.; Miyata, K.; Itaka, K.; Inoue, Y.; Takahashi, H.; Yanagi, Y.; Tamaki, Y.; Koyama, H.; Kataoka, K. Light-Induced Gene Transfer from Packaged DNA Enveloped in a Dendrimeric Photosensitizer. *Nat. Mater.* **2005,** *4,* 934–941.

Nunez, R.; Farras, P.; Teixidor, F.; Vinas, C.; Sillanpaa, R.; Kivekas, R. A Discrete P...I-I...P Assembly: The Large Influence of Weak Interactions on the 31P NMR Spectra of Phosphane–Diiodine Complexes. *Angew. Chemie Int. Ed.* **2006,** *45,* 1270–1272.

Nunez, R.; Tarres, M.; Ferrer-Ugalde, A.; de Biani, F. F.; Teixidor, F. Electrochemistry and Photoluminescence of Icosahedral Carboranes, Boranes, Metallacarboranes, and Their Derivatives. *Chem. Rev.* **2016,** *116,* 14307–14378.

O'Neill, M. J.; Lewis, J. A.; Noble, H. M.; Holland, S.; Mansat, C.; Farthing, J. E.; Foster, G.; Noble, D.; Lane, S. J.; Sidebottom, P. J.; Lynn, S. M.; Hayes, M. V.; Dix, C. J. Isolation of Translactone-Containing Triterpenes with Thrombin Inhibitory Activities from the Leaves of Lantana Camara. *J. Nat. Prod.* **1998,** *61,* 1328–1331.

Ohta, K.; Iijima, T.; Kawachi, E.; Kagechika, H.; Endo, Y. Novel Retinoid X Receptor (RXR) Antagonists Having a Dicarba-Closo-Dodecaborane as a Hydrophobic Moiety. *Bioorg. Med. Chem. Lett.* **2004,** *14,* 5913–5918.

Ohta, K.; Ogawa, T.; Endo, Y. Design and Synthesis of Iodocarborane-Containing Ligands with High Affinity and Selectivity toward ERβ. *Bioorg. Med. Chem. Lett.* **2017,** *27,* 4030–4033.

Ohta, K.; Ogawa, T.; Kaise, A.; Oda, A.; Endo, Y. Aliphatic Substitution of *o*-Carboranyl Phenols Enhances Estrogen Receptor Beta Selectivity. *Chem. Pharm. Bull.* **2014,** *62,* 386–391.

Olejniczak, A. B.; Adamska, A. M.; Paradowska, E.; Studzinska, M.; Suski, P.; Leśnikowski, Z. J. Modification of Selected Anti-HCMV Drugs with Lipophilic Boron Cluster Modulator. *Acta Pol. Pharm.* **2013,** *70,* 489–504.

Olejniczak, A. B.; Plesek, J.; Lesnikowski, Z. J. Nucleoside-Metallacarborane Conjugates for Base-Specific Metal-Labeling of DNA, *Chem. Eur. J.*, **2007a,** *13,* 311–318.

Olejniczak, A. B.; Wojtczak, B.; Lesnikowski, Z. J. 2′-Deoxyadenosine Bearing Hydrophobic Carborane Pharmacophore. *Nucleosides: Nucleotides Nucleic Acids* **2007b,** *26,* 1611–1613.

Page, M.; Jalisatgi, S.; Maderna, A.; Hawthorne, M. Design and Synthesis of a Candidate α-Human Thrombin Irreversible Inhibitor Containing a Hydrophobic Carborane Pharmacophore. *Synthesis-stuttgart* **2008,** *2008,* 555–563.

Papetti, S.; Heying, T. L. *P*-Carborane [1,12-Dicarbaclovododecaborane(12)]. *J. Am. Chem. Soc.* **1964,** *86,* 2295.

Poater, J.; Sola, M.; Vinas, C.; Teixidor, F. π-Aromaticity and Three-Dimensional Aromaticity: Two Sides of the Same Coin? *Angew. Chemie Int. Ed.* **2014,** *53,* 12191–12195.

Reynolds, R. C.; Campbell, S. R.; Fairchild, R. G.; Kisliuk, R. L.; Micca, P. L.; Queener, S. F.; Riordan, J. M.; Sedwick, W. D.; Waud, W. R.; Leung.; Dixon, R. W.; Suling, W.

J.; Borhani, D. W. Novel Boron-Containing, Nonclassical Antifolates: Synthesis and Preliminary Biological and Structural Evaluation. *J. Med. Chem.* **2007,** *50*, 3283–3289.

Scholz, M.; Blobaum, A. L.; Marnett, L. J.; Hey-Hawkins, E. Synthesis and Evaluation of Carbaborane Derivatives of Indomethacin as Cyclooxygenase Inhibitors. *Bioorg. Med. Chem.* **2011,** *19*, 3242–3248.

Scholz, M.; Hey-Hawkins, E. Carbaboranes as Pharmacophores: Properties, Synthesis, and Application Strategies. *Chem. Rev.* **2011,** *111*, 7035–7062.

Sivaev, I. B.; Bregadze, V. V. Polyhedral Boranes for Medical Applications: Current Status and Perspectives. *Eur. J. Inorg. Chem.* **2009,** *2009*, 1433–1450.

Srivastava, S.; Goswami, L. N.; Dikshit, D. K. Progress in the Design of Low Molecular Weight Thrombin Inhibitors. *Med. Res. Rev.* **2005,** *25*, 66–92.

Stubbs, M. T.; Bode, W. A Player of Many Parts: The Spotlight Falls on Thrombin's Structure. *Thromb. Res.* **1993,** *69*, 1–58.

Tarres, M.; Canetta, E.; Paul, E.; Forbes, J.; Azzouni, K.; Vinas, C.; Teixidor, F.; Harwood, A. J. Biological Interaction of Living Cells with COSAN-Based Synthetic Vesicles. *Sci. Rep.* **2015,** *5*, 7804.

Teixidor, F.; Barbera, G.; Vaca, A.; Kivekas, R.; Sillanpaa, R.; Oliva, J.; Vinas, C. Are Methyl Groups Electron-Donating or Electron-Withdrawing in Boron Clusters? Permethylation of o-Carborane. *J. Am. Chem. Soc.* **2005,** *127*, 10158–10159.

Weir, M. P.; Bethell, S. S.; Cleasby, A.; Campbell, C. J.; Dennis, R. J.; Dix, C. J.; Finch, H.; Jhoti, H.; Mooney, C. J.; Patel, S.; et al. Novel Natural Product *5*,5-Trans-Lactone Inhibitors of Human α-Thrombin: Mechanism of Action and Structural Studies. *Biochemistry* **1998,** *37*, 6645–6657.

Wiesboeck, R. A.; Hawthorne, M. F. Dicarbaundecaborane (13) and Derivatives. *J. Am. Chem. Soc.* **1964,** *86*, 1642–1643.

Wilkinson, S. M.; Gunosewoyo, H.; Barron, M. L.; Boucher, A.; McDonnell, M.; Turner, P.; Morrison, D. E.; Bennett, M. R.; McGregor, I. S.; Rendina, L. M.; Kassiou, M. The First CNS-Active Carborane: A Novel P2X7 Receptor Antagonist with Antidepressant Activity. *ACS Chem. Neurosci.* **2014,** *5*, 335–339.

Wilson, C. O.; Beale, J. M.; Block, J. H. *Wilson and Gisvold's Textbook of Organic Medicinal and Pharmaceutical Chemistry*, 12th ed.; Lippincott Williams & Wilkins: Baltimore, 2011.

Wojtczak, B. A.; Andrysiak, A.; Gruner, B.; Lesnikowski, Z. J. "Chemical Ligation" –A Versatile Method for Nucleoside Modification with Boron Clusters, *Chem. Eur. J.*, **2008,** *14*, 10675–10682.

Wojtczak, B. A.; Olejniczak, A. B.; Lesnikowski, Z. J. Nucleoside Modification with Boron Clusters and Their Metal Complexes. In *Current Protocols in Nucleic Acid Chemistry*; Beaucage, S. L., Ed.; John Wiley & Sons, Inc., 2009; Chapter 4, Unit 4.37, pp 1–26.

Wu, G.; Zhao, T.; Kang, D.; Zhang, J.; Song, Y.; Namasivayam, V.; Kongsted, J.; Pannecouque, C.; De Clercq, E.; Poongavanam, V.; Liu, X; Zhan, P. Overview of Recent Strategic Advances in Medicinal Chemistry. *J. Med. Chem.* **2019,** *62*, 9375–9414.

Wu, Y.; Carroll, P. J.; Kang, S. O.; Quintana, W. Synthesis, Characterization, and Reactivity of Isocyanato Dicarbaboranes Obtained from o-Carborane. *Inorg. Chem.* **1997,** *36*, 4753–4761.

Wu, Y.; Carroll, P. J.; Quintana, W. Synthesis and Characterization of a Protected Amino Alcohol Containing Ortho-Carborane. *Polyhedron* **1998,** *17*, 3391–3407.

Yamamoto, Y.; Seko, T.; Nakamura, H.; Nemoto, H.; Hojo, H.; Mukai, N.; Hashmioto, Y. Synthesis of Carboranes Containing Nucleoside Bases. Unexpectedly High Cytostatic and Cytocidal Toxicity towards Cancer Cells. *J. Chem. Soc. Chem. Commun.* **1992,** *2*, 157–158.

Zakharkin, L. I.; Kalinin, V. N.; Podvisotskaya, L. S. The Comparative Reactivity of Ortho-, Meta-, and Para-Carboranes. *Bull. Acad. Sci. USSR, Div. Chem. Sci.* **1970,** *19*, 1227–1231.

Zhao, X. B.; Bandyopadhyaya, A. K.; Johnsamuel, J.; Tiwari, R.; Golightly, D. W.; Patel, V.; Jehning, B. T.; Backer, M. V; Barth, R. F.; Lee, R. J.; et al. Receptor-Targeted Liposomal Delivery of Boron-Containing Cholesterol Mimics for Boron Neutron Capture Therapy (BNCT). *Bioconjug. Chem.* **2006,** *17*, 1141–1150.

# Index

## D



## E



## F

## G

## R



## S



## T



## U

## W



## X

Printed in the USA
CPSIA information can be obtained
at www.ICGtesting.com
CBHW062303241024
16389CB00003B/67

9 781774 91279